Administering Windows Server Hybrid Core Infrastructure AZ-800 Exam Guide

Design, implement, and manage Windows Server core infrastructure on-premises and in the cloud

Steve Miles

<packt>

BIRMINGHAM—MUMBAI

Administering Windows Server Hybrid Core Infrastructure AZ-800 Exam Guide

Group Product Manager: Mohd Riyan Khan
Publishing Product Manager: Mohd Riyan Khan
Senior Content Development Editor: Sayali Pingale
Technical Editor: Shruthi Shetty
Copy Editor: Safis Editing
Book Project Manager: Neil Dmello
Proofreader: Safis Editing
Indexer: Sejal Dsilva
Production Designer: Prashant Ghare
Marketing Coordinator: Ankita Bhonsle

First published: December 2022
Production reference: 1181122

Published by Packt Publishing Ltd.
Livery Place
35 Livery Street
Birmingham
B3 2PB, UK.

978-1-80323-920-0
www.packt.com

*This book is my contribution to the worldwide technical learning community and
I would like to thank all of you who are investing your valuable time in learning new skills and
committing to reading this book.*

Contributors

About the author

Steve Miles, aka *SMiles* or *Mr. Analogy*, is a Microsoft Azure MVP, MCT, and multi-cloud and hybrid technologies author and technical reviewer with over 20 years of experience in networking, data center infrastructure, managed hosting, and cloud solutions. His experience comes from working with end users and reseller channels and in vendor spaces and roles with a global network and app security vendor, with global telco hosters, in managed hosting, with colocation and data center service providers, and in hardware distribution. Currently, he is working for a multi-**Cloud Solution Provider** (**CSP**) distributor based in the UK and Dublin in a cloud and hybrid technology leadership role.

His current focus is on securing, protecting, and managing identities, Windows clients, and Windows Server workloads in hybrid and multi-cloud platform environments.

Most happy in front of a whiteboard, he prefers to communicate using illustrations. He is renowned for breaking down complex technologies with analogies and concepts into everyday, real-world scenarios.

His first Microsoft certification was on Windows NT and he is an MCP, MCITP, MCSA, and MCSE for Windows Server and many other Microsoft products. He also holds multiple Microsoft Fundamentals, Associate, Expert, and Specialty certifications in Azure Security, Identity, Network, and M365. He also holds multiple security and networking vendor certifications.

Finally, as part of the multi-cloud domain, he has experience with GCP, AWS, an Alibaba Cloud MVP, and is Alibaba Cloud-certified.

About the reviewer

Peter De Tender has more than 25 years of experience in architecting and deploying Microsoft solutions, starting with Windows NT4/Exchange5.5 in 1996. In early 2012, he started shifting to cloud technologies and quickly embraced Azure, working as a cloud architect and trainer. In September 2019, Peter joined Microsoft's prestigious Microsoft Technical Trainer team, providing Azure readiness workshops to their top customers and partners across the globe. Recently having relocated to Redmond, WA, Peter is continuing this role. Given his past status as Azure MVP and his passion for the community, Peter is still actively involved in public speaking, technical writing, and mentoring and coaching. You can follow Peter on Twitter at @pdtit.

Thanks Steve, for trusting my love for Azure. Also, a big thanks to my wife for supporting me in realizing my dreams.

Table of Contents

Part 1: Hybrid Identity

1

Implementing and
Managing Active Directory Domain Services   3

2

Implementing and Managing Azure Active Directory Domain Services 49

3

Managing Users and Computers with Group Policy 75

4

Implementing and Managing Hybrid Identities   105

Part 2: Hybrid Networking

5

Implementing and Managing On-Premises Network Infrastructure 139

6

Part 3: Hybrid Storage

7

8

Implementing a Hybrid File Server Infrastructure   267

Part 4: Hybrid Compute

9

10

11

Managing Windows Server Azure Virtual Machines   357

12

Managing Windows Server in a Hybrid Environment   379

13

Part 5: Exam Prep

14

Preface

Cloud computing is a technology commonly seen as a platform and mechanism to provide organizations with an enabler for digital transformation. However, the cloud computing journey need not be a *binary decision*; it need not be a broad sweeping decision to move *all* your workloads to the cloud and decommission *everything* on-premises.

While it may be appropriate and the best direction of travel for some, there are equally those that, for reasons such as data locality, compliance, control, performance, and so on, this will not be the case. There will be, for many, a need to remain with an appropriate amount of on-premises computing, storage networking, and other related resources to deliver the technology needs to their organizations. It is not for us to judge or rule on what is right or wrong for an organization but to provide options to meet and support all required outcomes that are mandated, or need to be delivered to an organization.

With that outlook and mindset on the changing face of cloud computing, we may look to extend the capabilities of cloud computing into our data centers to enhance and enrich the services operated from these facilities. It could be considered as much about *managing from the cloud* as it is about *moving to the cloud*.

We acknowledge that undeniably there are solid cases for digital transformation and, for many, being cloud native is the foremost approach; however, for many, the first step in the journey will be data center modernization through a hybrid cloud computing approach. Many a year's life is left in data centers, and Windows Server is a computing and core infrastructure platform that should be enriched and enhanced through cloud computing capabilities.

The content of this book intends to provide complete coverage of the exam requirements to prepare you for the *AZ-800: Administering Windows Server Hybrid Core Infrastructure* Microsoft Certification exam.

The exam is intended for candidates with extensive experience working with the Windows Server operating system in hybrid environments and implementing and managing the core hybrid infrastructure technologies of computing, storage, networking, identity, and management.

In addition, this book's added value is that it aims to go beyond the exam objectives, providing an extra depth of knowledge with practical hands-on skills to master, which will be of value in a day-to-day hybrid Windows Server environment role.

This book closes with exam preparation tests.

Who is this book for

This book is for those in technical implementation and administration roles looking to pass the *AZ-800: Administering Windows Server Hybrid Core Infrastructure* exam.

What this book covers

Chapter 1, Implementing and Managing Active Directory Domain Services, includes content on creating and managing **Active Directory Domain Services** (**AD DS**) in a hybrid environment.

Chapter 2, Implementing and Managing Azure Active Directory Domain Services, includes content on creating and managing **Azure Active Directory Domain Services** (**AAD DS**).

Chapter 3, Managing Users and Computers with Group Policy, includes content on implementing and managing group policy in AD DS and AAD DS.

Chapter 4, Implementing and Managing Hybrid Identities, includes content on implementing and managing hybrid identities using Azure AD Connect and Azure AD cloud sync.

Chapter 5, Implementing and Managing On-Premises Network Infrastructure, includes content on implementing and managing on-premises network infrastructure in a hybrid environment.

Chapter 6, Implementing and Managing Azure Network Infrastructure, includes content on implementing and managing Azure network infrastructure in a hybrid environment.

Chapter 7, Implementing Windows Server Storage Services, includes content on implementing Windows Server storage services in a hybrid environment.

Chapter 8, Implementing a Hybrid File Server Infrastructure, includes content on implementing Azure storage services in a hybrid environment.

Chapter 9, Implementing and Managing Hyper-V on Windows Server, includes content on implementing and managing Hyper-V on Windows Server in a hybrid environment.

Chapter 10, Implementing and Managing Windows Server Containers, includes content on implementing and managing containers in a hybrid environment.

Chapter 11, Managing Windows Server Azure Virtual Machines, includes content on managing Azure virtual machines.

Chapter 12, Managing Windows Server in a Hybrid Environment, includes content on managing Windows Server in a hybrid environment.

Chapter 13, Managing Windows Servers Using Azure Services, includes content on managing Windows Servers and workloads by using Azure services.

Chapter 14, Exam Preparation Practice Tests, provides practice tests for each skill section for *Exam AZ-800: Administering Windows Server Hybrid Core Infrastructure*.

To get the most out of this book

For this book, the following are required:

- A device with a browser, such as Edge or Chrome, to access the Azure portal: `https://portal.azure.com`

- An Azure AD tenancy and Azure subscription; you can use an existing account or sign up for free: `https://azure.microsoft.com/en-us/free`

- An Owner role for the Azure subscription

Software/hardware covered in the book	
Windows Server	Azure AD
Hyper-V	Azure Compute services
Active Directory	Azure Storage services
iSCSI storage	Azure Networking services
Storage Spaces and Storage Spaces Direct	Azure Security services
File Storage	Azure Management services

Download the color images

We also provide a PDF file that has color images of the screenshots/diagrams used in this book. You can download it here: `https://packt.link/Mc8to`.

Conventions used

There are a number of text conventions used throughout this book.

`Code in text`: Indicates code words in text, database table names, folder names, filenames, file extensions, pathnames, dummy URLs, user input, and Twitter handles. Here is an example: "The primary server is set with the `Delay configuration` attribute in the `Scope` properties on the secondary server."

A block of code is set as follows:

```
Install-WindowsFeature Web-Application-Proxy
-IncludeManagementTools
```

Bold: Indicates a new term, an important word, or words that you see on screen. For example, words in menus or dialog boxes appear in the text like this. Here is an example: "Launch **DNS Manager** from **Server Manager** under the **Tools** menu."

> **Tips or important notes**
> Appear like this.

Get in touch

Feedback from our readers is always welcome.

General feedback: If you have questions about any aspect of this book, mention the book title in the subject of your message and email us at customercare@packtpub.com.

Errata: Although we have taken every care to ensure the accuracy of our content, mistakes do happen. If you have found a mistake in this book, we would be grateful if you would report this to us. Please visit www.packtpub.com/support/errata, select your book, click on the **Errata Submission Form** link, and enter the details.

Piracy: If you come across any illegal copies of our works in any form on the Internet, we would be grateful if you would provide us with the location address or website name. Please contact us at copyright@packt.com with a link to the material.

If you are interested in becoming an author: If there is a topic that you have expertise in and you are interested in either writing or contributing to a book, please visit authors.packtpub.com.

Share Your Thoughts

Once you've read *Administering Windows Server Hybrid Core Infrastructure AZ-800 Exam Guide*, we'd love to hear your thoughts! Scan the QR code below to go straight to the Amazon review page for this book and share your feedback.

https://packt.link/r/1803239204

Your review is important to us and the tech community and will help us make sure we're delivering excellent quality content.

Download a free PDF copy of this book

Thanks for purchasing this book!

Do you like to read on the go but are unable to carry your print books everywhere? Is your eBook purchase not compatible with the device of your choice?

Don't worry, now with every Packt book you get a DRM-free PDF version of that book at no cost.

Read anywhere, any place, on any device. Search, copy, and paste code from your favorite technical books directly into your application.

The perks don't stop there, you can get exclusive access to discounts, newsletters, and great free content in your inbox daily

Follow these simple steps to get the benefits:

1. Scan the QR code or visit the link below

https://packt.link/free-ebook/9781803239200

2. Submit your proof of purchase

3. That's it! We'll send your free PDF and other benefits to your email directly

.

Part 1:
Hybrid Identity

This part will provide complete coverage of the knowledge and skills required for the skills measured in the *Deploy and manage AD DS in on-premises and cloud environments* section of the exam.

This part of the book comprises the following chapters:

- *Chapter 1, Implementing and Managing Active Directory Domain Services*
- *Chapter 2, Implementing and Managing Azure Active Directory Domain Services*
- *Chapter 3, Managing Users and Computers with Group Policies*
- *Chapter 4, Implementing and Managing Hybrid Identities*

Implementing and Managing Active Directory Domain Services

This chapter covers the AZ-800 Administering Windows Server Hybrid Core Infrastructure exam learning objective: *Deploy and manage AD DS in on-premises and cloud environments.*

We will start this first chapter with an understanding of Active Directory concepts and its portfolio of services. We will focus on **Active Directory Domain Services** (**AD DS**). We will look at each component of AD DS in more detail and then move on to understanding how to create and manage an instance of AD DS in on-premises environments. We will then conclude with a hands-on exercise to develop your skills further.

The following main topics will be covered in this chapter:

- What is Active Directory?
- What is Active Directory Domain Services?
- Active Directory Domain Services components
- Creating Active Directory Domain Services
- Managing Active Directory Domain Services
- Exercise – installing AD DS on Windows Server

This chapter aims to take your knowledge beyond the exam objectives to prepare you for a real-world, day-to-day hybrid environment-focused role.

What is Active Directory?

Before we dive into creating or configuring any services, we will look at some definitions and concepts to set a baseline and foundation of knowledge for you to build from. We will start this chapter by defining **Active Directory** (**AD**), which forms the basis of Windows Server identity, access management, and information protection services.

AD is part of Microsoft's identity, access, and information protection solutions. It runs as an installed service as part of Windows Server and was introduced in Windows 2000.

As its name suggests, AD is a **directory service** and an **identity provider** (**IDP**) whose primary function is to manage access to domain resources through *authentication* and *authorization*. It is used to control, centrally organize, locate, and secure access to these resources on a network.

At a simple level, you can think of it as an *identity store* and digital address book for resources on a network. It comprises a list of identities and their access rights to resources in the directory.

AD is not a single function service or solution; it is a collective or umbrella term for a portfolio of directory-based and identity-driven services, including **domain services**, **federation services**, **certificate services**, and **rights management services**. It provides capabilities such as **single sign-on** (**SSO**).

From a technical perspective, it is an *X.500 compatible directory service* and can be accessed using the **Lightweight Directory Access Protocol** (**LDAP**). It is based on a hierarchical, multi-master distributed database model that comprises partitions and an extensible schema.

This section introduced AD as the Microsoft solution for the foundation of identity, access, and information protection for on-premises and hybrid environments. In the next section, we will understand and define **Active Directory Domain Services** (**AD DS**).

What is Active Directory Domain Services?

AD DS is organized as a distributed and searchable hierarchical directory that controls access to network resources and allows settings and configurations to be applied through policies.

AD DS is a server role installed on Windows Server and is included with the **operating system** (**OS**); no software needs to be downloaded. A server with an installed AD DS role is referred to as a **domain controller** (**DC**).

AD DS provides access to resources by *authenticating* and *authorizing* domain object resources.

Accessing domain resources is based on a two-stage concept that consists of authenticating and then authorizing; in a nutshell, it involves identifying who you are and determining what you can do:

- **Authentication**, also referred to as **AuthN**, is the identity component; it is the process of establishing the identity of a person (or service) and proving they are who they say they are. This can be done by validating access credentials against stored or known identifying information.

- **Authorization**, also referred to as **AuthZ**, is the access component; it is the process of establishing what level of access the authenticated person (or service) has to the resource, what they can access, and what actions they may take.

The concept of authenticating and authorizing is shown in the following diagram:

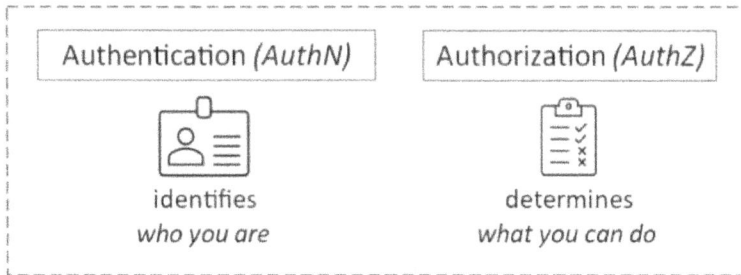

Figure 1.1 – Understanding authentication and authorization

The terms **Active Directory** and **Domain Services** (abbreviated to their short forms of **AD** and **DS**) are often used interchangeably to mean the same thing. When people refer to AD, they often mean just the DS component, mainly since it is the most common and foundational identity and access management service component to be implemented.

For this book and the exam, we will refer to AD in the context of the DS component only; we will refer to it simply as **AD DS** for brevity. For additional learning content on the other services that are part of AD, please refer to the *Further reading* section at the end of this chapter.

AD DS contains a list of objects, such as user accounts, along with their attributes, such as passwords and their assigned access rights to resources. This list can be queried to validate the identity and access rights to a resource in the domain that is created as an object in the information store database. This is shown in the following diagram:

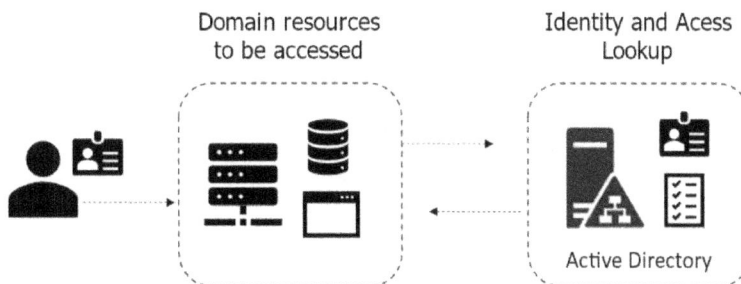

Figure 1.2 – Domain resource access

AD DS functions by classifying everything stored in its information store (**database**) as an object. Objects can be user accounts, computer accounts, groups (security principals), printers, network appliances, applications, or services. These objects are hierarchical, meaning that objects can contain other nested objects; these types of objects are called containers. We will look at these terms in more detail later in this chapter.

Each object stored in the database has a set of attributes that match the object's context; this is defined in a **schema**. The directories information data store is a database structure with a schema that is extensible and can store attributes that suit the business requirement. Being a directory service means every object can be searched, queried, access controlled, managed, and configured in a centralized and policy-driven manner.

The foundation of **identity and access management** (**IAM**) is that access to objects is controlled through **role-based access control** (**RBAC**) and the principle and practice of least privilege. This means providing the proper scope of control. Just enough access is given to perform a required task without unnecessarily providing elevated access rights that may give a broader scope than required and privileged lateral access if an account were to be compromised. The control mechanism is a group policy and provides group scoping for identities.

AD DS comprises a logical structure that often maps to the organization's *operating model* and a *physical structure* that should map to the *network topology*.

The following can be considered as the logical components:

- Domain
- Domain tree
- Forest
- OU
- Partition
- Schema
- Container

The following can be considered as the physical components:

- Data store
- Global catalog
- Domain controller
- Read-only domain controller
- Site
- Subnet

These components will be covered in more detail later in this chapter.

This section introduced us to **DS**, one of AD's services, which provides a centralized mechanism for authenticating and authorizing resources in a domain.

In summary, AD DS is built around a directory service that is a replicated information store (database) for all domain objects. It primarily provides authentication and authorization for accessing domain objects and configuration and management access.

Active Directory Domain Services components

In this section, we will look closer at the individual components of AD DS.

Objects

Objects refer to any element in the AD DS directory; examples of objects include printers, computers, users, groups, subnets, and so on. Objects have attributes that describe them; each will have required attributes and optional attributes. A name and an **object identifier** (**OID**) are two examples of required attributes.

In this section, we will look at the different classes of AD DS objects.

Container objects

These are *built-in objects* for logically grouping and holding other objects; think of these as hierarchical folder structures containing objects belonging to the same category or class.

When AD DS is installed, a set of default object containers are created; these will be visually represented as hierarchal folders. The following screenshot shows the default containers; some are hidden by default:

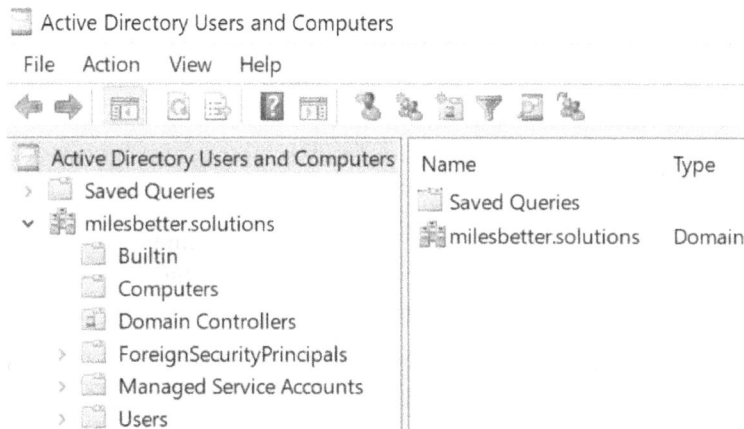

Figure 1.3 – Containers for objects

These default containers inherit a default set of permissions that are assigned to allow proper service administration and operations. These containers cannot be deleted, and you should not change the assigned permissions or delegate control of these; instead, you should create containers to meet your needs.

The essential containers to be aware of are as follows:

- The **Domain container** acts as the root container for the container hierarchy.

- The **Builtin container** provides a container that holds several default groups and default service admin accounts that are created when AD DS is installed.

- The **Computers container** provides a container that holds all computer account objects; it is the default location for all created computer accounts.

- The **Users container** provides a container that holds all user and group objects; it is also the default location for all created users and groups, some of which are automatically placed in this container location when AD DS is installed.

- The **Domain Controllers OU container** provides a container that holds the computer accounts for all DCs; it is the default location for all created DCs.

These default containers are automatically created when you install AD DS. They cannot be deleted, and assigned permissions and access control should not be modified. Still, custom containers can be created to meet your needs.

You can only apply policies to OU containers; to apply policies and delegate control over users or computers, you should create your own OU containers and move objects to those containers so that the policies and permissions can be applied.

The **domain controllers** container is an **organizational unit** (**OU**) container and is the only OU created by default when AD DS is installed. It is recommended that DCs not be moved out of their default location of the Domain Controllers OU container and not delegate control to non-service administrators.

Organizational unit

The OU functions as a container object and is used to collectively organize users, groups, and computers, providing a framework for targeting administrative control and policies on objects in the container.

An example of where OUs could be helpful is where you need to apply common administrative control to a group of resources that are part of the same workload (such as all Azure Virtual Desktop objects), environment (such as dev, test, prod, and so on), business unit, or region.

It is important to note that you can only link **group policy objects** (**GPOs**) to OU containers and not regular containers.

OUs can be created via the following methods:

- AD Users and Computers

- AD Admin Center

- PowerShell with the AD module

- **Windows Admin Center** (**WAC**), which is rapidly becoming the most widely adopted tool in hybrid environments

An OU structure can be hierarchal and map to organizational structures such as department, business group, or geography. These can be mapped to functions, projects, or resources such as workstations, servers, and so on.

OUs should group all those objects that need the same administration and policy settings and move them into an OU that can be controlled with Group Policy. If there are objects in an OU that you don't want to have the policy applied to, then you can create a new OU and move the objects out to that OU so that the policy does not apply.

Hierarchal grouping is supported; you can nest OUs within OUs (just like any hierarchal folder structure). However, you should carefully plan the object grouping and hierarchy so that it isn't too complex. Typically, five levels or less is optimum, and 10 is the recommended maximum to ease the admin burden and reduce complexity. You may also be constrained by a domain object resource that will only work with an OU structure of a supported method/configuration.

User objects

When a **user account** (an identity) is created in the directory, it's classed as an object. It is one of the core and foundation object classes in the directory, along with computer objects. Hence, we will look at an AD DS management tool known as **Active Directory Users and Computers** later in this chapter.

For a user to be able to authenticate to domain resources, they must use a user account that AD DS provides. AD DS stores user accounts in its database with object-specific identity information and attributes such as the login username and password; these are used for the user to sign in to access domain resources.

In addition, other attributes include any organization-specific information that needs to be stored, such as job title, department, office, email, phone number, and more.

Regarding access management, the groups the user is a member of are listed. These group memberships are used to determine the level of access to resources in the domain; this is the concept of who you are and what access you have.

The recommended practice is that access rights are not given to user objects directly but to group objects. Users are members of groups, so they inherit their access permissions from their group memberships; this simplifies administration and makes troubleshooting access issues much easier and quicker.

Every user account object has a critical and primary attribute known as the **User Principal Name** (**UPN**). This is in an internet communication standard format typically recognized and expressed as an email address format – `name@domain`. However, it is essential to understand that it is not an email address. We recognize that a user's email address could match their UPN and vice versa, but it's not an explicit rule.

The user's UPN is derived from a combination of the user's login name and the domain name. The @ symbol is a form of join or delimiter between these two names; for example, the user *Steve Miles* in the `milesbetter.solutions` AD domain may have a UPN of `smiles@milesbetter.solutions`, where `smiles` is the user login attribute for the user account of Steve Miles.

Two other terms are the **UPN prefix** and **UPN suffix**; the UPN prefix is the part of the name before the @ symbol, while the suffix is the part of the name after the @ symbol. The UPN prefix and suffix components are shown in the following screenshot:

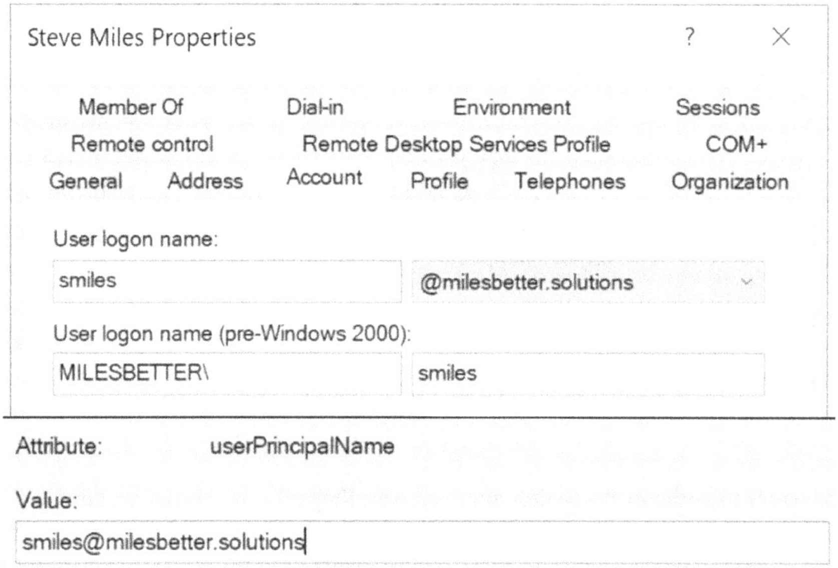

Figure 1.4 – UPN format

As in the previous user account example, the `smiles` login name attribute is then termed the UPN prefix, whereas the UPN suffix would be `milesbetter.solutions`, which refers to the domain part of the name.

Service objects

Service accounts are non-user-based account object classes used for applications and services that run in the background and require no user interaction. However, they still need to be authenticated to the directory service.

Managed Service Accounts (**MSAs**) are AD DS object classes that provide this service account capability; they benefit from lowering admin effort for **Service Principal Name** (**SPN**) management and service account password management.

There are two types of MSAs, as follows:

- **Standard Managed Service Accounts** (**sMSAs**) are *local to a computer*, service, or application and cannot be shared across other resources in the network. This limits their efficiency and adds an administrative burden. Needing to implement multiple local sMSAs increases the threat, security, and configuration drift surface area and exposure.

  Further information about sMSAs can be found at `https://docs.microsoft.com/en-us/azure/active-directory/fundamentals/service-accounts-standalone-managed`.

- **Group Managed Service Accounts** (**gMSAs**) have *domain scopes* and capabilities and can be used across multiple computers, services, and applications; NLB clusters and IIS are examples.

  gMSAs require you to create a **key distribution services** (**KDS**) root key on a DC to create gMSA accounts.

  The following are some AD PowerShell cmdlets that can be used to manage gMSAs:

  - `Get-ADServiceAccount`
  - `Install-ADServiceAccount`
  - `New-ADServiceAccount`
  - `Remove-ADServiceAccount`
  - `Set-ADServiceAccount`
  - `Test-ADServiceAccount`
  - `Uninstall-ADServiceAccount`

Further information about gMSAs can be found at `https://docs.microsoft.com/en-us/azure/active-directory/fundamentals/service-accounts-group-managed`.

Computer objects

Just as users have an account in the directory service so that we can search for and locate these users, we need the same mechanism to search for and locate computers in the domain. A **computer object** provides an account for a computer; it represents a computer resource on the network and allows it to uniquely identify itself and authenticate. Once a computer has a computer account, it can be managed as a domain resource and configured using Group Policy.

User account objects and **computer account** objects are the two foundation building block object classes in AD DS.

Group objects

Group object classes in the directory form the basis for access management.

Once you have been authenticated through your user account, your group object membership defines what you are authorized to access in the domain. The permissions assigned to the group define what actions you can carry out on that network resource.

Permissions shouldn't be set at the user account level; this is an admin burden, inefficient, prone to configuration drift, and becomes a governance and security posture issue.

All permissions should be assigned to a group, and users should be made members of the appropriate groups to give them the required permissions. This makes it easier to govern and control policies such as joiners, leaver policies, or where people change roles and now require different access levels. Rather than change individual resource permissions directly for what they can now do/not do based on their new role definition, we can add or remove them to the appropriate groups based on their new role's permissions requirements. This is much more efficient, less error-prone, and provides much better control and governance.

Adding user accounts to groups is an RBAC approach. The principle of least privilege should be adopted, meaning you only need just enough access to perform the tasks/activities required without the need for elevated permissions. This should be the cornerstone of protecting against lateral attacks following an account compromise.

In summary, by adopting the practice of creating groups, assigning permissions to groups, and then making user accounts members of those groups, we can efficiently manage access to resources for multiple users that all need the same access type.

Group types

There are two types of groups: **security groups** and **distribution groups**. Let's look at these in more detail:

- **Security groups** are used as *security-enabled* logical containers to collectively group users to assign access permission to resources. This is a much more efficient way to assign a set of permissions to resources than assigning them individually to each user. For example, to control access to a network resource, we would create any number of groups, each with different levels of access permissions. Then, users can be added to the appropriate group(s), depending on what level of access they should have. We should ensure we follow the least privilege approach when assigning users to groups – that is, if they only need read access, then don't make them members of a group with full-control permissions.

  Two default security groups are created automatically when AD DS is installed; these groups are located in the *Builtin* and *Users* containers.

- **Distribution groups** are used as *non-security enabled* logical containers, that is, no permissions can be assigned to these groups, and are used for grouping users for providing some other form of non-security related administration, and to target a subset of users, often a business unit, team, or project focus. Most commonly used for email applications, for sending emails to target all group members.

Group scopes

Groups also have a *scope*. This defines where they can exist/be effective, the types of members that can be added, and the capabilities and permissions that the group supports.

Four types of group scopes are supported in Windows Server: **local groups**, **domain-local groups**, **global groups**, and **universal groups**. Let's look at these in more detail:

- **Local groups** are effective only on the local machines they are created on and cannot be created on DCs; for clarity, these are not domain objects and exist only on local machines. The following are characteristics of local groups:

  - **Abilities and permissions**: Can be assigned to local (non-domain) resources only

  - **Members**: Any user account of the forest

- **Domain-local groups** only apply locally to the domain created within and exist on the AD DS DCs for that domain. They are the primary means of managing access rights and responsibilities to resources within the local domain. The following are characteristics of domain-local groups:

  - **Abilities and permissions**: Can only be assigned to domain-local resources

  - **Members**: Any user account of the forest

- **Global groups** are used to collectively group user accounts that share similar functions; this could be based on geographic location or business unit. The following are characteristics of global groups:

  - **Abilities and permissions**: Can be assigned to any forest resource

  - **Members**: Local domain only

- **Universal groups** are used in multi-domain environments and combine the functions of the domain-local and global groups. The following are characteristics of universal groups:

  - **Abilities and permissions**: Can be assigned to any forest resource

  - **Members**: Any user account of the AD DS forest

You set the group type and scope at the time of creation and choose based on your needs and requirements for each group; this must be given adequate planning.

This section looked at AD DS objects. The next section will look at the AD DS domain and domain tree topology.

Domains

As inferred by the name **Domain Services**, a domain is the core foundation component and the building block of AD DS. A domain is a logical container of objects in the directory for management purposes. AD DS directory objects could be user accounts, computer accounts, groups, printers, and so on; all objects within that domain are bound by and share the same admin and security policies.

The domain is an *administrative* and *replication* boundary for AD objects; security boundaries are only implemented at the forest level. We will explore this in the *Forests* section.

A domain has one or more DCs. This is a Windows Server role responsible for authenticating requests to access resources in the domain, such as authenticating and authorizing a login request to a computer or accessing a storage file share. The DCs hold a copy of the AD DS database that is writeable; we will look at the concept of RODC and its use cases later.

Domains have a hierarchal topology, referred to as a domain tree. It consists of domains in a parent/child relationship that share a parent root domain; they also share a contiguous namespace in DNS.

The AD DS domain and domain tree are shown in the following diagram:

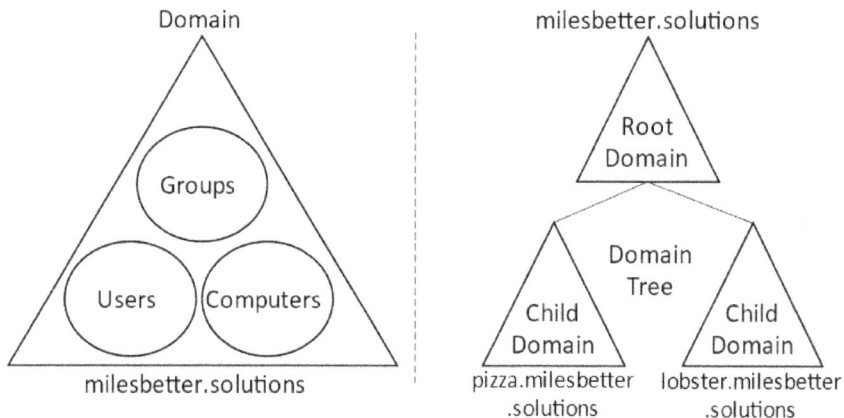

Figure 1.5 – Domain topology

AD DS uses a *multi-master replication model*, meaning that any DC can make changes to objects in the directory stored in the AD database. When changes to the domain are made on a DC, such as adding, removing, or changing an object, these changes are replicated to all other DCs in the domain. In a multi-domain case, only a subset of changes is replicated to all domains.

When the AD DS role is installed on Windows Server, a deployment option is provided for creating a new domain in an existing forest or a new domain in a new forest. The first domain in a new forest becomes what is known as the *forest root domain*; this is analogous to saying whether you would like to create a city in an existing state or whether you would like to create a new state and make this the first and parent/founding/capital city in that state – that is, the first domain in the new forest becomes the capital city, also known as the forest root domain.

This section looked at the AD DS domain and domain tree topology. The next section will look at the AD DS forest topology.

Forests

A **forest** is a top-level logical container definition. It is a collection of one or more domains that share a *common parent root domain*, *schema*, and *global catalog* that has a contiguous namespace; it is considered a *security* and *administrative* boundary.

To implement a security boundary for objects, they must be placed in different forests; all objects in a forest share the same security controls. It is a common misunderstanding that placing objects in different domains will isolate them between domains. It is essential to note this is not the case; only placing them in different forests can provide this isolation. The following diagram shows a visual representation of a forest:

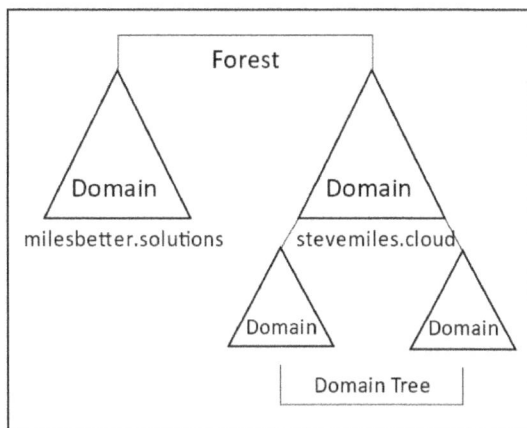

Figure 1.6 – Forest topology

The forest topology (structure or layout) means that a forest can contain one domain or multiple domains across multiple domain trees, much like a country can contain many states, which are collections of independent and autonomous cities. Each state would contain a state capital city; we refer to this as the forest root domain.

In this section, we looked closely at various AD DS components. In the next section, we will look at the function of AD DS trusts.

Trusts

Trusts are implemented to access resources when multiple domains and forests exist within an environment. Depending on the type of trust, these trusts can be one way or two way.

A trusted domain object is stored in the system container in AD DS and provides information on the trust types that have been created.

The most common trust types that can be implemented are as follows:

- **Parent and child trust**: This is created when a new domain is added to an existing tree. It is a transitive, two-way trust.

- **Tree-root trust**: This is created when a new tree is added to an existing forest. It is a transitive, two-way trust.

- **Forest trust**: This is a trust between forests and allows two forests to share resources. It is a transitive one-way or two-way trust.

- **Shortcut trust**: This is a trust created manually when authentication time between domains in different parts of the forest needs to be reduced. It is a nontransitive one-way or two-way trust.

- **External trust**: This is a trust that allows access to resources from a domain in another forest or an NT 4.0 domain. It is a nontransitive one-way or two-way trust.

- **Realm trust**: This is a trust between AD DS and a directory service other than AD DS that implements a Kerberos version 5 protocol realm. It is a transitive or nontransitive one-way or two-way trust.

These trust relationships are shown in the following diagram:

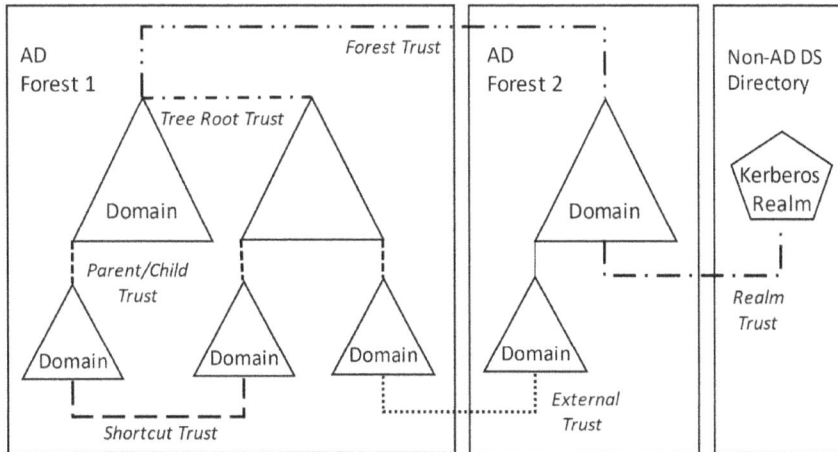

Figure 1.7 – Trusts

All automatically created two-way trusts in the forest are transitive. If domain B is trusted by domain A, and domain D is trusted by domain C, then domain A trusts domain C and D. This is a complex way of saying all the domains trust all the other domains in the forest.

However, in contrast, trusts between forests are not transitive, which means that if there are trusts between forests A and B and forests B and C, then C is not implicitly trusted by forest A.

Note that when a trust is set up between forests or domains, a function known as **SID filtering** is enabled by default. The purpose is to remove all foreign **security identifiers** (**SIDs**) from a user's access token when using a trust in the trusting domain to access resources. Although it should be disabled to allow access to resources via the trust, you can disable this on the outgoing trust of the trusting domain.

When it comes to SID filtering and resource access across trusts, it is important to understand the trust direction and its relationship to the direction of access and which way the trust relationship is set up – that is, outgoing or incoming. This can be unclear; the core concept is that the direction of trust will be the opposite of the direction of access. This is shown in the following diagram:

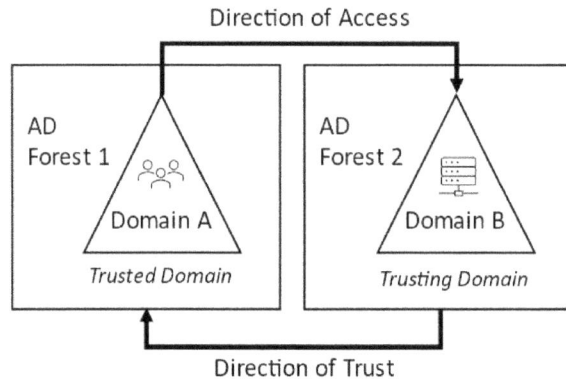

Figure 1.8 – Trust direction

This means that if you want to access resources in another forest, there must be an outgoing trust; conversely, an incoming trust allows outgoing access.

One final concept to note is that of the trust authentication types. External and forest trusts provide two modes of authentication: **forest-wide authentication** and **selective authentication**.

Further information about trusts can be found at `https://docs.microsoft.com/en-us/azure/active-directory-domain-services/concepts-forest-trust`.

In this section, we looked at AD DS trusts. Now, let's look at the AD DS data store.

Data store

A **directory information store** holds the AD database, which uses the **Extensible Storage Engine** (**ESE**) based on the Jet database engine.

The database is made up of two files – the database file itself (named `NTDS.DIT` and located in `C:\Windows\NTDS directory` by default) and the transaction log file (named `EDB.LOG` and also located in `C:\:Windows\NTDS directory` by default). These files are stored on each DC. The group policy files for the group policy objects are stored in the `SYSVOL` folder.

The AD DS database file (`NTDS.DIT`) contains four partitions (or naming contexts) that contain all the domain-related information. Different data is stored in each partition, and each partition is replicated between DCs with a dependent replication scope and schedule.

These partitions are shown in the following diagram:

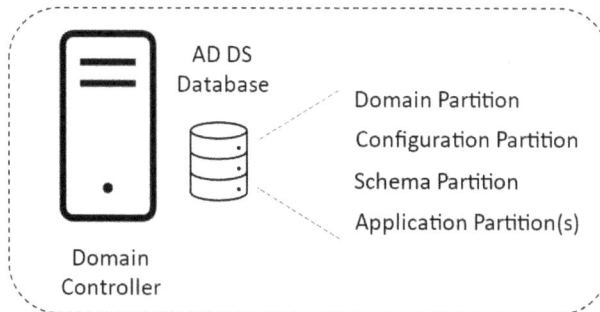

Figure 1.9 – AD DS database partitions

Let's look at these four directory partitions in more detail:

- The **schema partition** is where the AD DS schema is stored; this defines what can be stored in the AD database

- The **configuration partition** is where the configuration information for the forest and domain trees is stored, such as site and replication information

- The **domain partition** is where object information for the domain is stored – that is, information about the user and computer objects

- The **application partition** is where applications can store data in AD; multiple application partitions may exist

In this section, we looked at the AD DS data store, which contains the AD DS database. We also looked at the partitions for the AD database. In the next section, we will look at the AD DS schema.

Schema

The **schema** is a partition of the AD database; it provides the definitions for the object classes (categories) that can be stored in the directory and the object-describing attributes. The object classes and object attributes are schema objects; in other technology areas, they would be referred to as metadata – that is, data describing the data.

The schema can be edited and allows you to create, modify, and disable object categories and attributes of objects. The schema is forest-wide, so any change is replicated to every domain in the forest. Understand the impact before committing any changes to the schema. In addition, be aware that deletions are not supported in the schema.

The default built-in group schema administrators control access to the schema; you must be a member of this group to make changes to the schema.

In this section, we looked at the AD schema. In the next section, we will look at AD DS DCs.

Domain controllers

A DC is a server with the AD DS role installed; it holds a writeable copy of the AD database and functions through the multi-master replication model for this information store. It also provides a store for GPOs. It ensures that the configuration settings are applied to domain-managed objects.

The DC is also a Kerberos **Key Distribution Center** (**KDC**). Kerberos is the mechanism for all authentication and authorization within AD; a Kerberos key identifies an object (authentication – that is, who you are) and outlines what resources an object can access (authorization – that is, what you can do and access).

DCS must be highly available to process logon requests; they are required to authenticate requests for access to domain resources and apply policies for managing and configuring objects in the directory stored in the AD database.

All computers and users that wish to access objects in an AD DS domain must be authenticated by a DC, so their placement, operation, and availability are crucial.

Read-only domain controller

In Windows Server 2008, the **read-only domain controller** (**RODC**) concept was introduced as a deployment option. An RODC server holds a copy of the AD database and supports one-way, incoming replication. No direct changes can be made to the directory (such as LDAP writes); the only changes that can occur are those that are made via replication. A change should be made on a DC, allowing replication to push those changes to the RODC.

The RODC role was intended as a branch office for other locations that may not have adequate security to protect against Kerberos system components being compromised, such as the **KDC**, which is a database of keys, and the **Ticket Granting Service** (**TGS**), which is used to obtain Kerberos session keys. Unlike DCs, which share a Kerberos key, RODCs each have a unique key, minimizing the risk of exposing remote sites that may be isolated from secured and controlled corporate networks.

RODCs have two other use cases. The first is where locations, such as branch offices, don't have a lot of bandwidth for replication traffic. The other scenario is to support the principle of least privilege so that the local server admin account does not need to have domain admin access to function.

This section looked at DCs and their role in AD DS. Now, let's look at the global catalog and its purpose.

Global catalog

The **global catalog** (**GC**) runs as a service on DCs and provides an index that allows you to look up information for every object listed in the AD forest (such as group memberships).

The GC holds a list of every object in the directory, but not every attribute for those objects; the schema defines what attributes of the objects are stored in the GC. The schema can be modified to allow those object attributes you want to appear in the GC so that they can be looked up in the GC.

When a user logs in, a GC must be contactable to look up a user's universal group membership; however, when universal group caching is enabled, contacting a DC with a copy of the GC is less critical.

As part of the DC deployment, the first DC created in a forest holds a copy of the GC by default; this is for hybrid environments. GC placement should be designed for availability in each site, and perform local site searches to optimize network traffic. A DC that does not have a copy of the GC will have to send traffic over a WAN link to a site with a copy of the GC. GC placement is an important design decision; we will cover sites later in this chapter.

In this section, we looked at the purpose of the AD DS Global Catalog service. In the next section, we will look at the function of the operations master roles.

Operations master roles

Although AD DS runs on a multi-master model, some single master operation roles are known as **Flexible Single Master Operations** ((**FSMO**), pronounced *fizmo*) roles and run on DCs.

As the name implies, these are single instance roles. They are not replicated, so this does, to some extent, introduce a single point of failure. If a DC holding that role goes offline, then any operations that rely on being able to contact that role will not be available. This should be considered while managing expectations in any design.

The *five FSMO roles* are categorized as forest and domain operations scopes. These are as follows:

- Schema master (forest operations scope)
- Domain naming master (forest operations scope)
- RID master (domain operations scope)
- Infrastructure master (domain operations scope)
- PDC emulator master (domain operations scope)

Not all FSMO roles are created equal – some should be available for daily operations, while others are not required all the time. They each have different functions they perform, and some roles being available are more critical than others – that is, the schema master role could be offline. Only when you want to make schema changes would this impact you.

In this section, we looked at the function of AD DS operations master roles. Now, let's look at AD DS sites and subnets.

Sites and subnets

A **site** is an AD object representing a collection of IP subnets connected over low latency connections and supporting local **Remote Procedure Calls** (**RPCs**). You can think of a site as a physical location that maps to your network WAN topology.

A subnet object represents each subnet where a DC is located; a subnet object will be part of a site object and can only be created as a one-to-one mapping – that is, a subnet can only be part of a single site.

The site topology is shown in the following diagram:

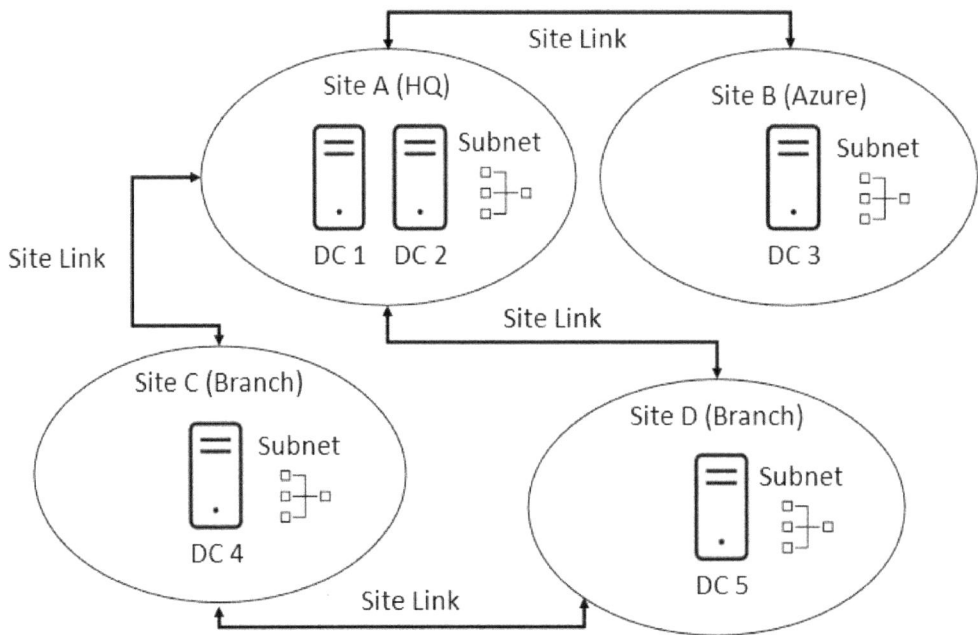

Figure 1.10 – Site topology

By default, there will be a single first site called *Default-First-Site-Name* (which can be renamed), containing all the subnets where a DC is located. However, in terms of network traffic for replication, user authentication, directory searches, and so on, this is often not an efficient implementation of sites and site design.

A site structure should map to the WAN topology of an organization, and you should create as many sites as needed to replicate this topology. This is to optimize replication so authentication can stay local to the site.

A DC determines its site by looking at the subnet object that matches the subnet that it is part of. After that, it uses the site associated with that subnet to determine the site it belongs to. This means it will control what site a DC belongs to; you must ensure that the DC is located in the subnet associated with that site or that the site contains that subnet.

Site links

Site links are the replication paths between sites; you determine which links the replication traffic goes over. When there is no DC in a client's local network, site links are used to direct client traffic to a site with a DC; you should consider the reasoning for not having a DC at each location to gain access to domain resources.

It is always recommended to put a DC close to clients who need access to a DC to authenticate resources, perform directory searches, and so on. You shouldn't have traffic pass over a WAN link but, instead, stay on the local network.

For this purpose, you should always create site design maps for your network topology. In a hybrid environment, this will often require you to go beyond the default and out-of-the-box setup of the single first site design; the default site link used to connect sites is called `DEFAULTIPSITELINK`.

This section concludes the first of the three skills areas of this chapter – *understanding the concepts and components of AD DS*. We looked at what AD is and looked at each component. In the next section, we will look at the second of this chapter's three skills areas: *creating AD DS*.

Creating Active Directory Domain Services

This section will cover some planning aspects and information about the components that will be created as part of a deployment. We will also cover deploying DCs on-premises and deploying Azure as **Infrastructure as a Service** (**IaaS**) resources and other considerations.

Planning for availability and performance

Before diving straight into deploying, let's cover some planning steps.

The **availability** of the AD DS database is provided by the built-in native capability of the **multi-master replication** model, where the data is replicated across multiple DCs; there is no master-slave.

Deploying at least two DCs is a recommended best practice to provide the necessary services and AD database availability through this native replication capability.

This can be a double-edged sword in that it has pros and cons; multiple DCs can be made available to process authentication and authorization for access to domain resources, especially in the case of many multiple concurrent requests, such as login storms to a virtual desktop infrastructure, or a **line of business** (**LoB**) app or service. However, the downside is that this data replication can cause network congestion when data is transferred across the network. We can address this through sites and links, which help control this replication traffic. With that said, to address most enterprises' availability and performance needs, you should have at least two DCs per geographical region. Here, each geographical region represents a site that improves performance and reduces replication and auth traffic over a WAN link to remote DCs; instead, this traffic can stay local to the region's network.

Planning for data protection

In this section, you will learn how to protect your AD DS data store, which holds your database. We will also look at how to restore AD DS.

When we talk about data protection in general terms, the industry focuses on backing up data, which is an important task to discuss. However, the focus should always be on the desired outcome rather than the action – that is, we must focus on restoring data rather than backing it up. We always hear about *backup strategies*, but I encourage you to consider your restore strategy.

With AD DS, two levels of data protection are provided and, more importantly, data recovery; these are the AD recycle bin and being able to restore from a traditional backup copy of the data (this must have the included system state). Let's look at each of those methods in more detail.

Recycle bin restore

The **recycle bin** is the first level of object restore and allows you to quickly and efficiently restore objects that may have been accidentally deleted. It only allows you to restore any deleted objects in the directory. It can't roll back any changes made to an object by recovering to an earlier time; this can only be achieved with a traditional data backup method.

The recycle bin's primary benefit is its **Recovery Time Objective** (**RTO**) – the time it takes to bring a deleted object back into operation. This can be achieved by requiring no downtime on the server, which would be required from the more traditional restore from the backup data approach.

For simplicity and speed, which are needed in a data recovery crisis, this restore can be performed from within the UI of the server itself. No backup software or hardware components need to be installed, and no license needs to be configured or paid for. However, the feature must be enabled, so it is important to do this before putting any DC into operation.

Once enabled, you will see it referenced as the *Deleted Objects* container in the **AD Administrative Center** (**ADAC**). Objects have a lifetime of 180 days by default, though this can be changed. The original location or alternate location can be set as the restore location.

Backup copy restore

The traditional **Windows Server backup** method is the next level of object restore; however, it has a much longer RTO due to the temporary offline nature of the traditional restore method when using a backup copy of the data to restore from. The system state must be explicitly included for the backups jobs to restore AD DS using the traditional Windows Server backup method.

The `NTDSUtil`, `replmon`, and `repadmin` command-line tools can also be used in the data protection operations to run validations, restore, seize FSMO roles, and so on.

Windows Server has a safe mode boot for DCs; this is called **Directory Services Restore Mode** (**DSRM**), allowing you to restore the directory database. Two types of restoration can be carried out: an authoritative restore and a non-authoritative restore. Let's briefly look at each.

Authoritative restore

This type of restore replaces objects in the directory with a copy of the objects stored in the backup.

An **authoritative restore** means you wish to set the restored objects so that they persist in the directory and replace all the other copies of the objects through replication. That is, all the other DC copies will be updated and made consistent with the copy of the database stored in the restored DC.

Non-authoritative restore

This type of restore also replaces objects in the directory with a copy of the objects stored in the backup. However, the objects are not persisted, and the DC requests a pull replication from other DCs to ensure it has the latest copies of the database, including any changes that were made since the backup occurred; the restored objects are not restored to other DCs.

In this section, we looked at AD DS data protection. Now, let's learn how to deploy AD DS DCs.

Deploying domain controllers

In this first section of this top-level topic, we will collectively look at deploying AD DS DCs in several scenarios, including deploying to on-premises platforms and deploying to the Microsoft Azure public cloud platform using IaaS **virtual machines** (**VMs**). However, note that deploying domain controllers for Azure AD DS won't be covered here; this will be covered in *Chapter 2, Implementing and Managing Azure Active Directory Domain Services*.

Deploying a DC on Windows Server using Server Manager GUI

A DC is a server with the AD DS role installed on it. This can be installed and configured in many ways, such as via *PowerShell*, *Server Manager*, or *WAC*. As we continue our hybrid journey, we will see that WAC, in many cases, will often be the preferred and most optimal deployment and management tool for many hybrid scenarios.

A critical part of AD DS is **DNS**; AD DS requires a DNS server. The DNS server role is typically installed on the DC as part of the deployment configuration and will use AD-integrated DNS for a seamless experience.

Regarding availability, to ensure users can still log on, servers and other resources can authenticate to a DC if a DC can't be reached. A minimum of two should be deployed for each domain in the forest. DCs should be placed as close to where the resources that require authentication are located, and ideally, in each physical location that requires local authentication to take place to reduce authentication traffic that would have to occur over a **Wide Area Network** (**WAN**) connection. We covered this in more detail in the *Sites* and *Site links* sections.

When using Server Manager to install the AD DS role, the process will start by installing some AD DS components and tools on the server. Once completed, you will be prompted to complete the post-deployment configuration, which will step you through promoting this server to a DC.

The configuration wizard will prompt you with several deployment configuration selections; we will look at these in this section so that you understand these when it comes to the hands-on exercise.

For the deployment configuration, you can select **Add a domain controller to an existing domain**, **Add a new domain to an existing forest**, or **Add a new forest**. If this is a new deployment of AD DS rather than deploying DCs to an existing AD DS instance, then you need to use the third option – **add a new forest**.

When you implement AD DS and deploy a domain and promote a server to be a DC that won't be deployed into an existing AD forest, then, by default, a new forest will be created as the top-level container in the topology or hierarchy for the new AD DS instance. The first domain that's created will be created as the *forest root domain*:

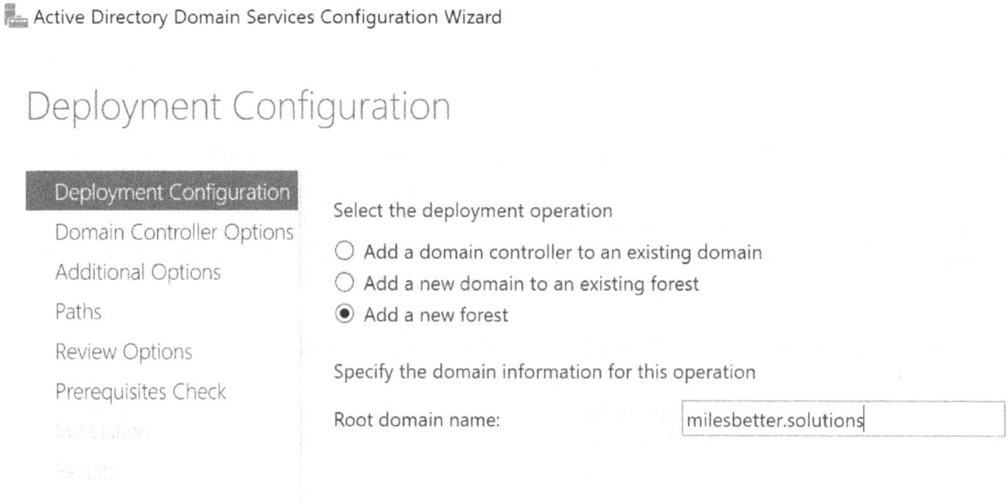

Active Directory Domain Services Configuration Wizard

Deployment Configuration

Deployment Configuration
Domain Controller Options
Additional Options
Paths
Review Options
Prerequisites Check

Select the deployment operation

○ Add a domain controller to an existing domain
○ Add a new domain to an existing forest
◉ Add a new forest

Specify the domain information for this operation

Root domain name: milesbetter.solutions

Figure 1.11 – Deployment operation options

Here, you specify a domain name for the root domain name; all subsequent child domains created within the forest will share the same contiguous DNS namespace. To define a domain with a new DNS namespace within the same forest, you would need to use the **Add a new domain to an existing forest** option, creating a new domain tree. This is shown in the following diagram:

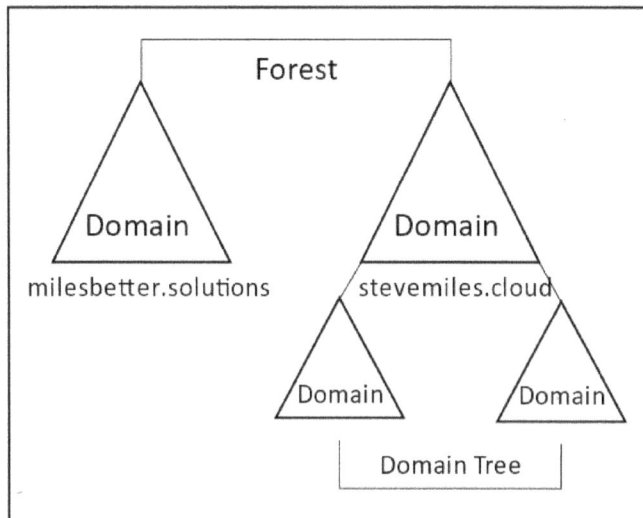

Figure 1.12 – Domain tree

Once the deployment configuration step is complete, you will be presented with the **Domain Controller Options** screen. From here, you can set the forest's functional level. Note that this is where you get to set the DC's capabilities.

Note that at the bottom of this screen, there is a hyperlink for **More about domain controller options**. This will take you to the Microsoft documentation. If you haven't done so already, it is recommended that you follow this link and study the information in the documentation before proceeding so that you are aware of all the information we do not have space to cover in this book:

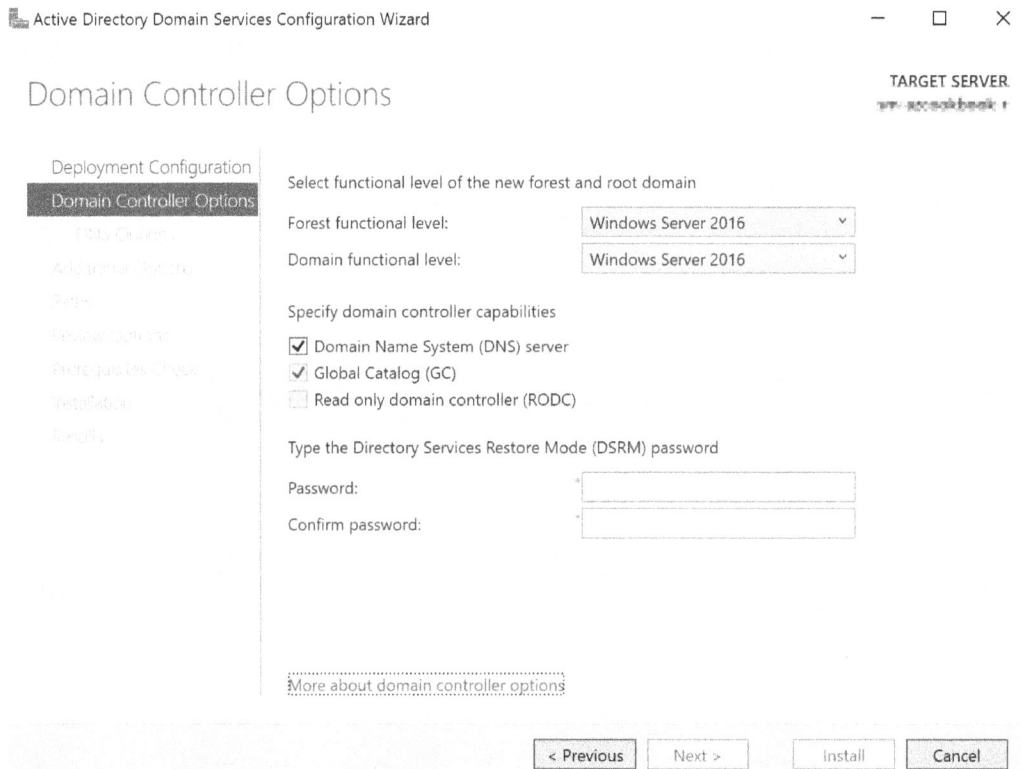

Figure 1.13 – Domain controller options

As shown in the preceding screenshot, you can choose whether the DC should be a DNS server. Note that the **Global Catalog (GC)** option is automatically assigned as the first DC in a new domain and that the **Read-only domain controller (RODC)** option is grayed out; again, this is because this is the first DC in the domain. When you choose the **Add a domain controller to an existing domain** option on the previous **Deployment Configuration** screen, these options can be selected; we will look at installing an RODC later.

On the following screens, you have the option to create **DNS delegation records** and verify and change the generated **NetBIOS name** if necessary.

Under the **Paths** section of the configuration wizard, you can specify different locations for the *AD DS database*, *log files*, and *SYSVOL*:

Active Directory Domain Services Configuration Wizard   — □ ×

Paths

Deployment Configuration

Domain Controller Options

DNS Options

Additional Options

Paths

Review Options

Prerequisites Check

Installation

Results

Specify the location of the AD DS database, log files, and SYSVOL

Database folder:	C:\Windows\NTDS
Log files folder:	C:\Windows\NTDS
SYSVOL folder:	C:\Windows\SYSVOL

More about Active Directory paths

< Previous   Next >   Install   Cancel

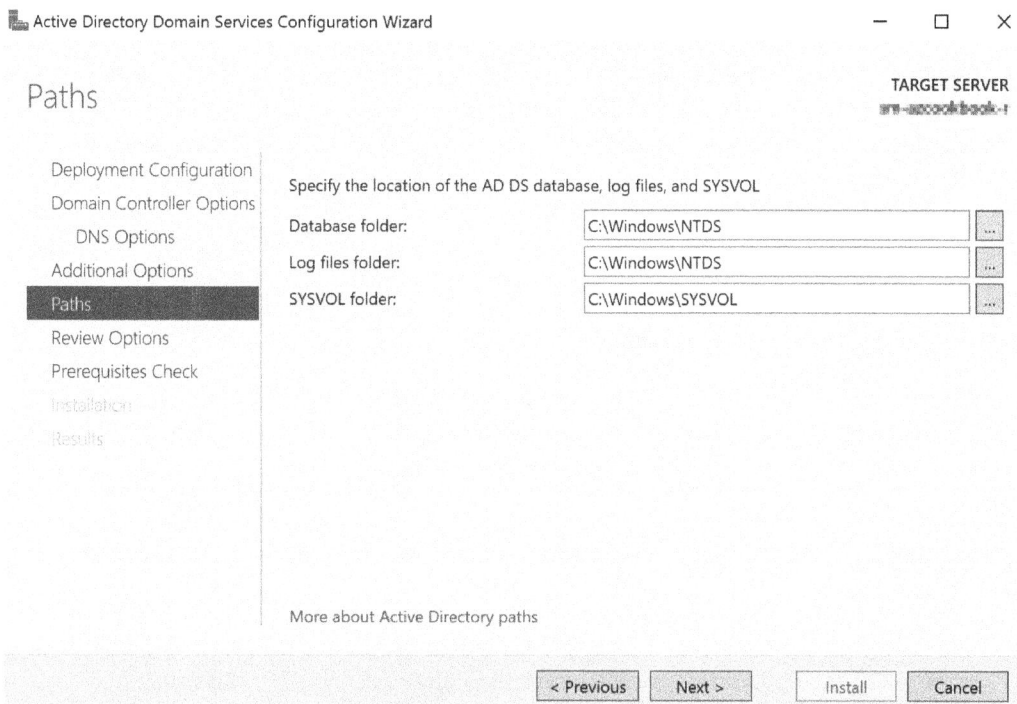

Figure 1.14 – Install paths

You should specify the location that meets your needs; the component's paths can impact the performance of AD DS. This is more critical when installing AD DS on IaaS VM DCs, which we will look at later in this chapter, due to disk caching.

Deploying a DC on Windows Server Core

There is no GUI on Windows Server Core. So, to install AD DS, you need an alternative method such as Windows PowerShell with the AD module; alternatively, you can use the **Remote Server Administration Tools** (**RSAT**), Windows Admin Center, Server Manager, and PowerShell remoting tools from a remote machine that can connect to the server's core machine.

Deploying a DC from media

This is only for deploying additional DCs to an existing AD DS environment. This deployment approach creates a backup of AD DS and then transfers it to removable media such as a USB drive. You start by deploying AD DS on the server that will be the DC via the Server Manager GUI and select the **Install from media** option. This is very similar to a non-authoritative recovery – that is, the AD DS database and its objects will only be as current as the date when the backup was taken. So, to get the latest changes to the directory since the backup, pull replication must occur. This will ensure that the new DC is consistent with all the other DCs and has the same copy of the directory objects.

Deploying a DC in Azure as IaaS

As we learned earlier in this chapter, as well as deploying DCs to on-premises platforms, we can also deploy DCs to the Microsoft Azure public cloud platform as IaaS VMs. However, note that deploying AD DS is not covered here and will be covered in *Chapter 2, Implementing and Managing Azure Active Directory Domain Services*.

When deploying a DC in Azure or the AD DS role in an IaaS VM, there are some considerations you need to be aware of:

- **Networks**: To add DCs to an existing AD DS environment, you should consider network connectivity. Here, you must have a line of sight to those existing DCs. This will require your networks that have DCs placed on them to be extended into Azure. This is typically done over a VPN or, in some scenarios, ExpressRoute.

- **Sites/Subnets**: These should be created to reflect the address space(s) defined in your Azure VNet.

- **DNS**: Azure AD DNS cannot be used for AD DS; you should use the DNS role within Windows Server or Azure private DNS zones.

- **VMs**: Due to the burstable CPU capability in the B Series VMs, they are an excellent fit to be used for DCs.

- **Disks**: Since implementing AD DS will install a database on the VM; we should follow some guidance on that scenario – that is, we shouldn't install on any disks that have caching enabled – especially write caching. To meet this guidance, you should not place the `NTDS.DIT` and `SYSVOL` files in the default path offered to you during the configuration wizard; instead, you should attach a data disk to the VM (which has to write caching off by default) and change the path to store these files in the data drive.

Deploying an RODC

To deploy an RODC, we can use a pre-staged account to perform the AD DS install tasks. This will ensure that a user does not need to be a member of the domain's admin group or enterprise admins group.

A new site and subnet should be created to control this traffic replication to the RODCs. The replication schedule can be set on the default site link.

Upgrading the OS of an existing DC

The underlying OS of a DC from Windows Server 2012 R2 to Windows Server 2022 can be upgraded as if it did not have the AD DS role installed. There are no differences, known issues, or limitations placed on it because it performs as a DC.

This section concludes the second of this chapter's three skills areas: *creating AD DS*, where we looked at planning and deployment information. In the next section, we will look at the last of this chapter's three skills areas: *managing AD DS*.

Managing Active Directory Domain Services

This section will introduce managing AD DS and the tools that are used. We will look at **Active Directory Administrative Center**, **Remote Server Administration Tools**, **Windows Admin Center**, and **PowerShell**, along with the **AD module** and other additional management tools.

Active Directory Administrative Center (**ADAC**) is a PowerShell-based GUI available in Windows Server (not in Windows Server Core).

The following tasks can be carried out with this tool:

- Manage multiple domains through a single tool instance
- Search the directory for objects
- Create and manage directory objects, such as users, groups, computers, and OUs
- Manage Dynamic Access Control
- Create and manage fine-grained password policies
- AD recycle bin operations

This tool replaces the functionality previously provided through the **Microsoft Management Console** (**MMC**) snap-in tool known as **Active Directory Users and Computers**.

Further information about ADAC can be found at `https://docs.microsoft.com/en-us/windows-server/identity/ad-ds/get-started/adac/active-directory-administrative-center`.

Now, let's look at the management tools that can be used for the Azure AD DS managed domain.

RSAT

RSAT allows you to manage servers remotely via a GUI; a set of AD DS tools is included. This was the primary tool console until the introduction of WAC, which we will look at in the next section.

The consoles for these tools are available on Windows 10/11 and Windows Server. With Windows 10/11, these tools are now included within the OS rather than a separate download, which was added through the **Optional features** setting.

Further information about RSAT can be found at `https://docs.microsoft.com/en-us/troubleshoot/windows-server/system-management-components/remote-server-administration-tools`.

WAC

This browser-based admin tool can be downloaded and installed locally on Windows 10/11 and Windows Server. It can also be accessed directly via the Azure portal, so no download or local install is required, much like CloudShell has to install PowerShell locally.

For a local install of WAC, you must ensure your network allows the required ports; the default is port `6516` for standalone mode in Windows 10. The gateway mode in Windows Server is TCP `443`. Both can be changed.

Further information about WAC can be found at `https://docs.microsoft.com/en-us/windows-server/manage/windows-admin-center/overview`.

PowerShell with the AD module

This is an alternative to using a GUI to manage AD DS. You can use PowerShell commands via an AD module that provides a collection of cmdlets.

If you wish to use the module on a local install of PowerShell on a client/desktop OS such as Windows 10/11, then the module is part of RSAT, which you will need to download and install.

Further information about the AD module can be found at `https://docs.microsoft.com/en-us/powershell/module/activedirectory/?view=windowsserver2022-ps`.

MMC snap-in tools

MMC is a GUI console that contains a collection of tools called snap-ins. The following snap-in tools are available for managing AD DS, most of which are self-explanatory:

- **Active Directory Users and Computers** allows you to carry out everyday tasks to manage objects such as users, groups, and computers; this is replaced by ADAC and provides additional capabilities
- **Active Directory Sites and Services** allows you to create and manage sites, subnets, replication, and associated services
- **Active Directory Domains and Trusts** allow you to create and manage domain and forest trusts
- **Active Directory Schema snap-in** allows you to view and modify the schema

Further information about MMC can be found at `https://docs.microsoft.com/en-us/troubleshoot/windows-server/system-management-components/what-is-microsoft-management-console`.

This section looked at a variety of AD DS management tools. In the next section, we will look at some of AD DS's monitoring and troubleshooting tools.

Monitoring and troubleshooting tools

In this section, we will look at some of AD DS's monitoring and troubleshooting tools.

Performance monitoring tools

Windows Server contains the following built-in native tools for monitoring performance and analyzing service operations:

- Performance monitor – **Directory Replication Agent (DRA)** counters

- Resource Monitor

- Task Manager

- Event Viewer

These tools can help you analyze and identify any overutilization and depletion of these system resources. They will help you find the root cause and the source of any system performance issues caused by a bottleneck. A system can only suffer from one bottleneck at a time; this could lie in the CPU, memory, disk, or networking. You should address each in turn and then move on to the next.

Repadmin

This tool helps you view the service's health and diagnose replication problems between DCs. It allows you to view the replication topology, manually create a replication topology, and force replication. It is available when the AD DS role is installed on a server and is also included as part of the AD DS tools in the RSAT tools.

Further information and syntax about Repadmin can be found at `https://docs.microsoft.com/en-us/previous-versions/windows/it-pro/windows-server-2012-r2-and-2012/cc770963(v=ws.11)`.

dcdiag

This tool will analyze the state of the health of AD DS DCs. It is available when the AD DS role is installed on a server and is also included as part of the AD DS tools in the RSAT tools.

Further information and syntax about dcdiag can be found at `https://docs.microsoft.com/en-us/previous-versions/windows/it-pro/windows-server-2012-r2-and-2012/cc731968(v=ws.11)`.

netdom

This tool allows you to manage AD DS trusts; it can also join a computer to a domain, manage computer accounts, query for domain information such as which DCs hold the FSMO roles, and more. It is available when the AD DS role is installed on a server and is also included as part of the AD DS tools in the RSAT tools.

Further information and syntax about netdom can be found at `https://docs.microsoft.com/en-us/previous-versions/windows/it-pro/windows-server-2012-r2-and-2012/cc772217(v=ws.11)`.

In this section, we looked at some of AD DS's monitoring and troubleshooting tools. In the next section, we will complete a hands-on exercise to reinforce some of the concepts covered in this chapter.

Hands-on exercise

To support your learning with practical skills, let's learn how to create some of the services we looked at in this chapter. You will learn how to install AD DS on Windows Server.

Getting started

To start this hands-on exercise, you will need access to a physical or virtual machine running *Windows Server 2012 Standard/Datacenter* or later.

For this exercise, we could use a nested virtualization environment in Azure, meaning no on-premises hardware is required. However, to help you with your learning and demo purposes, we will complete this exercise using IaaS VMs in an Azure environment where we have the correct level of access to create the required resources.

You can create a free Azure account at `https://azure.microsoft.com/free`. This free Azure account provides the following:

- 12 months of free services
- $200 credit to explore Azure for 30 days
- 25+ services that are always free

If you will be using Azure IaaS VMs for DCs, then recommended practice is that each VM should have a data disk attached to store the AD DS database, log files, and SYSVOL. Alternatively, you could install them on the default paths provided for learning purposes. However, this should not be done in a production scenario.

Let's move on to the exercise.

Exercise – installing AD DS on Windows Server

This section will teach you how to install AD DS on Windows Server.

The following steps must be carried out on the local OS of a machine you have admin access to. We will install the AD DS role directly on the server we wish to be our *first DC* in our *new domain*, in a *new forest*; we will not use remoting.

The Server Manager *Add Roles wizard* and the *AD DS Configuration wizard* are used to install and configure AD DS.

Follow these steps:

> **Note**
>
> The dcpromo.exe AD DS Installation Wizard has been deprecated as a deployment method starting with Windows Server 2012.

1. Log in to your server. Then, from **Manage** in **Server Manager**, click **Add Roles and Features**:

Figure 1.15 – Server Manager

2. On the **Before you begin** page, click **Next**.

3. On the **Select installation type** page, leave **Role-based or feature-based installation** set and click **Next**.

4. On the **Select a destination server** page, leave **Select a server from the server pool** set and ensure the server where you want to install AD DS is selected. Then, click **Next**.

5. From the list of available roles on the **Select server roles** page, select the box for the **Active Directory Domain Services** role:

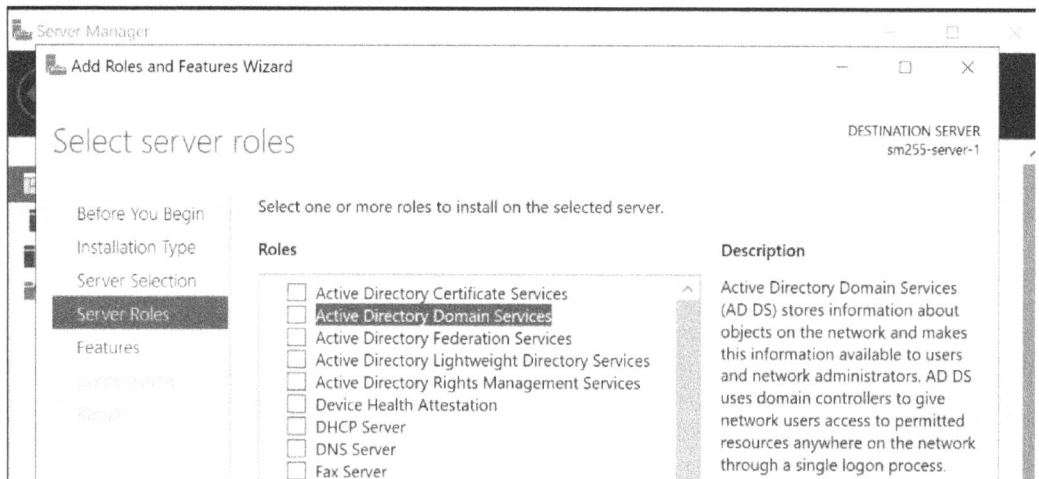

Figure 1.16 – The Select server roles screen

6.   From the **Add Roles and Features Wizard** pop-up screen, click **Add Features**:

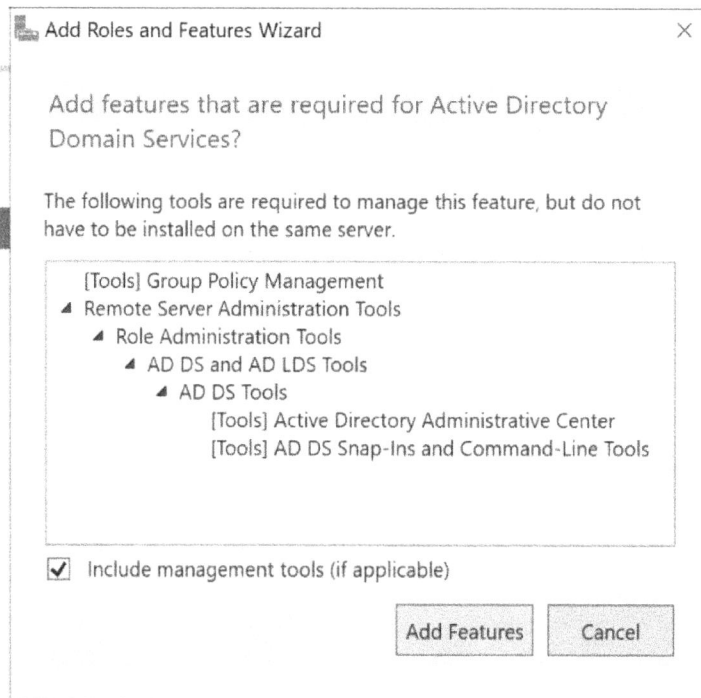

Figure 1.17 – The Add Roles and Feature Wizard pop-up screen

7. Back on the **Select server roles** page, click **Next**.

8. On the **Features** page, leave all the defaults as-is, review the features selected as a reference, and click **Next**.

9. On the **AD DS** page, review the information and click **Next**.

10. If required, select **Restart the destination server automatically** on the Confirmation page. Then, review the selections installed as a reference and click **Install**.

11. On the **Results** page, observe and monitor the installation progress; click **Close** when you see a message stating the **Installation succeeded on [YourServerName]**:

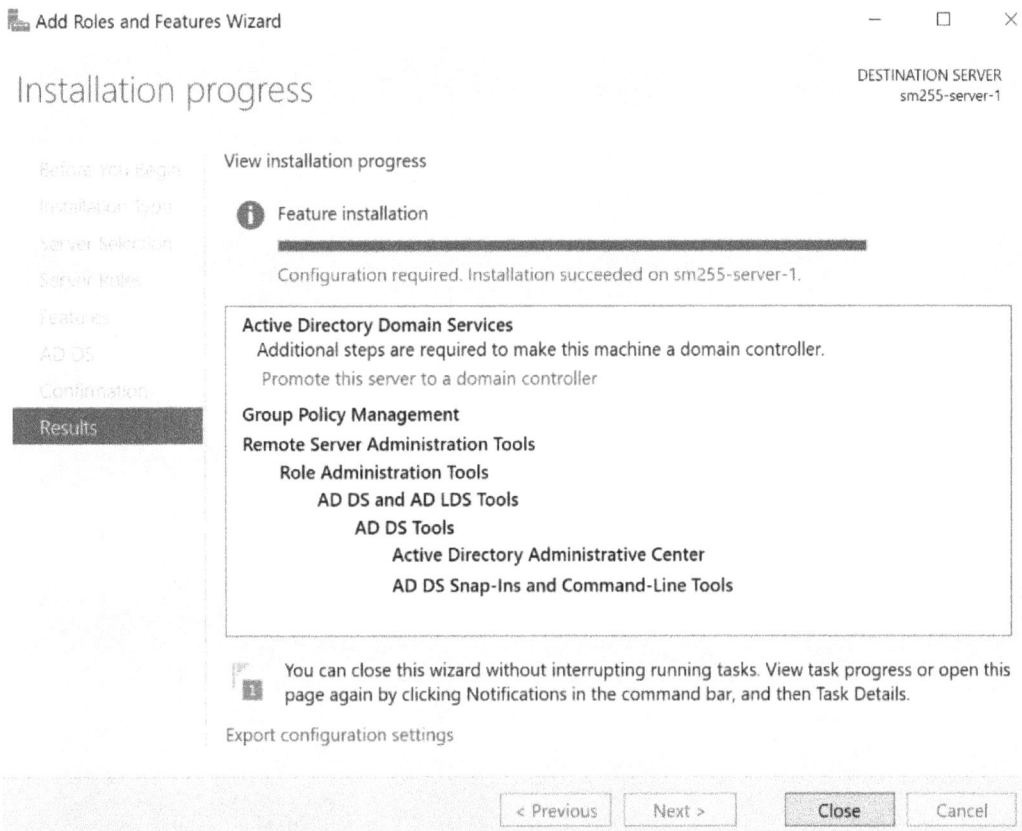

Figure 1.18 – The Installation progress screen

12. From **Server Manager**, click on **Notifications**, then **Promote this server to a domain controller**:

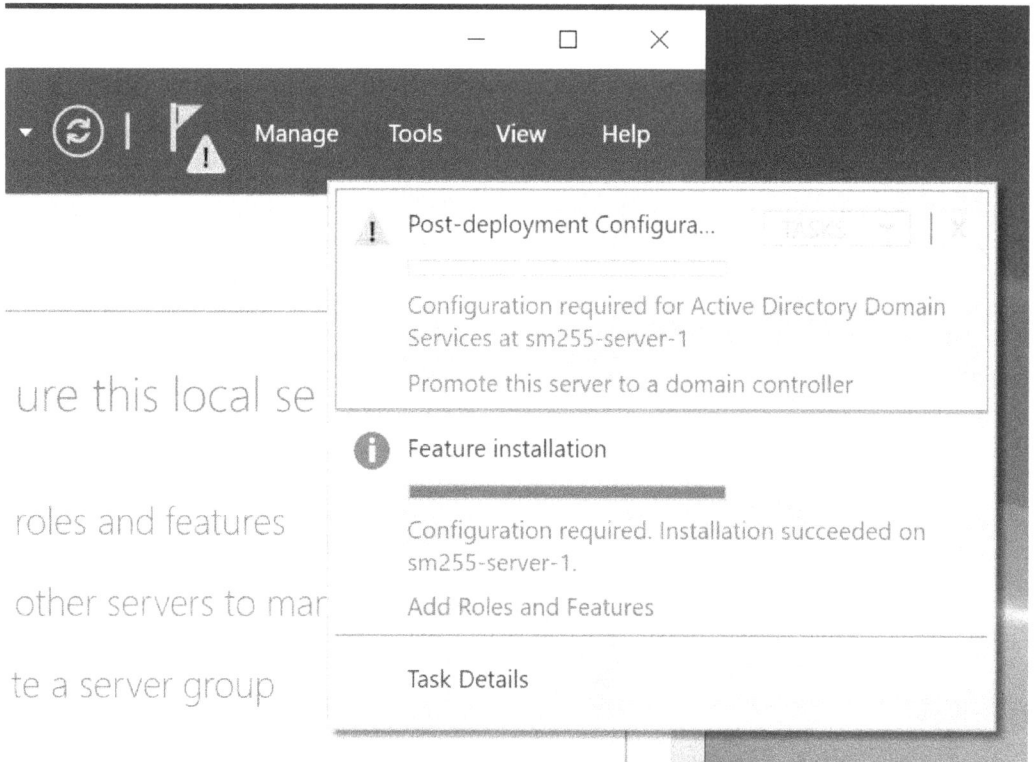

Figure 1.19 – The Promote this server to a domain controller notification screen

13. On the **AD DS Configuration Wizard** pop-up screen, note the deployment operation options on the **Deployment Configuration** screen. Select **Add a new forest** for this exercise:

Figure 1.20 – The Deployment Configuration screen

14. Enter a domain name under the **Root domain name** and click **Next**:

For further reference, click the **More about deployment configurations** hyperlink at the bottom of the page:

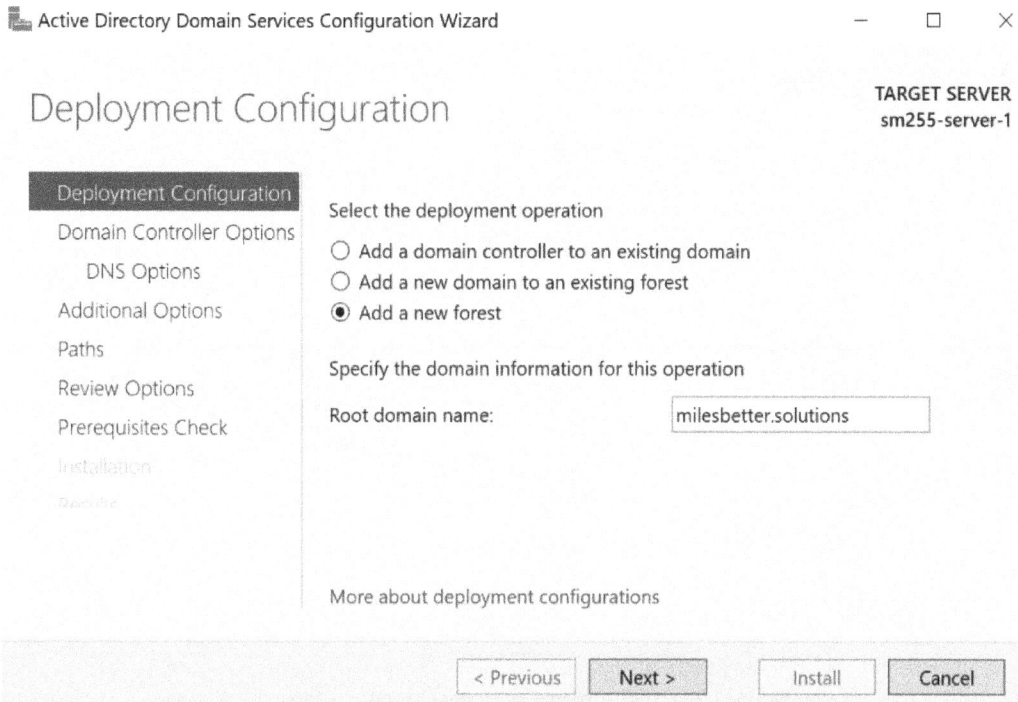

Figure 1.21 – The Deployment Configuration screen

15. On the **Domain Controller Options** page, note that the **Domain Name System (DNS) server** and **Global Catalog (GC)** options are selected by default but that the **Read only domain controller (RODC)** option is not available; leave all the defaults as-is and enter the DSRM password and confirm. Then, click **Install**.

For further reference, click the **More about domain controller options** hyperlink at the bottom of the page:

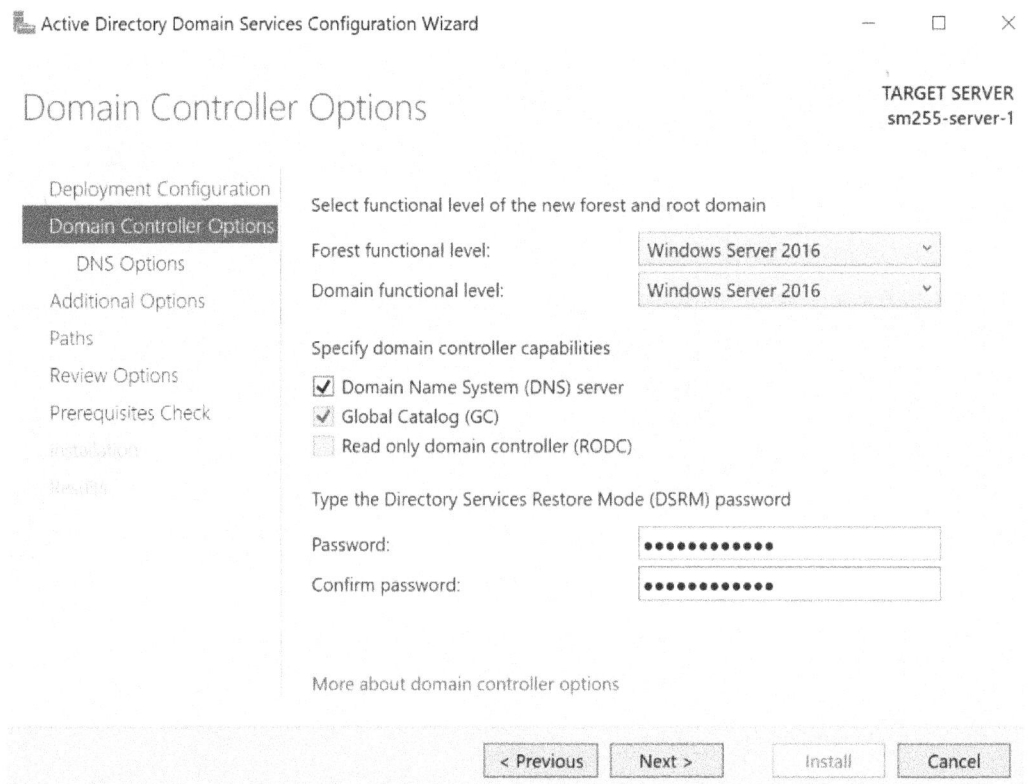

Figure 1.22 – The Domain Controller Options screen

16. On the **DNS Options** page, ignore the message that states that delegation for this DNS server cannot be created and click **Next**.

17. Wait while the NETBIOS domain name is auto-populated on the Additional Options page. Then, click **Next**:

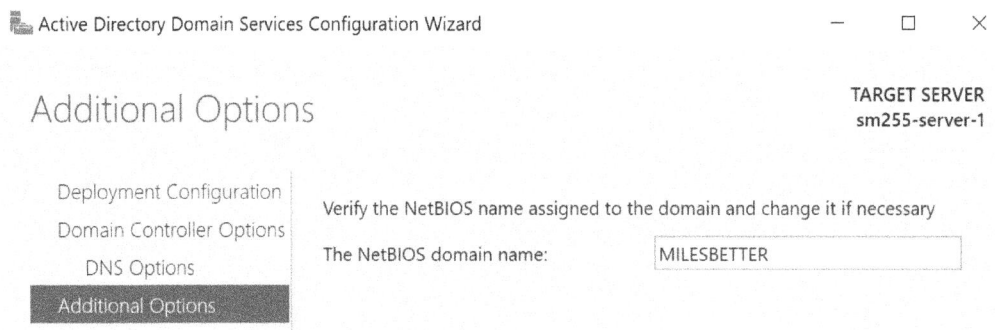

Figure 1.23 – The Additional Options screen

18. From the **Paths** page, specify the location for the AD DS database, log files, and SYSVOL. For this exercise, leave the defaults as-is. Then, click **Next**.

For further reference, click the **More about Active Directory paths** hyperlink at the bottom of the page:

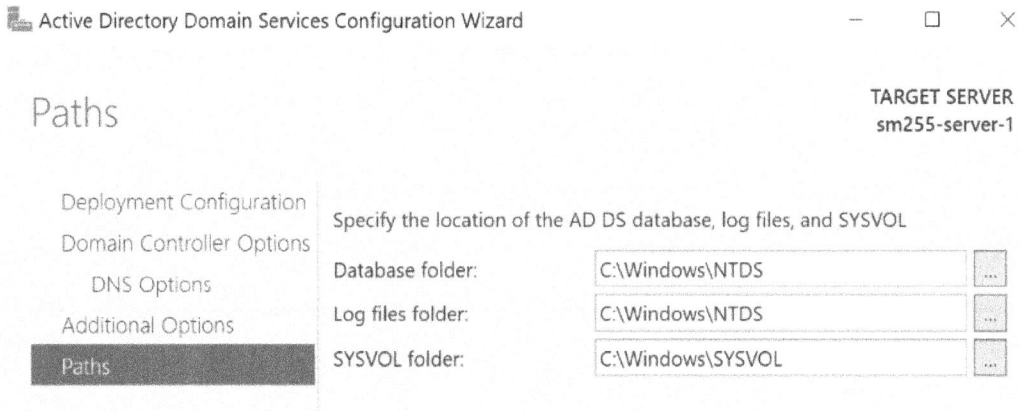

Active Directory Domain Services Configuration Wizard   —   □   ✕

Paths

TARGET SERVER
sm255-server-1

Deployment Configuration
Domain Controller Options
DNS Options
Additional Options
Paths

Specify the location of the AD DS database, log files, and SYSVOL

Database folder:	C:\Windows\NTDS
Log files folder:	C:\Windows\NTDS
SYSVOL folder:	C:\Windows\SYSVOL

Figure 1.24 – The Paths screen

19. From the **Review Options** page, review the selections and click **Next**.

20. The **Prerequisites Check** page confirms the status of all prerequisite checks. A green tick should appear with a message stating **All prerequisite checks passed successfully. Click 'Install' to begin the installation**. Then, click **Install**:

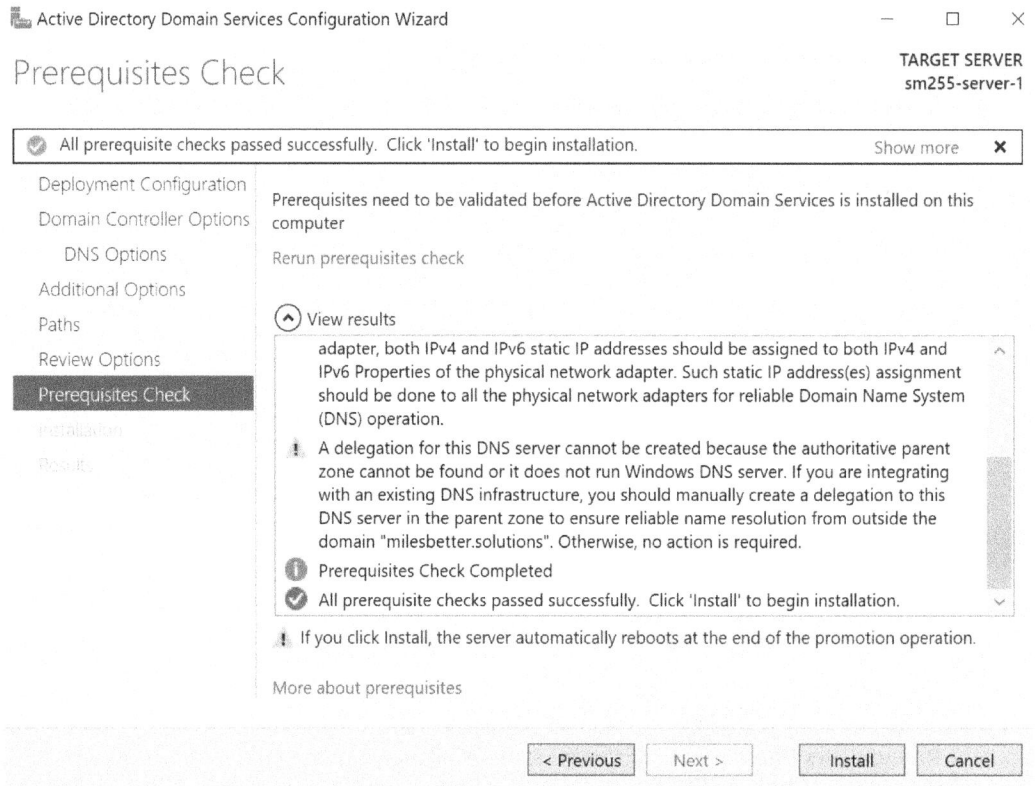

Figure 1.25 – The Prerequisites Check screen

21. From the **Installation** page, observe and monitor the installation progress; your server will automatically restart:

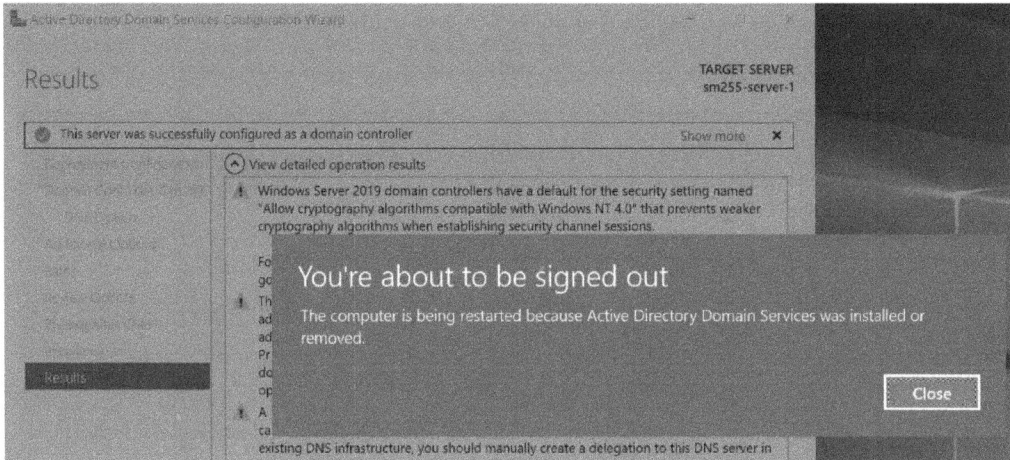

Figure 1.26 – Installation progress screen

22. From the login screen of the server, you will need to use the *domain account* for your user, *not* the local account; this will be in UPN format – that is, `UserName@domain.name`:

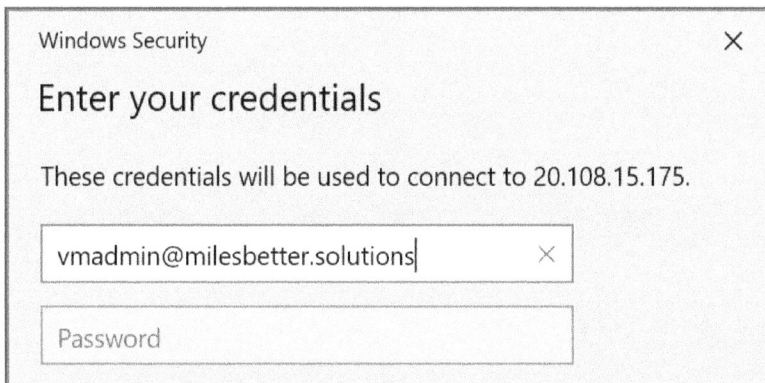

Figure 1.27 – Server login screen

23. From **Server Manager**, you will see that the AD DS role has been installed, as well as the DNS:

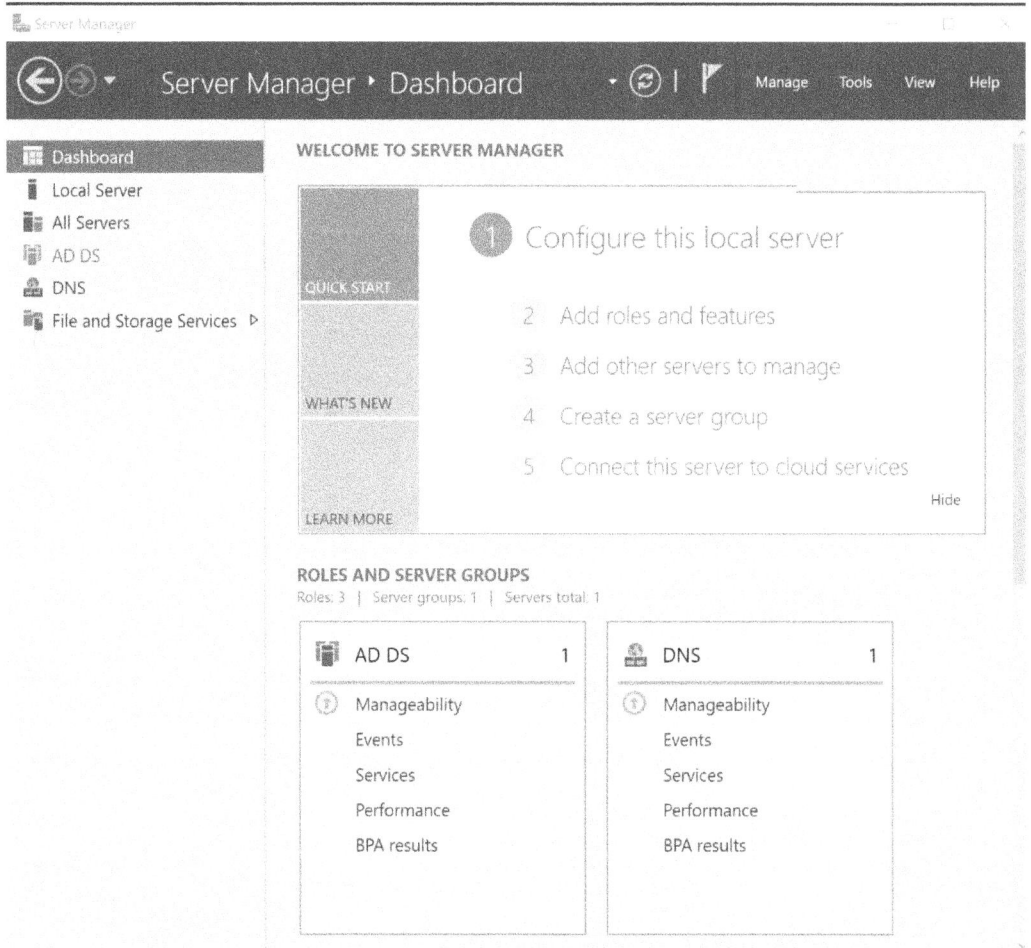

Figure 1.28 – Server Manager

24. From **Server Manager**, click **Tools**, then **Active Directory Administrative Center**:

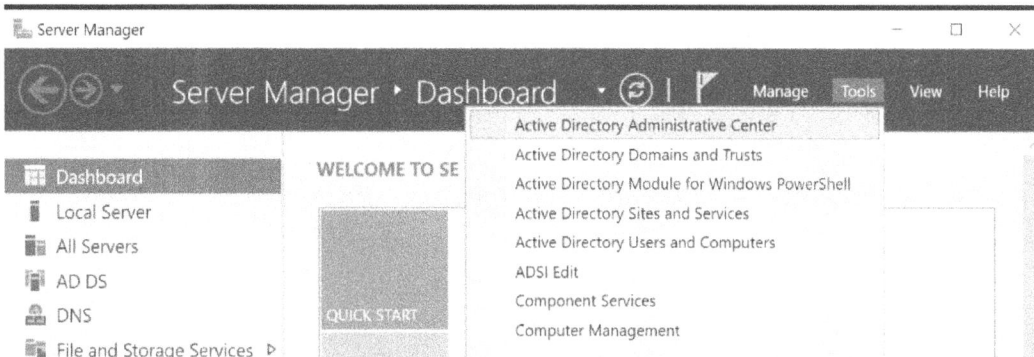

Figure 1.29 – AD Administrative Center

Congratulations! You have completed this exercise and installed AD DS on Windows Server.

In this exercise, you installed AD DS on Windows Server and accessed it via the ADAC. This helped you reinforce this chapter's theory, along with some practical skills.

Now, let's summarize this chapter.

Summary

This chapter has provided coverage for the AZ-800 Administering Windows Server Hybrid Core Infrastructure exam learning objective: *Deploy and manage AD DS in on-premises and cloud environments*.

We learned about AD concepts, its services, such as AD DS and its components. We looked at the creation and management of AD DS instances in on-premises environments. A hands-on exercise then finished the chapter to provide you with additional practical skills.

This chapter aimed to take your knowledge beyond the exam objectives; we added new skills and learning with the content provided. This further develops your knowledge and skills for on-premises network infrastructure services and enables you to be prepared for a real-world, day-to-day hybrid environment-focused role.

In the next chapter, you will learn about some of the concepts and components of Azure AD DS and learn how to create and manage an Azure AD.

Further reading

This section provides links to additional exam information and study references:

- *Microsoft Certified: Windows Server Hybrid Administrator Associate*: `https://docs.microsoft.com/en-us/learn/certifications/windows-server-hybrid-administrator/`

- *Exam AZ-800: Administering Windows Server Hybrid Core Infrastructure*: `https://docs.microsoft.com/en-us/learn/certifications/exams/az-800`

- *Exam AZ-800: skills outline*: `https://query.prod.cms.rt.microsoft.com/cms/api/am/binary/RWKI0r`

- *Microsoft Learn: Deploy and manage identity infrastructure*: `https://docs.microsoft.com/en-us/learn/paths/deploy-manage-identity-infrastructure/`

Skills check

Challenge yourself regarding what you have learned in this chapter:

1. What is Active Directory?
2. List three services that form part of Active Directory.
3. What is Active Directory Domain Services?
4. Explain the difference between AD and Azure AD.
5. What is Hybrid Identity?
6. What is the difference between Authentication and Authorization?
7. List seven logical components of AD DS.
8. List six physical components of AD DS.
9. What are AD DS objects?
10. What is an OU?
11. What is a service object, and how does it differ from a user object?
12. What is the difference between Standard and Group Managed Service Accounts?
13. What is a UPN?
14. Explain the group types and scopes.
15. Explain domains and forests.
16. List the trust types.
17. Explain trust directions.

18. What is a domain controller?

19. What are the five FSMO roles?

20. How can replication traffic be minimized?

21. List three considerations for deploying AD DS in Azure with IaaS VMs.

22. What are the two levels of AD DS data protection?

23. Name five AD DS management tools.

24. What is the `dcdiag` tool used for?

25. What is the `netdom` tool used for?

2

Implementing and Managing Azure Active Directory Domain Services

This chapter covers the AZ-800 Administering Windows Server Hybrid Core Infrastructure exam learning objective: *Deploy and manage AD DS in on-premises and cloud environments.*

In the previous chapter, we added skills to understand AD DS's concepts, creation, and management in on-premises environments.

We will start this chapter with some concepts and an understanding of Azure AD and then how it compares to Active Directory. We will then look at creating and managing an Azure AD DS managed domain. We will then conclude with a hands-on exercise to develop your skills further.

The following topics will be covered in this chapter:

- What is Azure AD?
- What is Azure AD DS?
- Creating an Azure AD DS managed domain
- Managing an Azure AD DS managed domain
- Exercise – Installing Azure AD DS

Before we dive into any service creation or management, we will look at some definitions and concepts to set a baseline and foundation of knowledge to build upon. We start this chapter by defining Azure AD.

What is Azure AD?

When is AD not AD? When it is Azure AD!

We will take a brief moment to introduce the tale of two directories.

Azure AD is often shrouded in the misconception that it is purely the cloud equivalent of the traditional Windows Server-based AD; in some respects, it's quite close. In others, it's not, so it is essential to understand that this is not the case, and we should not think of this as AD in the cloud, as it were; if only it were that simple.

While both AD and Azure AD are **identity providers** (**IDPs**) and share AD in the name, they function very differently.

At the moment, at least at the time of publishing, Azure AD *cannot fully replace* the functionality and capabilities of traditional Windows Server AD implementations. However, this may not be a bad thing, and the solution you need may not require the full functionality and capabilities of what we can refer to as *traditional AD*.

Microsoft provides Azure AD as a fully managed IDP platform, provided as **software as a service** (**SaaS**).

AD can be connected to Azure AD using *Azure AD Connect*, a free download tool to allow an organization to establish a hybrid identity. This tool synchronizes user identities, attributes, and objects between both IDPs.

Unlike AD, there is nothing to install for Azure AD, and no DCs are required while still allowing devices, users, and groups to appear as objects in the directory.

Some of the most common services offered by AD and Azure AD are shown in the following figure:

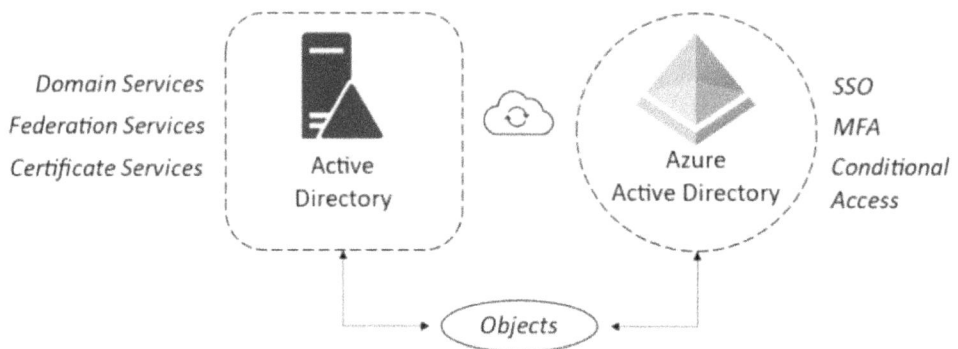

Figure 2.1 – AD and Azure AD

A *hybrid identity* approach allows users to access resources in Azure AD using their AD identity; the same username and password are used to access resources accessed in both *IDP environments*. Further reading can be found at this URL if you wish to continue learning more about hybrid identities and Azure AD Connect: `https://docs.microsoft.com/azure/active-directory/hybrid/whatis-azure-ad-connect`.

What is Azure AD DS?

Azure AD DS is a *managed Azure identity service* provided as a **Platform as a Service** (**PaaS**); in simple terms, it provides *Domain Services as a service*.

When you implement Azure AD DS as part of an Azure AD tenant, you create a *Microsoft managed domain*, which is a managed implementation of AD DS. This provides the functions of Kerberos/NTLM authentication, **lightweight directory access protocol** (**LDAP**), domain join, and group policy.

The use case for Azure AD DS is where workloads running in Azure depend on Domain Services functions and where apps or services cannot be modified or rewritten to utilize Azure AD and modern authentication, such as OAuth, SAML, and REST.

The following diagram outlines the question of how to provide the same on-premises domain functions for workloads once moved into Azure without having to provide an instance of AD DS that you manage:

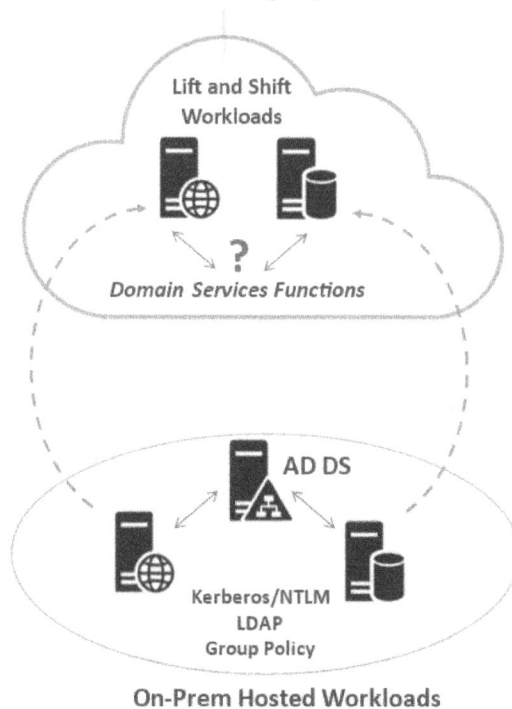

Figure 2.2 – Azure AD DS use case scenario

An Azure AD DS *managed domain* could be a solution for scenarios such as **Azure Virtual Desktop** (**AVD**), where you may have cloud-only identities but still depend on Domain Services not yet supported by Azure AD. This allows you to utilize the PaaS capabilities of Azure AD DS rather than self-managed DCs. Combined with Microsoft 365, this approach can provide a tightly integrated solution.

The advantage of providing AD DS DCs as a PaaS service includes not considering the physical infrastructure or logical components, such as data store, schema, trusts, and partitions, and AD DS's availability and protection.

The *managed domain* instance is created by Microsoft with two DCs for high availability, with additional region redundancy across two availability zones if available in that region. Microsoft will automatically configure the distribution of DCs across the zones.

The following are some characteristics of a managed domain:

- Microsoft-managed, not self-managed.

- A standalone domain–no possibility of extension or joining the managed domain to an existing AD DS forest.

- The trusts are one-way outbound forest trusts only.

- No domain or enterprise administrator privileges.

- Supports domain join, LDAP, and Kerberos/NTLM.

- Two built-in OU containers for computers and users are provided, with an associated built-in **Group Policy Object** (**GPO**).

- Supports custom OU structure and group policy via a flat structure.

- No schema extensions.

- DCs are provided as a service and managed by Microsoft.

- Azure AD DS is integrated into the Azure AD tenant for cloud-only identity scenarios and can be combined with AD DS for hybrid identity scenarios.

- Azure AD DS can be managed similarly to AD DS through the RSAT tools.

- Object sync is one-way, from Azure AD to Azure AD DS; any changes made directly to Azure AD DS will be overwritten at the next sync. These should be considered read-only DCs with no direct creation or changes to objects with DCs.

- Passwords cannot be reset in the managed domain; they must be reset in AD/Azure AD and synchronized into the managed domain.

You can review the complete list of differences between Azure AD DS and AD DS in the links provided in the *Further reading* section at the end of this chapter.

In the cloud-only identity scenario, users, passwords, and groups are created and managed in the Azure AD tenant, which is the authoritative directory service and IDP; these objects are synced to the Azure AD DS managed domain.

There are no direct changes to objects in the managed domain. Any changes should be made in the authoritative IDP, which is Azure AD in this scenario, and those changes must then be synchronized to the managed domain to become effective.

This is done in Azure AD if a user's password needs to be reset. This change is then synchronized to the managed domain DCs; enabling **self-service password reset** (**SSPR**) in the AD tenant is recommended.

The following figure shows the relationship between Azure AD and the Azure AD DS managed domain in the **cloud-only identity** scenario:

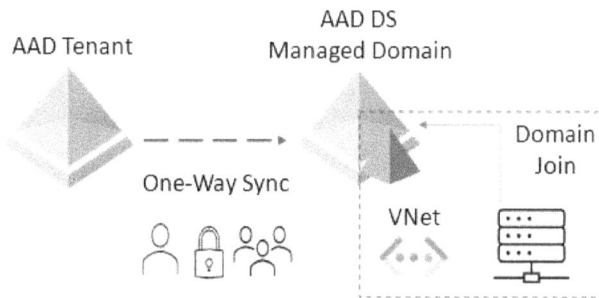

Figure 2.3 – Azure AD cloud-only identity scenario

If a user's password needs to be reset, this is reset in Azure AD. This change is then synchronized to the managed domain DCs; enabling SSPR in the AD tenant is recommended.

The following figure shows the relationship between AD, Azure AD, and the Azure AD DS managed domain in the **hybrid identity** scenario:

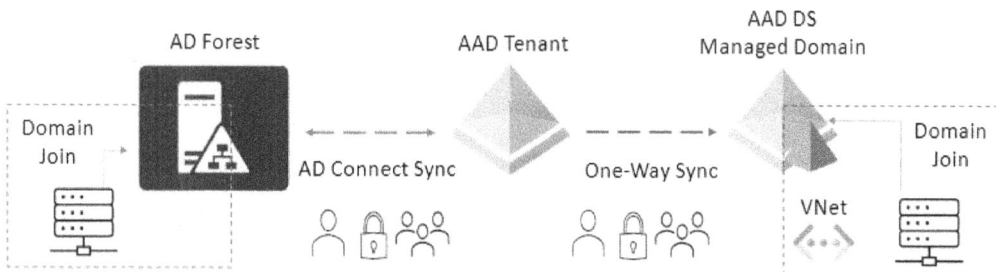

Figure 2.4 – Azure AD hybrid identity scenario

In the hybrid identity scenario, users, passwords, and groups are created and managed in the AD forest, which is the authoritative directory service and IDP.

These objects are then synchronized two-way into the Azure AD tenant using **AD Connect**, and then the objects are synchronized one-way to the Azure AD DS managed domain. There are no direct changes to objects to be made in the managed domain; any changes should be made in the authoritative IDP, which in this case is the AD forest. These changes must then be synchronized to the Azure AD tenant using AD Connect and then synchronized from the Azure AD tenant to the managed domain to become effective.

In summary, no changes will be made in the Azure AD DS managed domain in the previous cloud-only and hybrid identities scenarios. All objects are synchronized from the Azure AD tenant, so the only query is how the objects get created in Azure AD in the first instance. This will be done by creating objects directly in the Azure AD tenant in the case of cloud-only accounts or by allowing these objects to be created in the Azure AD tenant from the AD Connect sync from the AD domain controllers.

This section concludes the first of this chapter's three skills areas: *What is Azure AD DS?* We have looked at how it provides managed domain services. In the next section, we will look at the second of this chapter's three skills areas: *Creating Azure AD DS.*

Creating an Azure AD DS managed domain

This section will cover some planning aspects and information needed for creating Azure AD DS managed domains, and we will finish this chapter with a hands-on exercise section.

The installation of Azure AD DS is carried out in the Azure portal and requires an Azure AD tenant and an Azure subscription. The objects can be cloud-only or synchronized from AD for a hybrid identity scenario.

Two enterprise applications are created in the Azure AD tenant to support the operation of the domain; these should not be removed or edited and consist of the following:

- Domain Controller Services
- Azure AD Domain Controller Service

The managed domain instance is created by Microsoft, which will automatically configure the distribution of the managed DCs across the zones.

The following section will examine the planning aspects required before creating an Azure AD DS managed domain.

Azure AD tenant and subscription

To create Azure resources, you will need an **Azure subscription**. This will act as a logical container, access control, and billing mechanism for the resources created and consumed. You must associate an Azure AD tenant with this subscription; this tenant can be synchronized with AD for a hybrid identity scenario or a cloud-only identity approach.

The Azure AD tenant can only have one Azure AD DS managed domain. Ensure you plan adequately first, as once you have created the managed domain; you cannot change any of the following fields: *tenant, subscription, region, resource group, VNet,* or *subnet.*

Privileges to create an Azure AD DS managed domain

You should ensure you have the following privileges for an account you wish to use to create Azure AD DS managed domains:

An account with `Owner` or `Contributor` rights at the subscription level or an account with *Domain Services Contributor, Groups Administrator,* and *Application Administrator* Azure AD roles.

Network and DNS

The default **virtual network** (**VNet**) created through the deployment wizard for Azure AD DS uses a *dedicated Azure AD DS subnet* named `aadds-vnet`. A single subnet is only created within the VNet named `aadds-subnet`. This subnet should not be used for anything other than the managed domain resources; do not add any workloads to this subnet.

VMs or resources you create should be in their own *workload subnet* in the `aadds-vnet`; alternatively, you can create another *workload VNet* and connect it to the Azure AD DS VNet using VNet peering.

The following figure outlines both approaches to the network topology:

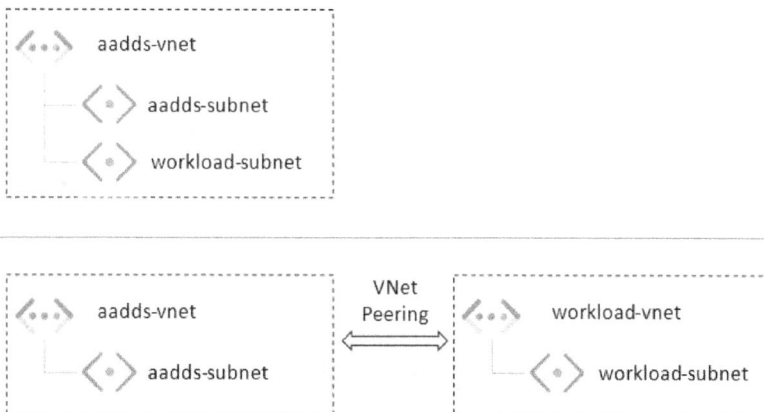

Figure 2.5 – Azure AD DS network topology

When Azure AD DS is created, you will be given *two IP addresses*, the Microsoft-managed DCs; these act as the *DNS servers* for the managed domain.

You must configure the DNS on the peered VNets to allow resources to talk to the Azure AD DS VNet. This can be done using the IP address of the managed domain DCs or existing DNS servers configured in the peered VNets. DNS should be configured to direct queries to the managed domain by allowing conditional DNS forwarding.

Domain name

You must specify a *DNS name* for the domain name of the managed domain. It is essential to ensure that you choose the correct DNS domain name, as once created, you cannot change it without deleting the managed domain and recreating it.

> Note
> You should plan the domain names to avoid conflict with any existing DNS namespace or planned namespace.

The *prefix* for the managed domain cannot be longer than **15 characters**; this is because that is the maximum supported for the *NetBIOS hostname*.

For the *suffix*, you should not use a domain name with a *non-routable suffix*, such as `.local`; this can cause DNS-related issues. Instead, you can use the default built-in domain name for the tenant directory; this will use the `.onmicrosoft.com` suffix. The problem with this approach is that the **Certificate Authority** (**CA**) won't issue you a certificate, so you cannot secure this domain, as Microsoft owns that domain.

To address this, you can use a *routable custom domain* that you own and have registered in DNS, where you have access to create and edit the DNS records and generate a certificate if needed. It is recommended to keep this separate from any existing domain name in the organization for a custom domain name. It must not conflict with any current namespace in a VNet or connected via a VPN or ExpressRoute to on-premises.

In the case of the existing domain name, `milesbetter.solutions`, a recommended naming approach could be to use a *subdomain*, `aadds.milesbetter.solutions`, to avoid conflict with the existing `root` namespace.

Location

You should deploy the Azure AD DS managed domain in the same region where the implemented workloads need to use the managed domain. An Azure availability zone can be selected for additional redundancy in that region if available.

You should ensure to deploy to the correct region, as once created, you cannot change without deleting and recreating.

SKU

The purpose of the SKU is to choose the *performance* and *backup frequency* of the managed domain. All SKUs provide the same *unlimited object count*, although a suggested object count is given as a guideline.

The SKUs at the time of publishing are as follows:

- **Standard**:

  - Suggested auth load (peak, per hour): 0 to 3,000

  - Suggested object count: 0 to 25,000

  - Backup frequency: every 5 days

  - User forest only

- **Enterprise**:

  - Suggested auth load (peak, per hour): 3,000 to 10,000

  - Suggested object count: 25,000 to 100,000

  - Backup frequency: every 3 days

  - User forest and resource forest

  - Five resource forest trusts

- **Premium**:

  - Suggested auth load (peak, per hour): 10,000 to 70,000

  - Suggested object count: 100,000 to 500,000

  - Backup frequency: every 5 days

  - User forest and resource forest

  - Ten resource forest trusts

You can change between SKUs after they are created without deleting and recreating the managed domain.

Forest type

The default forest type created is the **user forest**; this synchronizes objects from Azure AD into the managed domain. These objects can also be those created and synchronized from AD DS forests. A **resource forest** synchronizes only users and groups created in Azure AD.

This section concludes the second of this chapter's three skills areas: *Creating Azure AD DS*. We looked at planning and deployment information. In the next section, we will look at the last of this chapter's three skills areas: *Managing Azure AD DS*.

Managing an Azure AD DS managed domain

This section will cover information for managing an Azure AD DS managed domain instance.

As this is a Microsoft-managed service, there is a limit on what administrative tasks can be carried out. You don't have access to some administrative capabilities in the managed domain; we will cover these areas in this section. We will finish this chapter with a hands-on exercise section.

Management tools

The same set of management tools used for managing AD DS can be used for managing an instance of Azure AD DS, such as **Active Directory Administrative Center (ADAC)**, **Active Directory Users and Computers snap-in**, and **AD PowerShell**. AAD DC Administrators group members can use these tools to administer managed domains.

As these are Microsoft-managed DCs, these are locked down, and there is no direct access to the DCs for the managed domain, such as by **Remote Desktop Protocol (RDP)**. All administrative tasks are intended to be run remotely from a domain-joined management machine on the network.

These remote management tools can be installed on a domain-joined VM (server or client) as a management machine using the **Remote Server Administration Tools (RSAT)** feature.

Managing Azure AD DS administrative privileges

The managed domain will *not* provide access to the Domain Administrator or Enterprise Administrator permissions. Instead, when Azure AD DS is created, an *administrative group* for managing the managed domain is created, and this group is named **Azure AD DC Administrators**.

User accounts that are members of this group are granted *admin permissions* on domain-joined VMs; this group is added to the **local administrator's group**. This group also provides **remote desktop** rights to connect to a domain-joined VM.

You should add users from the Azure AD directory that you wish to be a member of this group.

The following are the administrative tasks that you can carry out in the managed domain:

- Join computers in the managed domain
- Manage the *built-in GPO* for the containers **AADDC Computers** and **AADDC Users**
- Create and manage custom OUs and GPOs
- Manage DNS

The following are the capabilities that you *cannot* carry out on the managed domain:

- Connect to DCs via remote desktop
- Add DCs to the managed domain
- Extend an existing AD DS forest to Azure AD DS; the managed domain is a standalone forest instance
- No *domain administrator* or *enterprise administrator* privileges
- Extend the schema

The previous points should be carefully considered to aid in the decision process of whether Azure AD is the most appropriate identity solution for your needs.

Configuring VNets to use the managed domain

By default, the managed domain VNet only contains one subnet. This is only for the use of the managed domain service, and you should not add any other resources or VMs into this subnet. If you wish to create resources or VMs within the managed domain VNet, you should create additional subnets for these.

However, you may use an existing VNet or create a new VNet and use network segmentation to separate off the managed domain VNet; this recommended approach will vary based on your needs.

To allow resources in other VNets to use the managed domain, the first step is to ensure that those VNets have connectivity to the managed domain VNet (through VNet peering). The DNS server settings for the VNet should be updated to reflect those that can perform lookups against managed domain DCs.

VMs may be required to be restarted to receive new DNS settings; you can run an `ipconfig /all` CLI command on the VM to check what DNS settings are being used before and after the restart.

Managing user accounts

User accounts are not created directly in the managed domain; they are synchronized from Azure AD.

Azure AD DS requires password hashes to be in the format required by NTLM or Kerberos, and Azure AD doesn't provide this format until after the managed domain is enabled on the tenant. Existing users won't be able to authenticate to the managed domain immediately; their password hashes will not have been synchronized to Azure AD DS in the format it uses. Therefore, we need to generate usable password hashes in Azure AD so they can be synchronized to the Azure AD DS managed domain.

The user's password must be changed for this generation of the new password hash to occur; this can be forced so that a user must change it at the next sign-in, or they can change it manually. This change causes the password to be recreated in the correct format for Azure AD DS support and correct Kerberos and NTLM authentication.

Once this password is changed, Azure AD stores this new password hash, and the account will be synchronized to the managed domain. Users can now successfully sign in to the managed domain with their Azure AD account credentials.

It should be noted that *Azure AD Connect Cloud Sync* is not supported with Azure AD DS; AD users need to be synchronized using *Azure AD Connect*.

This *password hash regeneration* scenario does not occur for accounts created after the managed domain is created in the tenant; for these accounts, Azure AD will create password hashes in the correct format that Azure AD DS requires.

SSPR must be implemented in the Azure AD tenant to allow users to reset their passwords.

There is a default password policy in the managed domain, where complexity, age, and lockout are defined; custom password policies can be created to override this.

Managing OUs

There are two built-in **Organizational Units (OUs)** created for the managed domain; these are as follows:

- **AADDC Users**: This is the default location for all user and group objects synchronized into the managed domain from the Azure AD tenant. Includes users and groups synchronized in from the Azure AD tenant.

- **AADDC Computers**: This is the default location for all computer objects joined to the managed domain.

In a hybrid identity scenario where user accounts and groups are synchronized from AD DS, the OU objects are *not synchronized* because the managed domain only supports a *flat OU* structure. Creating custom OUs and associated GPOs to meet your needs is recommended.

Managing VM domain join

Azure AD DS provides *managed domain services* so that you can provide VM join capabilities without the need for self-managing IaaS VM DCs. You can authenticate to this managed domain using your *Azure AD credentials*. User accounts can be cloud-only or synchronized from AD into Azure AD via Azure AD Connect.

This gives us a *common identity* that can be used across AD DS, Azure AD, and Azure AD DS. With user account password hashes synchronized from AD into the managed domain via Azure AD, this gives us a lot of choice as to what is then possible.

The account to join the computer to the domain must be synchronized to the managed domain or from the Azure AD tenant; *external directory accounts* cannot be used.

If the VM cannot join the managed domain, it is often because of three common causes:

- **Connectivity issues**: Ensure you can reach the managed domain DC's IP addresses; you may need to check that the VNets are peered if you have workloads in a different VNet to the VNet where Azure AD was created. Ensure that there is no network security filtering traffic for the managed domain.

- **DNS issues**: Ensure the VM has the correct DNS servers listed to look up the managed domain; check that the IPs of the managed domain DCs have been set as the DNS servers to use for the VNet where the VM is attempting to do the *Domain join*.

- **Credentials**: Make sure the account used belongs to the Azure AD tenant or has successfully synchronized to the managed domain and that the password hashes are the correct format for Azure AD DS. We learned earlier in this section that the correct password hash may not have been synchronized into the managed domain and that a password reset will generate the correct password hash format; you should wait for this to be synchronized to the managed domain.

It is recommended that the account credentials are in the *UPN suffix* format; for example, for a `domainjoin` user, you would use a `domainjoin@aadds.milesbetter.onmicrosoft.com` UPN suffix. Alternatively, the *SAMAccountName* format can be used; for example, for a `domainjoin` user, you would use an `AADDS\domainjoin` *SAMAccountName* format.

In this section, we looked at managing Azure AD DS. In the next section, we will complete a hands-on exercise to reinforce some of the concepts covered in this chapter.

Hands-on exercise

To support your learning with some practical skills, we will learn how to create Azure AD DS, which we looked at in this chapter.

We will look at the following exercise:

- Exercise – Installing Azure AD DS

Getting started

To get started with this hands-on exercise section, if you do not already have an Azure subscription, you can create a free Azure account at `https://Azure.microsoft.com/free`. This free Azure account provides the following:

- 12 months of free services

- $200 credit to explore Azure for 30 days

- 25+ services that are always free

Let's move on to the exercise for this chapter.

Exercise – Installing Azure AD DS

In this section, we will learn how to install Azure AD DS. The steps in the exercise will be carried out in the Azure portal: `https://portal.Azure.com/`.

Follow these steps to create an *Azure AD DS* instance:

1. Log in to the Azure portal at `https://portal.Azure.com`. Alternatively, you can use the Azure desktop app: `https://portal.Azure.com/App/Download`.

2. Type `Azure AD Domain Services` in the search bar and click on the **Azure AD Domain Services** blade from the list of services shown.

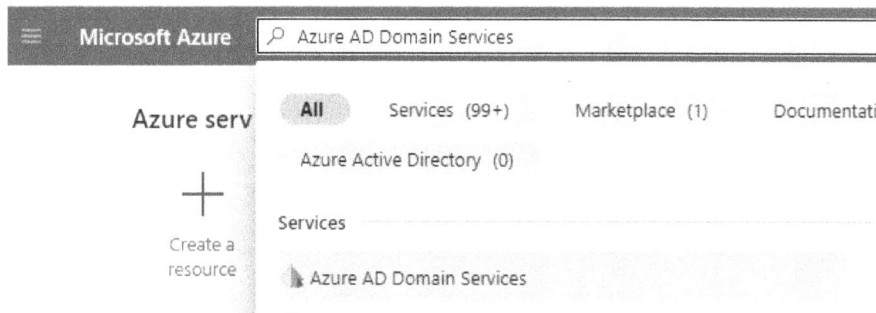

Figure 2.6 – Search for the service in the Azure portal

3. On the **Azure AD Domain Services** blade, click the + **Create** resource. It should be at the top left of the blade.

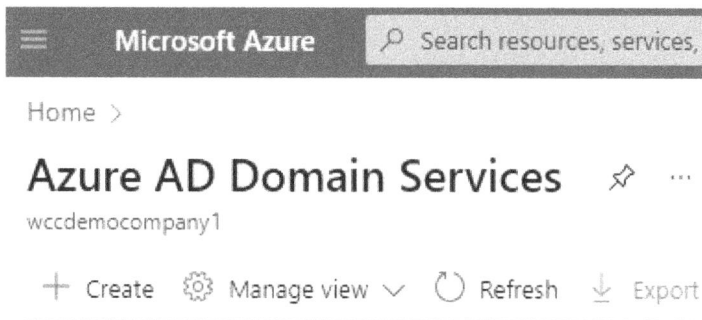

Figure 2.7 – Azure AD Domain Services blade

4. For reference purposes, review the information provided about the functionality of Azure AD Domain Services and the information in the **Project details** section.

Figure 2.8 – Review creation information

5. Review the information from the **Help me choose the subscription and resource group** helper link; then select **Subscription** and **Resource group** or create a new Resource group as required.

Figure 2.9 – Choose subscription and resource group

6. Review the information from the **Help me choose the DNS name** helper link; then enter a *domain name*. For this exercise, we will modify the default to use a subdomain to avoid conflict with the default namespace; you should set it as required for your scenario.

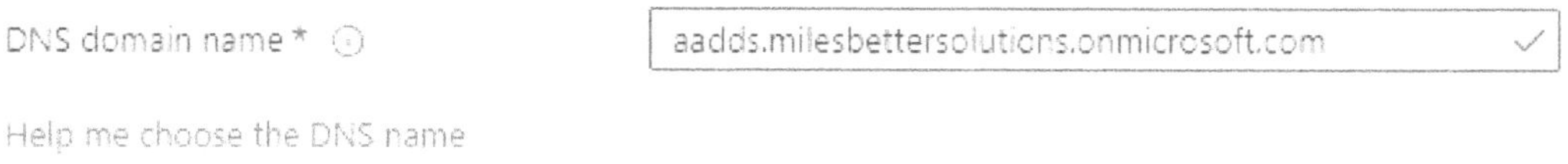

DNS domain name * ⓘ

aadds.milesbettersolutions.onmicrosoft.com   ✓

Help me choose the DNS name

Figure 2.10 – Choose DNS domain name

7. Set **Region**.

Region * ⓘ

East US   ⌄

Figure 2.11 – Choose the region

8. Review the information from the **Help me choose a SKU** helper link; then set **SKU**.

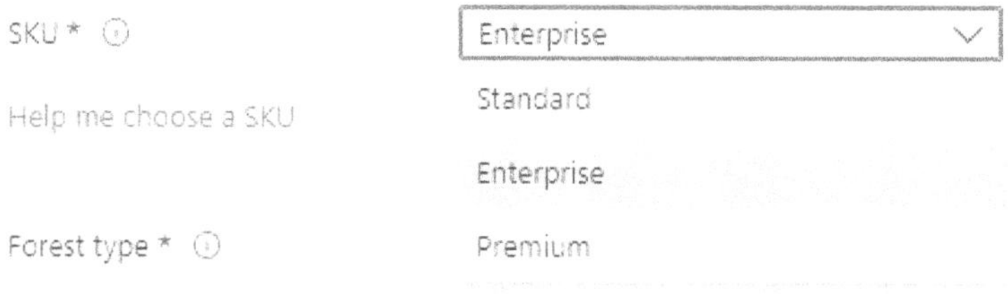

SKU * ⓘ

Enterprise   ⌄

Help me choose a SKU

Standard

Enterprise

Forest type * ⓘ

Premium

Figure 2.12 – Choose SKU

9. Review the information from the **Help me choose a forest type** helper link; then set **Forest type**.

Forest type * ⓘ

(**User**   Resource)

Help me choose a forest type

Figure 2.13 – Setting Forest type

10. Click **Next**.

11. On the **Networking** tab, review the reference information about using an existing network.

˟Basics   ˟Networking   Administration   Synchronization   Security Settings   Tags   Review + create

Azure AD Domain Services uses a dedicated subnet within a virtual network to hold all of its resources. If using an existing network, ensure that the network configuration does not block the ports required for Azure AD Domain Services to run. Learn more

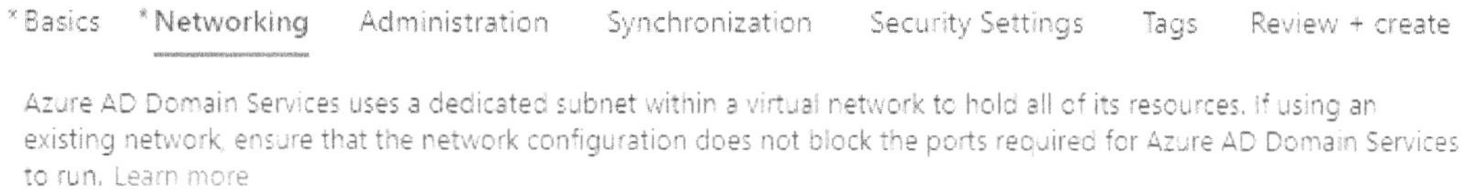

Figure 2.14 – Review network information

12. Review the information from the **Help me choose the virtual network and address** helper link; then set **Virtual network** or click **Create new** as required.

Virtual network * ⓘ   (new) aadds-vnet   ˅
                             Create new

Help me choose the virtual network and address

Figure 2.15 – Choose a virtual network

13. Review the information from the **Help me choose the subnet and NSG** helper link; then, set **Subnet**.

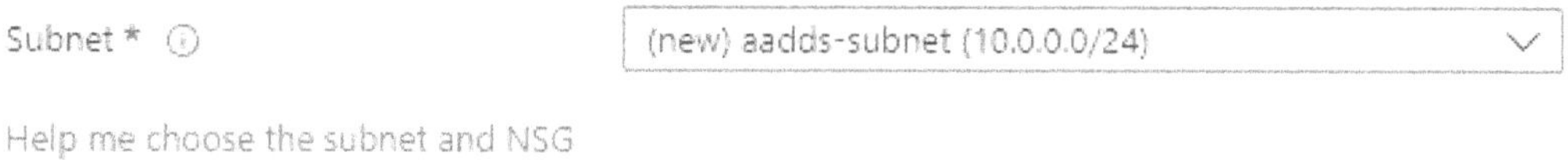

Subnet * ⓘ   (new) aadds-subnet (10.0.0.0/24)   ˅

Help me choose the subnet and NSG

Figure 2.16 – Choose the subnet

14. Click **Next**.

15. On the **Administration** tab, review the information from the **Help me choose AAD DC Admins** helper link; complete the **AAD DC Administrators** group membership as required, or skip and manage later.

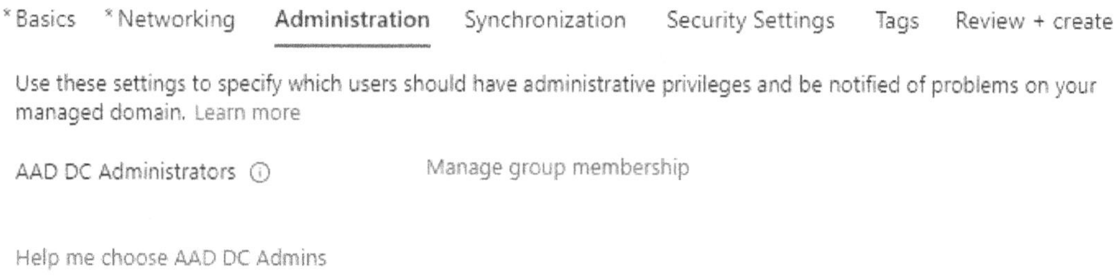

˟Basics   ˟Networking   Administration   Synchronization   Security Settings   Tags   Review + create

Use these settings to specify which users should have administrative privileges and be notified of problems on your managed domain. Learn more

AAD DC Administrators ⓘ   Manage group membership

Help me choose AAD DC Admins

Figure 2.17 – Choose admin group membership

16. Review the information from the **Help me choose who gets notifications** helper link; set as required.

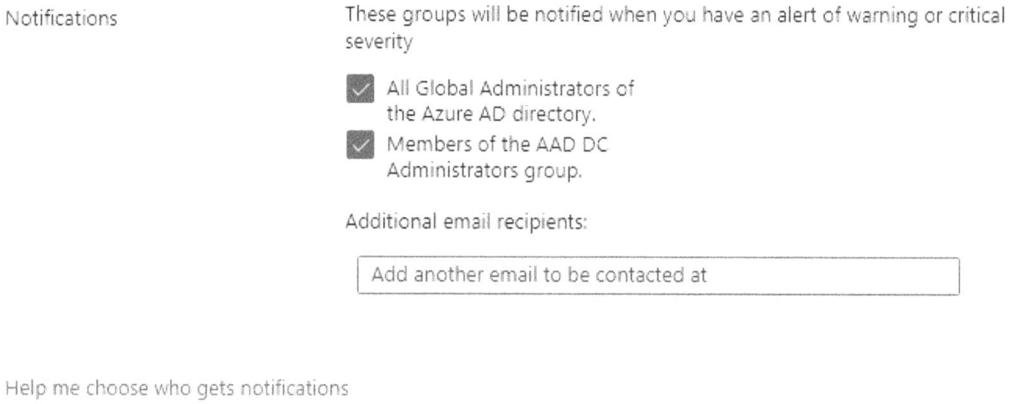

Notifications

These groups will be notified when you have an alert of warning or critical severity

☑ All Global Administrators of the Azure AD directory.
☑ Members of the AAD DC Administrators group.

Additional email recipients:

Add another email to be contacted at

Help me choose who gets notifications

Figure 2.18 – Choose notifications

17. Click **Next**.

18. Review the **Synchronization** tab's reference information about synchronization and scoped synchronization.

˟Basics   ˟Networking   Administration   Synchronization   Security Settings   Tags   Review + create

Azure AD Domain Services provides a one-way synchronization from Azure Active Directory to the managed domain. In addition, only certain attributes are synchronized down to the managed domain, along with groups, group memberships, and passwords. Learn more

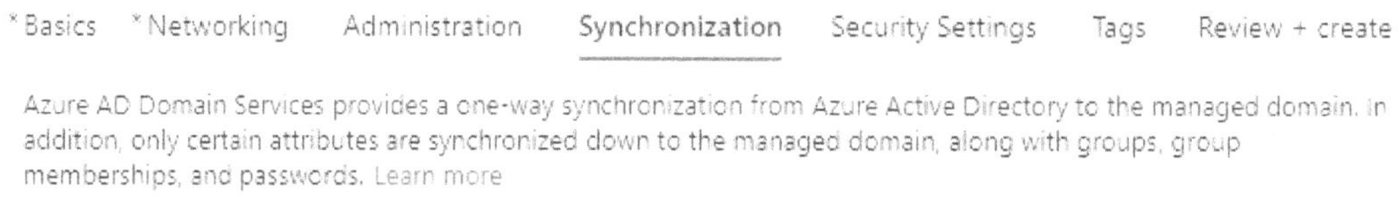

Figure 2.19 – Review synchronization information

19. Review the information from the **Help me choose the synchronization type** helper link; set it as **All** or **Scoped** as required.

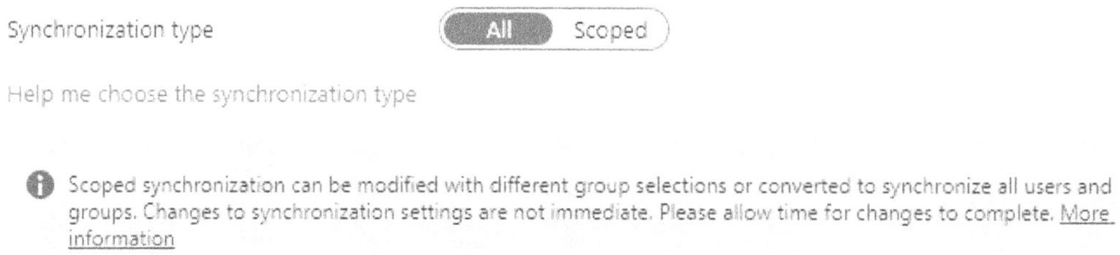

Synchronization type   (**All**   Scoped)

Help me choose the synchronization type

ⓘ Scoped synchronization can be modified with different group selections or converted to synchronize all users and groups. Changes to synchronization settings are not immediate. Please allow time for changes to complete. More information

Figure 2.20 – Choose synchronization

20. Click **Next**.

21. On the **Security Settings** tab, review the information from the three helper links, then leave the set defaults or change as required.

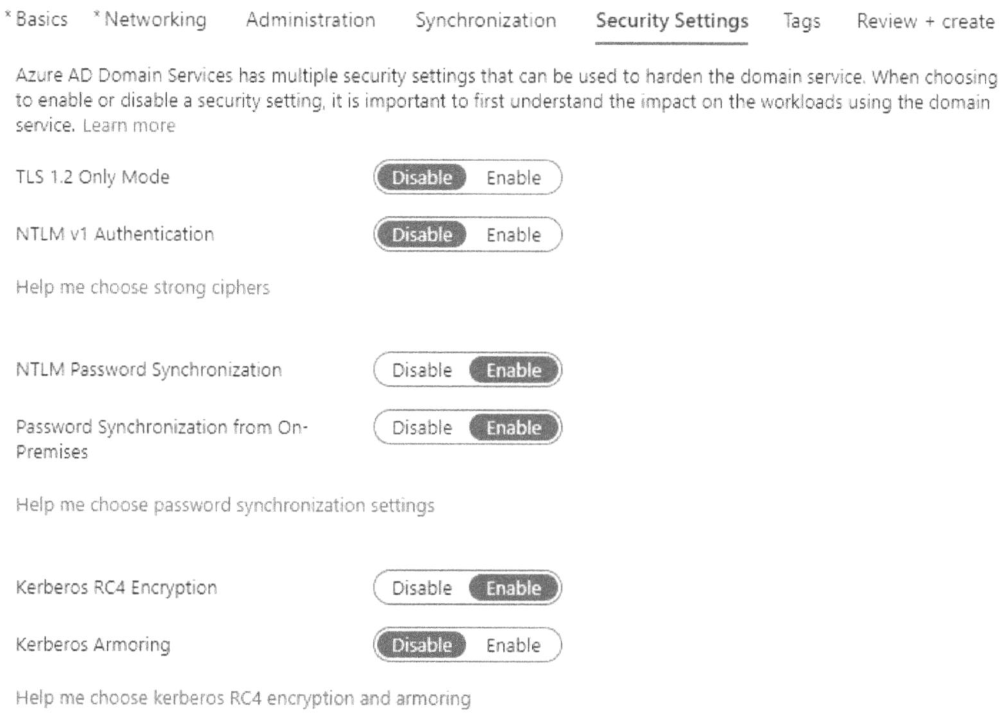

Basics   Networking   Administration   Synchronization   **Security Settings**   Tags   Review + create

Azure AD Domain Services has multiple security settings that can be used to harden the domain service. When choosing to enable or disable a security setting, it is important to first understand the impact on the workloads using the domain service. Learn more

TLS 1.2 Only Mode   (Disable   Enable)

NTLM v1 Authentication   (Disable   Enable)

Help me choose strong ciphers

NTLM Password Synchronization   (Disable   Enable)

Password Synchronization from On-   (Disable   Enable)
Premises

Help me choose password synchronization settings

Kerberos RC4 Encryption   (Disable   Enable)

Kerberos Armoring   (Disable   Enable)

Help me choose kerberos RC4 encryption and armoring

Figure 2.21 – Choose security settings

22. Click **Next**.

23. On the **Tags** tab, leave the set defaults or change as required. Then, click **Next**: **Review + create**.

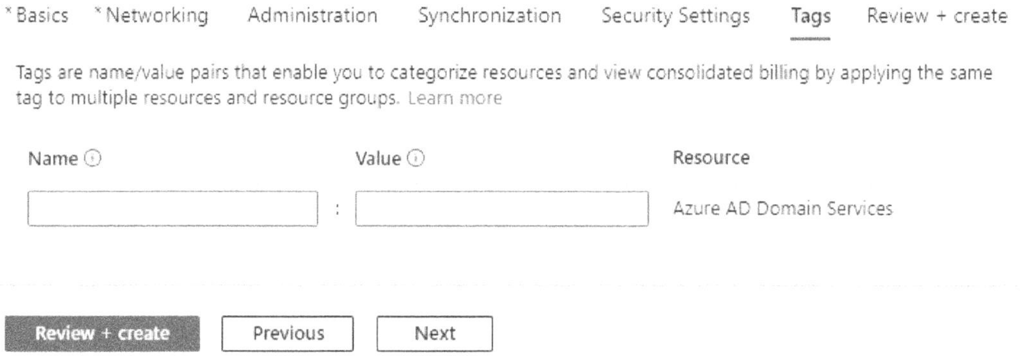

ˣ Basics	ˣ Networking	Administration	Synchronization	Security Settings	Tags	Review + create

Tags are name/value pairs that enable you to categorize resources and view consolidated billing by applying the same tag to multiple resources and resource groups. Learn more

Name ⓘ	Value ⓘ	Resource
	:	Azure AD Domain Services

Review + create   Previous   Next

Figure 2.22 – Add tags

24. On the **Review + create** tab, review your settings; you may go back to the previous tabs and make any edits if required. Once you have confirmed your settings, click **Create**.

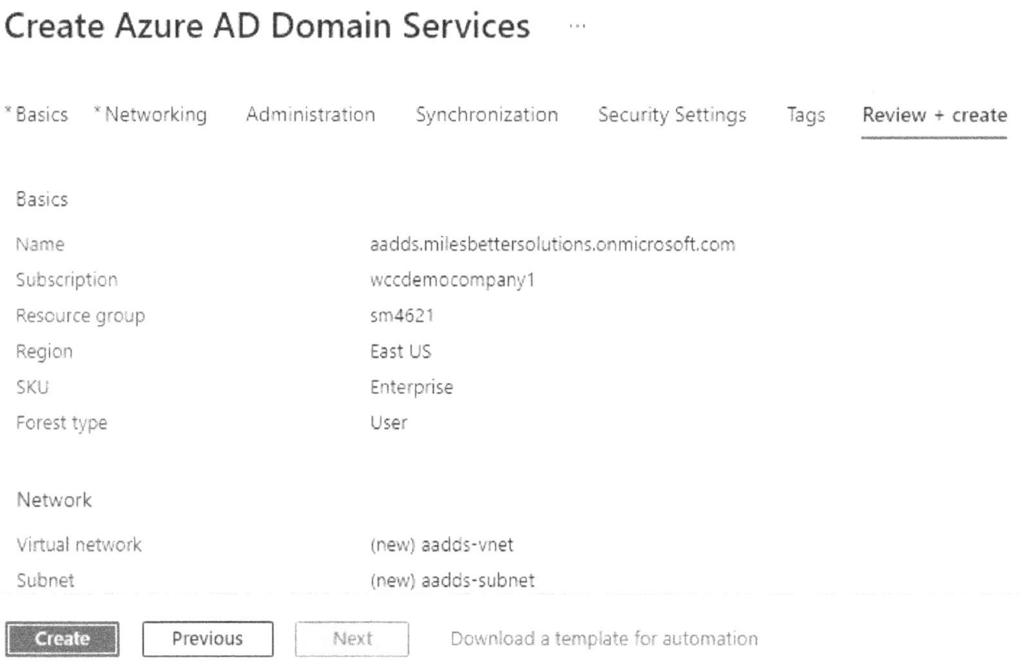

Create Azure AD Domain Services   ...

ˣ Basics	ˣ Networking	Administration	Synchronization	Security Settings	Tags	Review + create

Basics

Name	aadds.milesbettersolutions.onmicrosoft.com
Subscription	wccdemocompany1
Resource group	sm4621
Region	East US
SKU	Enterprise
Forest type	User

Network

Virtual network	(new) aadds-vnet
Subnet	(new) aadds-subnet

Create   Previous   Next   Download a template for automation

Figure 2.23 – Review creation options

25. You will see a pop-up screen asking you to acknowledge your choices, as these cannot be changed after the managed domain has been created. Review this information, click **OK** to continue, or go back and review the change as required.

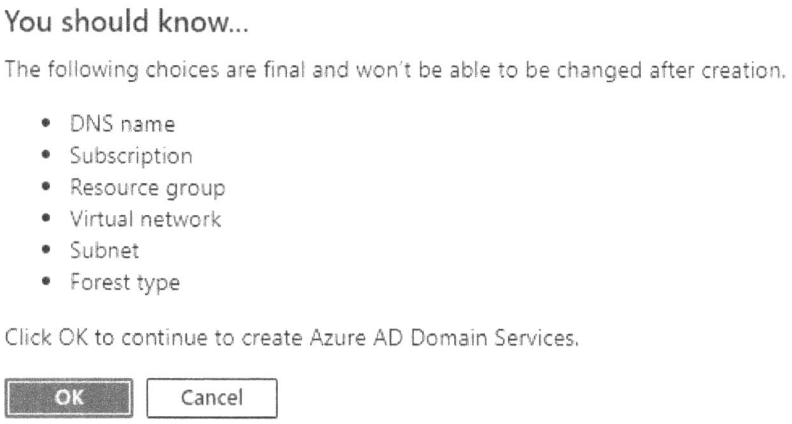

You should know...

The following choices are final and won't be able to be changed after creation.

- DNS name
- Subscription
- Resource group
- Virtual network
- Subnet
- Forest type

Click OK to continue to create Azure AD Domain Services.

OK	Cancel

Figure 2.24 – Review final choices

26. You will now see a **Deployment is in progress** screen; this will take approximately an hour to complete.

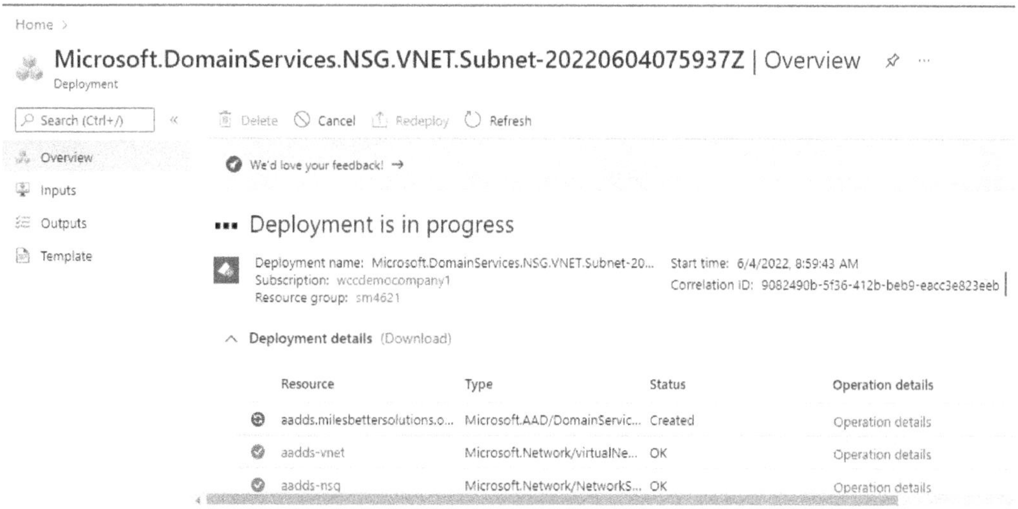

Home >

🌸 **Microsoft.DomainServices.NSG.VNET.Subnet-20220604075937Z** | Overview ✧ ⋯
    Deployment

🔍 Search (Ctrl+/) «	🗑 Delete ⊘ Cancel ⬆ Redeploy ↻ Refresh
🏭 Overview	✅ We'd love your feedback! →
🖥 Inputs	
⊟ Outputs	••• Deployment is in progress
📄 Template	

■ Deployment name: Microsoft.DomainServices.NSG.VNET.Subnet-20...   Start time: 6/4/2022, 8:59:43 AM
   Subscription: wccdemocompany1   Correlation ID: 9082490b-5f36-412b-beb9-eacc3e823eeb
   Resource group: sm4621

∧ Deployment details (Download)

	Resource	Type	Status	Operation details
⊕	aadds.milesbettersolutions.o...	Microsoft.AAD/DomainServic...	Created	Operation details
✅	aadds-vnet	Microsoft.Network/virtualNe...	OK	Operation details
✅	aadds-nsg	Microsoft.Network/NetworkS...	OK	Operation details

Figure 2.25 – Deployment progress

27. You will see a **Your deployment is complete** message and will receive a **Deployment succeeded** notification.

ervices.NSG.VNET.Subnet-202206040

🗑 Delete   ⊘ Cancel   ⬆ Redeploy   ↻ Refresh

✅ We'd love your feedback! →

✅ **Your deployment is complete**

🔷 Deployment name: Microsoft.DomainServices.NSG.VNET.S
Subscription: wccdemocompany1
Resource group: sm4621

⌄ Deployment details (Download)

⌃ Next steps

Go to resource

Notifications   ✕

More events in the activity log →   Dismiss all ⌄

✅ Deployment succeeded   ✕
Deployment 'Microsoft.DomainServices.NSG.VNET.Subnet-20220604075937Z' to
resource group 'sm4621' was successful.

Go to resource   ⭐ Pin to dashboard

46 minutes ago

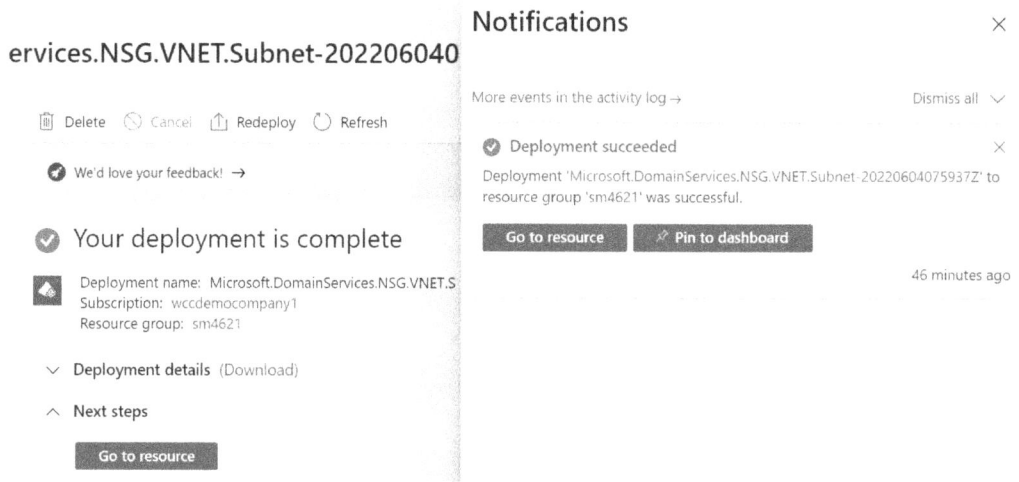

Figure 2.26 – Deployment completion

28. Click **Go to resource** or navigate to the **Azure AD Domain Services** blade; the created Azure AD Domain Services instance can be seen.

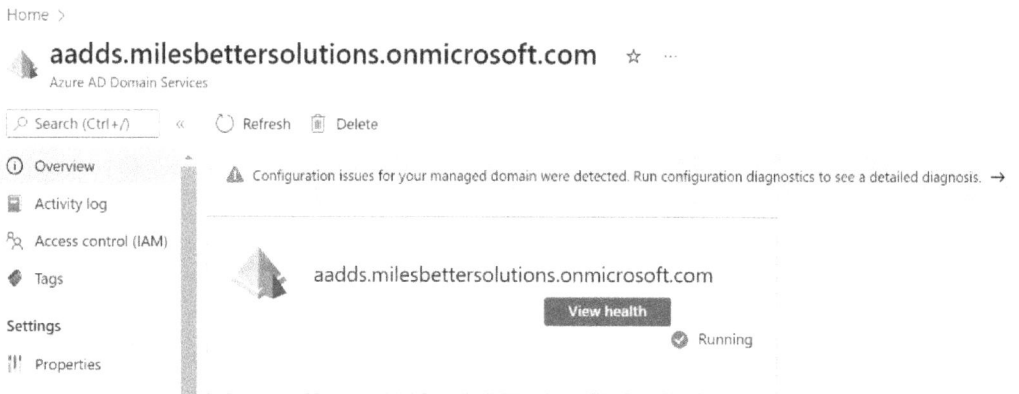

Home >

🔷 **aadds.milesbettersolutions.onmicrosoft.com**   ☆ ⋯
Azure AD Domain Services

🔍 Search (Ctrl+/)   «   ↻ Refresh   🗑 Delete

ⓘ Overview   ⚠ Configuration issues for your managed domain were detected. Run configuration diagnostics to see a detailed diagnosis. →
🗄 Activity log
Rₐ Access control (IAM)
🏷 Tags   🔷   aadds.milesbettersolutions.onmicrosoft.com
Settings   View health
⫴ Properties   ✅ Running

Figure 2.27 – Azure AD Domain Services blade

29. From the **Overview** page, review the information in the **Required configuration steps** section to complete the configuration.

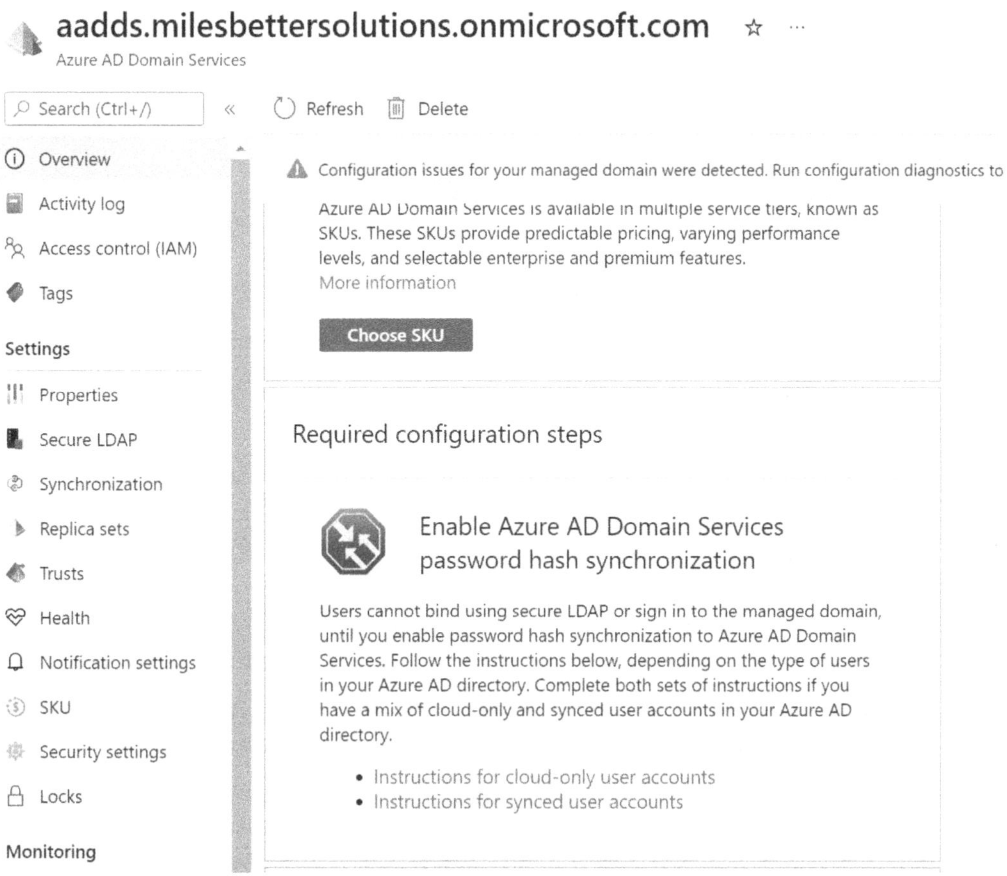

aadds.milesbettersolutions.onmicrosoft.com ☆ ...
Azure AD Domain Services

🔍 Search (Ctrl+/)   «   ↻ Refresh   🗑 Delete

ⓘ Overview

▤ Activity log

👥 Access control (IAM)

🏷 Tags

Settings

⫶|⫶ Properties

▮ Secure LDAP

🔁 Synchronization

▶ Replica sets

🛡 Trusts

💟 Health

🔔 Notification settings

💲 SKU

⚙ Security settings

🔒 Locks

Monitoring

⚠ Configuration issues for your managed domain were detected. Run configuration diagnostics to

Azure AD Domain Services is available in multiple service tiers, known as SKUs. These SKUs provide predictable pricing, varying performance levels, and selectable enterprise and premium features.
More information

Choose SKU

Required configuration steps

Enable Azure AD Domain Services password hash synchronization

Users cannot bind using secure LDAP or sign in to the managed domain, until you enable password hash synchronization to Azure AD Domain Services. Follow the instructions below, depending on the type of users in your Azure AD directory. Complete both sets of instructions if you have a mix of cloud-only and synced user accounts in your Azure AD directory.

- Instructions for cloud-only user accounts
- Instructions for synced user accounts

Figure 2.28 – Final configuration steps

With this, we have finished the creation of AD DS.

In this exercise, we looked at creating an Azure AD Domain Services instance.

Summary

This chapter has provided coverage for the AZ-800 Administering Windows Server Hybrid Core Infrastructure exam learning objective: *Deploy and manage AD DS in on-premises and cloud environments.*

We learned about the concepts of Azure AD and how it compares with AD. We looked at creating and managing an Azure AD DS managed domain. A hands-on exercise then finished the chapter to provide you with additional practical skills.

This chapter aimed to take your knowledge beyond the exam objectives; we added new skills and learning with the content provided. This further develops your knowledge and skills for on-premises network infrastructure services and enables you to be prepared for a real-world, day-to-day hybrid environment-focused role. In the next chapter, we learn the essential skills of managing users and computers with group policy.

Further reading

This section provides links to additional exam information and study references:

- *Microsoft Certified: Windows Server Hybrid Administrator Associate*: `https://docs.microsoft.com/en-us/learn/certifications/windows-server-hybrid-administrator/`

- *Exam AZ-800: Administering Windows Server Hybrid Core Infrastructure*: `https://docs.microsoft.com/en-us/learn/certifications/exams/az-800`

- *Exam AZ-800: skills outline*: `https://query.prod.cms.rt.microsoft.com/cms/api/am/binary/RWKI0r`

- *Microsoft Learn: Deploy and manage identity infrastructure*: `https://docs.microsoft.com/en-us/learn/paths/deploy-manage-identity-infrastructure/`

- *What is Azure Active Directory Domain Services*: `https://docs.microsoft.com/en-gb/Azure/active-directory-domain-services/overview`

- *Compare self-managed Active Directory Domain Services, Azure Active Directory, and managed Azure Active Directory Domain Services*: `https://docs.microsoft.com/en-gb/Azure/active-directory-domain-services/compare-identity-solutions`

- *Tutorial: Configure virtual networking for an Azure Active Directory Domain Services managed domain*: `https://docs.microsoft.com/en-us/Azure/active-directory-domain-services/tutorial-configure-networking`

- *Tutorial: Create and configure an Azure Active Directory Domain Services managed domain*: `https://docs.microsoft.com/en-us/Azure/active-directory-domain-services/tutorial-create-instance`

Skills check

Challenge yourself with what you have learned in this chapter by answering the following questions:

1. What is Azure AD DS?

2. Explain the differences between Azure AD DS and AD DS.

3. What are the use cases for Azure AD DS?

4. What is the relationship between Azure AD and Azure AD DS in the cloud-only identity scenario?

5. What is the relationship between AD, Azure AD, and Azure AD DS in the hybrid identity scenario?

6. List five crucial characteristics of Azure AD DS.

7. What are the two enterprise applications created in the Azure AD tenant when the managed domain is created, and what are their purposes?

8. What privileges are required to create an instance of Azure AD DS in the Azure AD tenant?

9. What are the two approaches to the VNets used for the Azure AD DS and for workloads that will need to use Azure AD DS?

10. Name some crucial considerations for choosing the domain name used for Azure AD DS.

11. What are the SKUs for Azure AD DS, and what are their differences?

12. Why is it essential to select the correct name, location, and SKU for Azure AD DS?

13. What is the default forest type for Azure AD DS?

14. What management tools can be used with Azure AD DS?

15. List some of the limitations of the administrative aspects of a managed domain.

Managing Users and Computers with Group Policy

This chapter covers the AZ-800 Administering Windows Server Hybrid Core Infrastructure exam learning objective: *Manage Windows Server by using domain-based Group Policies*.

In the previous chapter, we added skills to understand the concepts, creation, and management of Azure AD DS in Microsoft Azure-hosted environments.

We will start this chapter with some concepts and an understanding of Group Policy. We will then look at the Group Policy components, their capabilities and operations, management, and troubleshooting. We will then conclude with a hands-on exercise to develop your skills further. The following topics will be covered in this chapter:

- What is Group Policy?
- What is a GPO?
- GPO components
- GPO administrative templates
- GPO scope
- GPOs in Azure AD DS
- GPO management and troubleshooting tools
- Exercise – Installing Group Policy Management tools
- Exercise – Implementing Group Policy

Before we dive into any policy creation or management, we will look at some definitions and concepts to set a baseline and foundation of knowledge to build from. We start this chapter by defining **Group Policy** and **Group Policy Objects** (**GPOs**).

What is Group Policy?

Group Policy is a *Microsoft governance* feature and *configuration management framework* built as part of the Windows OS; it provides a method for the centralized configuration and management of users' and computers' OS and applications.

It is used to force a baseline of standardized configuration settings on *user* and *computer* objects that are part of an AD domain. It is a powerful management capability and can manage nearly all configurable settings.

Group Policy can be used with *Azure AD DS* to manage user and computer object settings in much the same way as for AD DS, with some limitations. We will cover this in more detail in the *GPOs in the Azure AD DS* section.

There are two configuration settings nodes, as follows:

- **User Configuration**: These apply to a user. These settings can restrict a user from accessing certain functionality of the OS, such as removing items from the **Start Menu** and **Taskbar**. These settings are applied to a user on whatever domain computer they access.

- **Computer Configuration**: These apply to a computer. These settings can be for configuration, such as **Remote Desktop Session** settings. These settings apply to all users who log on to this computer.

The following screenshot shows these two nodes:

Figure 3.1 – GPO configuration nodes

The policies exist on the domain controllers for a domain. Part of the management aspect is ensuring a copy of the latest policies is passed through the replication process. With Azure AD DS, this is taken care of for you by Microsoft as one of the benefits of having a *managed domain*.

Group Policy uses built-in management tools to create and manage the policy settings. The following screenshot shows the **Group Policy Management Console (GPMC)** tool:

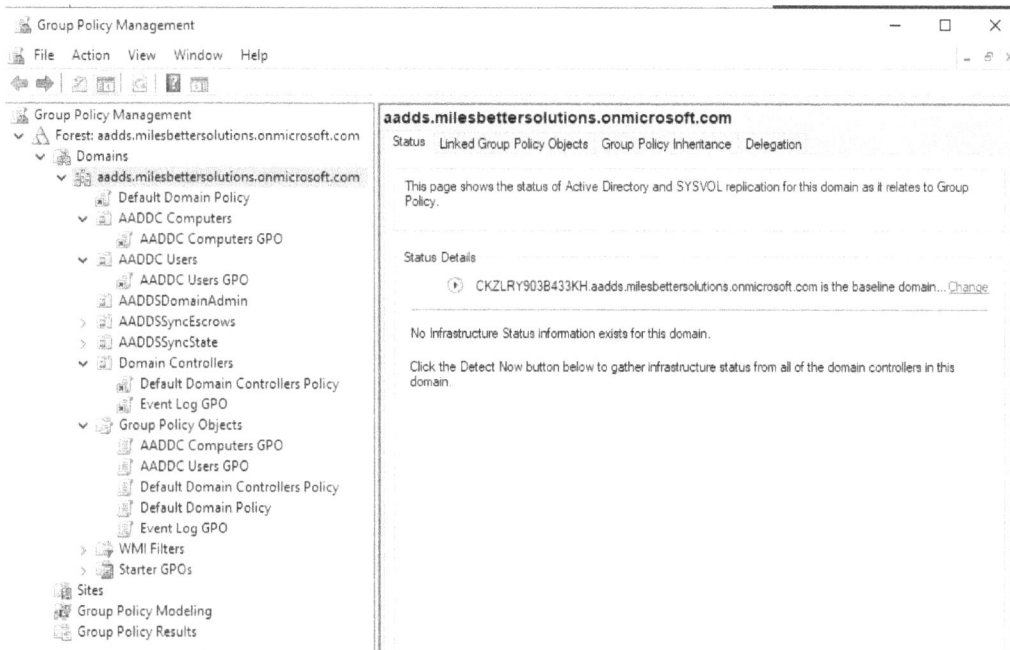

Figure 3.2 – The GPMC tool

Before we move on to the next section, it would be helpful for us to recap the first chapter on OUs. An **Organizational Unit (OU)** functions as a container object. It is used to collectively organize users, groups, and computers, providing a framework for targeting administrative control and policies on objects in the container.

In this section, we introduced Group Policy concepts and a recap of OUs. In the next section, we will look at GPOs, which form the basis and building blocks of Group Policy.

What is a GPO?

GPOs can be considered a *store* of the collection of individual policy settings that can be applied to the *user* and *computer* domain objects. These objects are much like a *manifest document* that holds and defines the individual policy settings that should be applied to objects for management and configuration in an automated, governed, and controlled manner.

GPOs can be one of the following two types:

- **Local GPO**: Windows OS computers have policies whose settings only apply to the *local computer* and cannot be linked to any other object; they are applicable for computers that are *not AD joined*.

- **Domain GPO**: These are created and stored on AD domain controllers. These GPOs can then be linked to a *site*, *domain*, and *OU container* objects (not groups) and provide central management of configuration settings that wish to be applied to user and computer objects in the domain.

There are two default policies created when AD DS is installed on Windows Server, and these are as follows:

- **Default Domain Policy**: This is linked to the *domain container* and affects all computers. Security filtering settings can be applied to a specific security group; the default security filtering will apply to the *Authenticated Users* group.

- **Default Domain Controllers Policy**: This is linked to the *domain controllers OU* container for the exclusive purpose of applying domain-controller-specific settings.

You should create additional GPOs and link those to the site, domain, or OU container objects as required if you need additional settings that are unavailable in these GPOs.

In this section, we defined GPOs. In the next section, we will look at the GPO components.

GPO components

An AD domain GPO consists of the following two components:

- **GPO Container**: This is an AD object and is a metadata store for the GPOs. The **Directory Replication Agent** (**DRA**) provides replication between DCs.

- **GPO Template**: This is a store for a set of files for the GPO. **Distributed File System Replication** (**DFSR**) provides replication between DCs.

For a GPO to be applied to a user or a computer, it must be linked to a site, domain, or OU container that contains computer or user objects that require to be affected by the policy; a GPO cannot be applied directly to a group. GPOs provide two groups of configuration settings, one for computers and one for users; there are no configuration settings for groups. The configuration settings nodes in any policy are as follows:

- **Computer Configuration**: The policies defined in this section will only affect computer objects. These settings are evaluated and applied at computer boot time.

- **User Configuration**: The policies defined in this section will only affect the user object's login sessions. These settings are evaluated and applied at user logon and only apply to the session while it lasts.

The following screenshot shows these configuration node sections:

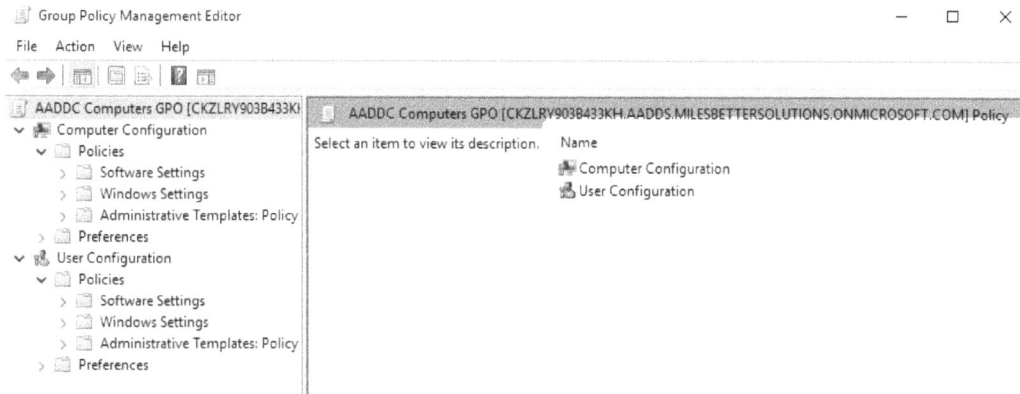

Figure 3.3 – GPO configuration nodes

You should plan whether a policy setting should be applied in the computer or user settings:

- Computer settings will affect all users that access that computer
- User settings will affect the user no matter what computer they access

The GPO's user or computer configuration section can also be disabled independently. This is useful if you have no settings configured in one of these, as every section will need to be processed if enabled; disabling the section with no settings can result in faster policy processing times.

The individual policy setting is the most granular level and is the most basic building block of the GPO hierarchy; the specific setting configured for affecting the computer and user objects will be defined here. This setting could prevent a user from shutting down the computer, configuring the remote desktop session settings, or setting a registry entry.

The following screenshot shows an individual group policy setting:

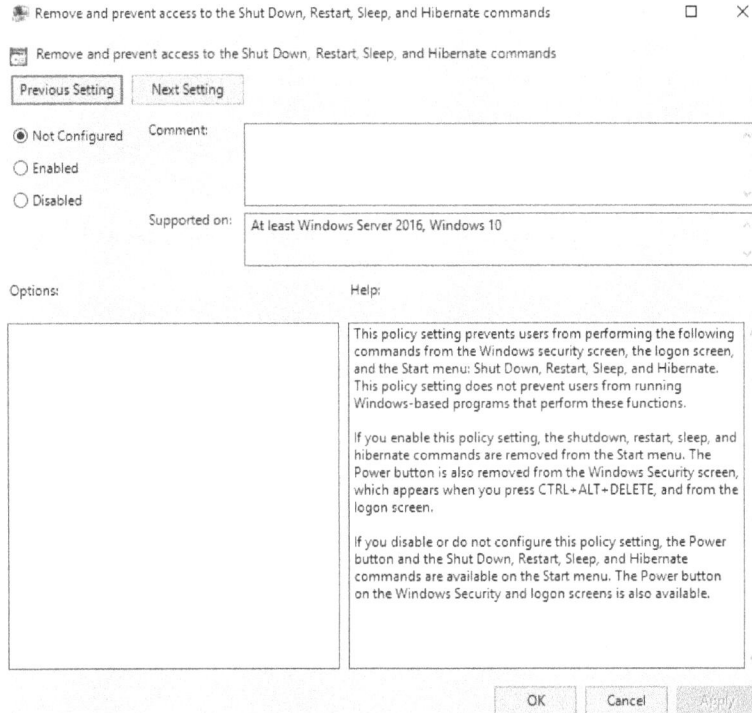

Figure 3.4 – Group policy setting

With Group Policy's extensible framework, nearly any configurable setting can be applied with GPOs.

In this section, we looked at the GPO components. In the next section, we will look at administrative templates and their purpose.

GPO administrative templates

These provide the basis for *registry-based policy*. As the name implies, the settings you configure are used to make changes to the registry; deploying configurations, apps, and software in this way is often the most efficient approach for centralized deployments.

Templates have the following settings that can be configured:

- **Computer-related**: The settings defined here will edit the HKEY_LOCAL_MACHINE registry hive
- **User-related**: The settings defined here will edit the HKEY_CURRENT_USER registry hive

If settings conflict, the *computer settings* registry changes take precedence.

Templates are located under the **Administrative Templates** node of the GPOs **Computer Configuration** and **User Configuration** nodes, as seen in the following screenshot:

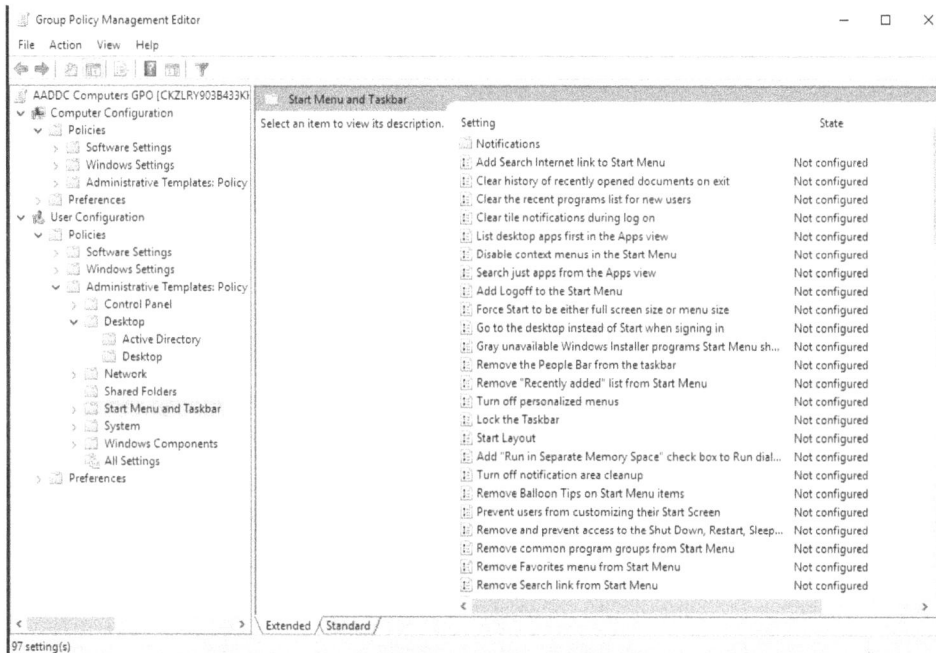

Figure 3.5 – Administrative templates

The administrative template is made up of two file types, as follows:

- **.admx file**: This specifies the group policy settings and is *language-neutral*. These should be copied into the `%SYSTEMROT%\PolicyDefinitions` folder or a central location such as *SYSVOL*; for Azure AD DS, you will not have access to SYSVOL.

- **.adml file**: This specifies the language used with the `.admx` file. These are commonly referred to as *language files*. These should be copied into the `%SYSTEMROT%\PolicyDefinitions\`
`[Langauage-Country Region]` folder. Each language has its own `\PolicyDefintions` subfolder; that is, the `en-US` folder should be used to store the English files, and so on. The default `.adml` language file present is the language of the installed OS of the computer.

The following screenshot shows the locations for these files:

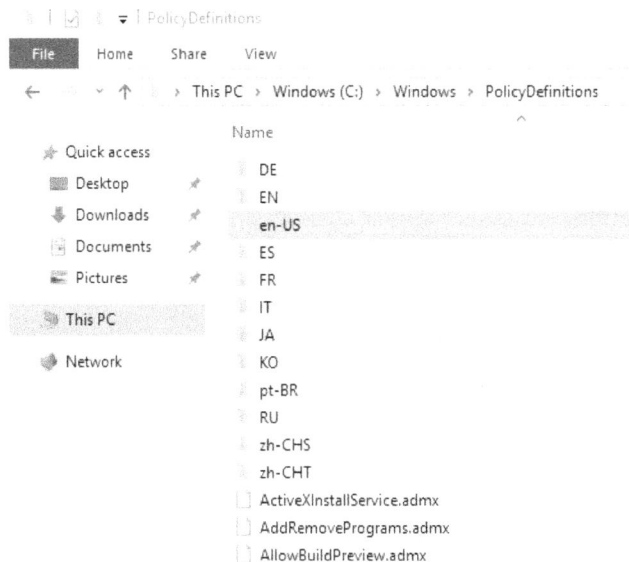

Figure 3.6 – Administrative templates location

This section looked at using administrative templates to configure registry settings centrally. In the next section, we will look at the GPO scope.

GPO scope

A GPO contains configuration settings that you specify that could apply to computer and user objects. Still, you must define the domain objects you want to be affected by the policies. This is the *scope*, or, defined another way, the blast radius/area of impact of where the policy settings will be applied.

A GPO can have a scope of the following three domain object levels:

- **Site scope**: All users and computers within this site will be affected by the linked GPO with a scope set at this level

- **Domain scope**: All users and computers within this domain will be affected by the linked GPO with a scope set at this level

- **OU scope**: Only users and computers within this OU will be affected by the linked GPO with a scope set at this level

By default, all users and computers will be affected by the policy linked to the container they are part of; a GPO can be linked to multiple sites, domains, or OU container objects.

> **Note**
> GPOs can only apply/affect users and computers and cannot be applied directly to groups.

However, we can use security group filtering to ensure the right users get the policy applied; by default, security filtering is set to the **Authenticated Users** group.

You should consider the effect of linking GPOs to multiple sites in multiple domain forests, as this can have an impact on performance; to process the GPOs, they may need to be accessed over what could be a slow WAN link.

GPOs can also be *unlinked*, so their policy settings are not affecting the user or computer objects they were linked to; GPOs can also be *disabled* or *deleted*.

This section looked at the scopes for a GPO. The following section will look at the GPO processing order and inheritance.

The order of processing for GPOs

As we learned in the previous section, GPOs can be linked to an AD site, domain, or OU and will be applied to any user or computer object in that container.

Each GPO will be processed in the following order of inheritance:

1. **Local policy**
2. **Site-level linked GPO**
3. **Domain-level linked GPO**
4. **OU-level linked GPO**

The following illustration aims to outline the order of linked GPO processing:

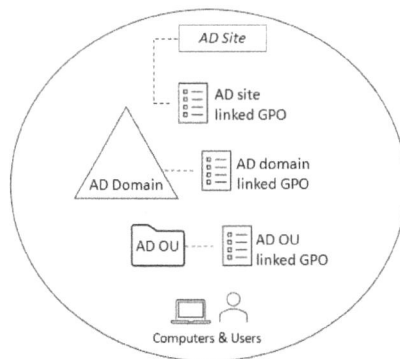

Figure 3.7 – Domain GPO processing order

The settings applied to the user and computer objects will be the combined effect of policies linked at each level, creating a **Resultant Set of Policies** (**RSoP**). There can be multiple GPOs linked at each container level.

When one GPO has the same policy setting as a different value to another linked GPO, there will be a conflict; for example, *prevent shutdown* may be *enabled* in a site GPO, but the same setting might have a value of *disabled* in a GPO linked to a lower container such as an OU GPO. By default, lower-level containers override the GPOs set at the higher levels; in this case, the result would be that a computer can be shut down, as the policy setting to prevent shutdown has been disabled through the order of precedence.

It is essential to consider that settings you intended to be applied by a GPO at one level are not *overridden* by another GPO processed afterward.

The default behavior, however, can be changed so that you can set **block inheritance** of policy settings on a domain or OU container. As its name suggests, this blocking action prevents the policy settings from GPOs linked at higher levels in the hierarchy from being applied.

In addition to blocking inheritance, one final control is the ability to *enforce* a linked GPO. This GPO's settings will take precedence over any conflicting settings, maybe from a GPO linked at a higher level. An enforce overrides a block, meaning the settings will be enforced even if set to block inheritance.

The challenge with the Group Policy framework is that it can become a governance issue. Many GPOs can be created and linked to different containers. With inheritance, blocking, and enforcement, it can become very complex to understand the resulting or effective settings that would be applied to a user or computer; the use of inheritance blocking should be limited.

When we consider blocking and enforcement, this resulting precedence can be quite hard to track; the **Group Policy Inheritance** tab can simplify this, showing the resulting precedence of GPOs considering all factors, so you do not need to figure it out. This is shown in the following screenshot:

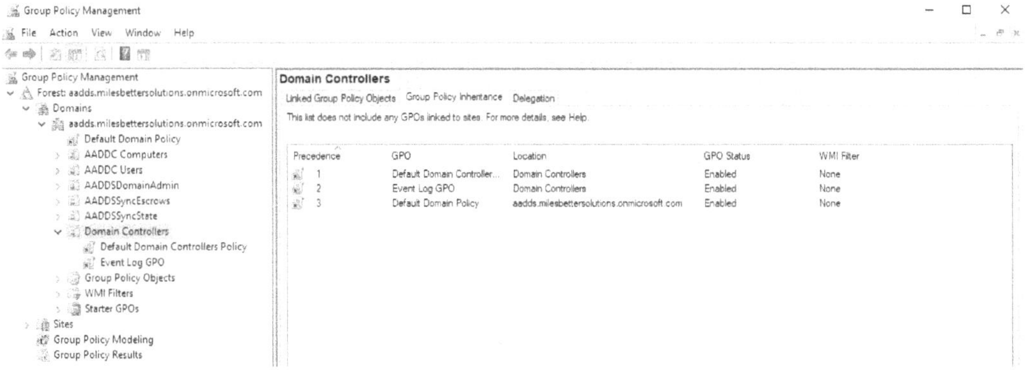

Figure 3.8 – Group Policy inheritance precedence evaluation

This section looked at the GPO order of processing and inheritance. The following section will look at GPOs specific to Azure AD DS.

GPOs in Azure AD DS

Azure AD DS manages Group Policy in much the same way as AD to manage settings for the user and computer objects, but with restrictions on what settings can be accessed and configured in the *managed domain*.

When Azure AD DS is implemented in an Azure AD tenant, two built-in GPOs are created for managing the configuration of users and computers. These are as follows:

- **AADDC Users GPO**

- **AADDC Computers GPO**

These are shown in the following screenshot:

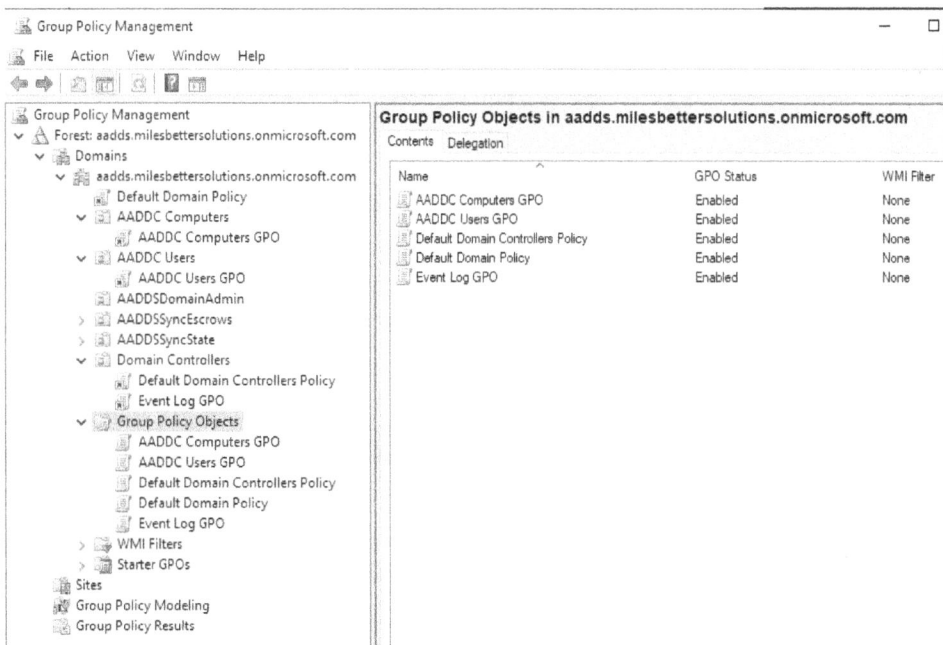

Figure 3.9 – Default Azure AD DS GPOs

You can create custom GPOs if you need additional settings not available in these GPOs. These can only be linked to OUs, and you can create additional custom OUs as required to meet your needs. For Group Policy administration privileges in the managed domain, you need to be a member of the **AAD DC administrators** group.

In a hybrid identity environment where both AD DS and Azure AD DS are utilized, group policies from the AD DS DCs will not synchronize to the Azure AD DS managed domain. You must manually create these settings in Azure AD DS to work around this. You should be aware that Azure AD DS only supports a flat GPO structure, so you may have to rethink how you will provide the same user and computer configuration if your existing GPO design does not translate directly into what is possible in Group Policy on Azure AD DS.

In this section, we defined GPOs in Azure AD DS. The following section will look at the tools to create and manage GPOs.

GPO management and troubleshooting tools

GPOs can be managed primarily through a GUI. The GPMC is the built-in group policy management tool in the Windows OS. This has a built-in editor feature, **Group Policy Management Editor**. The GPMC can be launched with the `gpedit.msc` command.

This tool allows you to create, link, edit, and delete GPOs, provide reports, and model how they will affect users and the computers they are applied to.

The following command-line tools can also be used to manage and troubleshoot GPOs:

- **GPMC**
- **GPRESULT**
- **GPUPDATE**
- **LDIFDE**
- **PowerShell**

When you create a new GPO or edit a GPO, the GPO files are replicated to every member server every *90 minutes* (DC servers are replicated every 5 minutes). You don't need to have a user log off or reboot to apply the policy. You can also force an application if you don't want to wait by using the `gpupdate/force` command.

The **Group Policy Management Editor** pane in the following screenshot shows the configuration settings nodes of **Computer Configuration** and **User Configuration** in the GPO settings hierarchy and how they are visually represented:

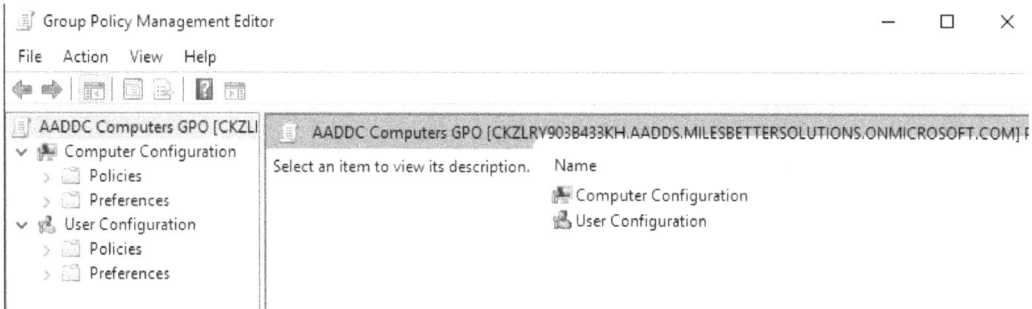

Figure 3.10 – GPO configuration nodes

The **Computer Configuration** and **User Configuration** nodes in the GPO settings hierarchy are then broken down into sections with folders for **Policies** and **Preferences**, and beneath those are further configuration areas. The **Group Policy Management Editor** pane shows how the **Policies** and **Preferences** folders and subconfiguration group folders are visually represented in the following screenshot:

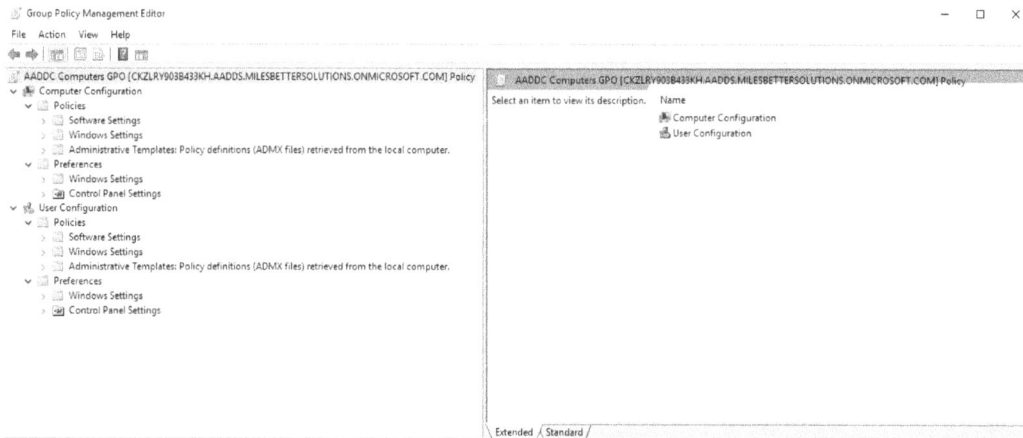

Figure 3.11 – GPO configuration settings

GPOs linked to an object container will have the following attributes to be aware of for managing GPOs, including the following:

- *The link order*
- *The GPO name*
- *Whether the GPO is enforced or not*
- *Whether the link is enabled*
- *Whether the GPO is enabled*
- *Whether there is a WMI filter*

The following screenshot shows the GPOs linked to a *domain object container* and its attributes:

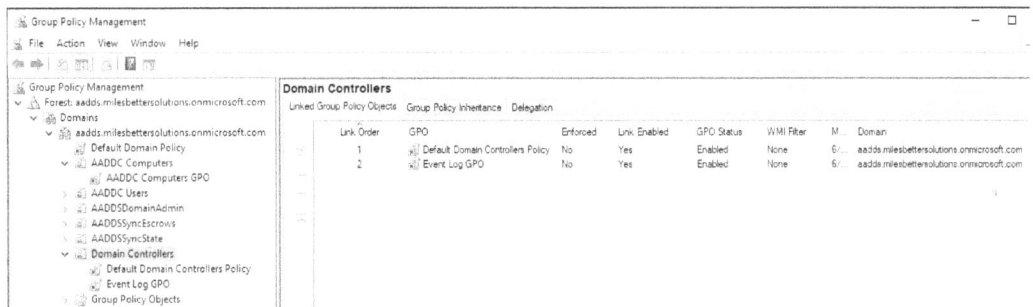

Figure 3.12 – Linked GPOs

The Group Policy framework's strong configuration capabilities mean it can manage nearly all configurable settings. The most granular level of the Group Policy framework is the individual policy setting. The policies can be one of three states:

- **Not Configured**
- **Enabled**
- **Disabled**

The following screenshot shows these individual Group Policy states:

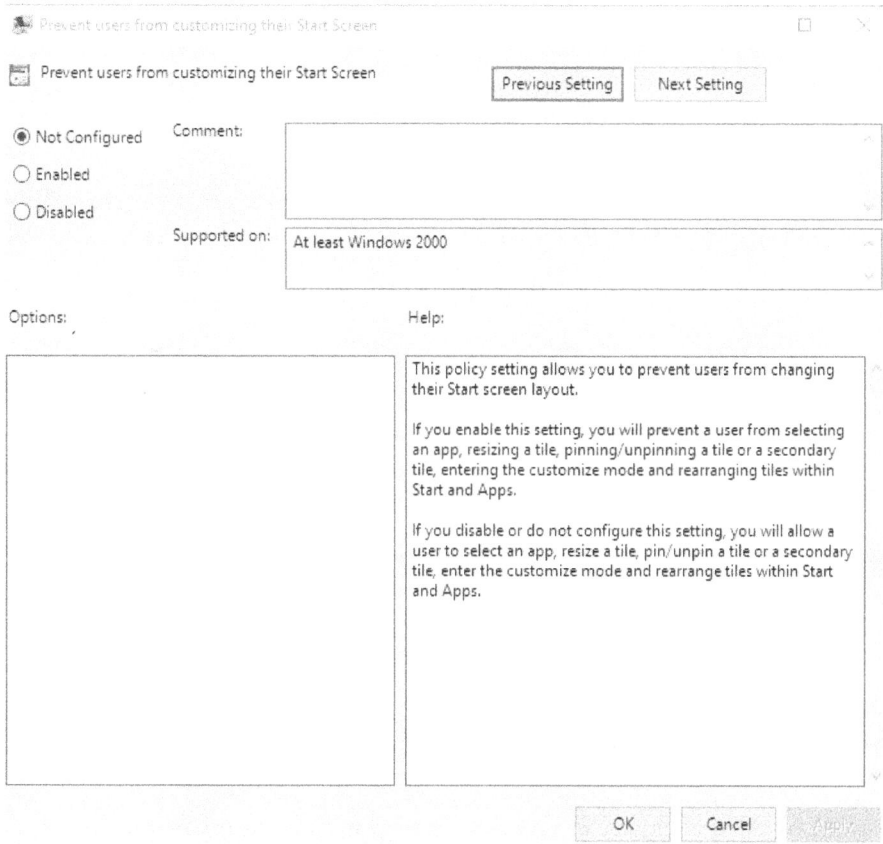

Figure 3.13 – Group Policy setting states

When we create a new policy, the default is that all settings are set to **Not Configured**. When a policy state is changed between **Enabled** and **Disabled**, the changes are applied to the computer and user objects affected by the GPO.

This section looked at the tools for creating and managing GPOs. In the next section, we will complete a hands-on exercise section to re-enforce some of the concepts covered in this chapter.

Hands-on exercise

To support your learning with some practical skills, we will utilize the concepts and understanding gained from the earlier sections of this chapter and put them into practical application.

We will look at the following exercises:

- Exercise – Installing Group Policy Management tools

- Exercise – Implementing Group Policy

Getting started

To get started with this section, if you do not already have an Azure subscription, you can create a free Azure account at `https://Azure.microsoft.com/free`. This free Azure account provides the following:

- 12 months of free services

- $200 credit to explore Azure for 30 days

- 25+ services that are always free

Let's move on to the first exercise for this chapter.

Exercise – Installing Group Policy Management tools

This section will look at installing the *Group Policy Management tools* on a management server VM. This will allow you to create and configure GPOs and custom OUs in the managed domain.

For this exercise, we will use the Azure AD DS managed domain we created in the *Azure AD tenant* exercise in the previous chapter. If needed, please complete this exercise and then return. Alternatively, please use your own environment so that you can follow this exercise.

You will need the following additional requirements in place to complete this exercise:

- For this Azure AD DS managed domain, a user account that is a member of the **Azure AD DC administrators** group in the Azure AD tenant

- A Windows Server management VM joined to the Azure AD DS managed domain; if needed, please complete the steps in the following Microsoft article and then return: `https://docs.microsoft.com/en-us/azure/active-directory-domain-services/join-windows-vm`

Follow these steps to install the tools:

1. Log in to your server with a domain account and click **Add Roles and Features** from **Manage** in **Server Manager**.

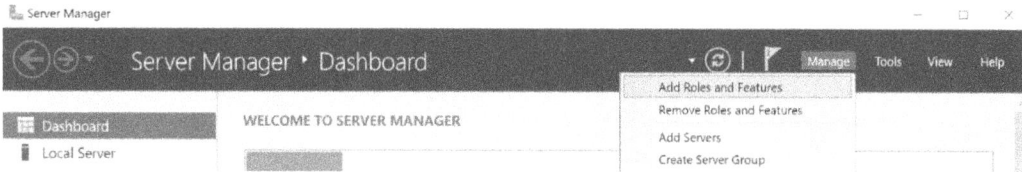

Figure 3.14 – Server Manager

2. On the **Before you begin** page, click **Next**.

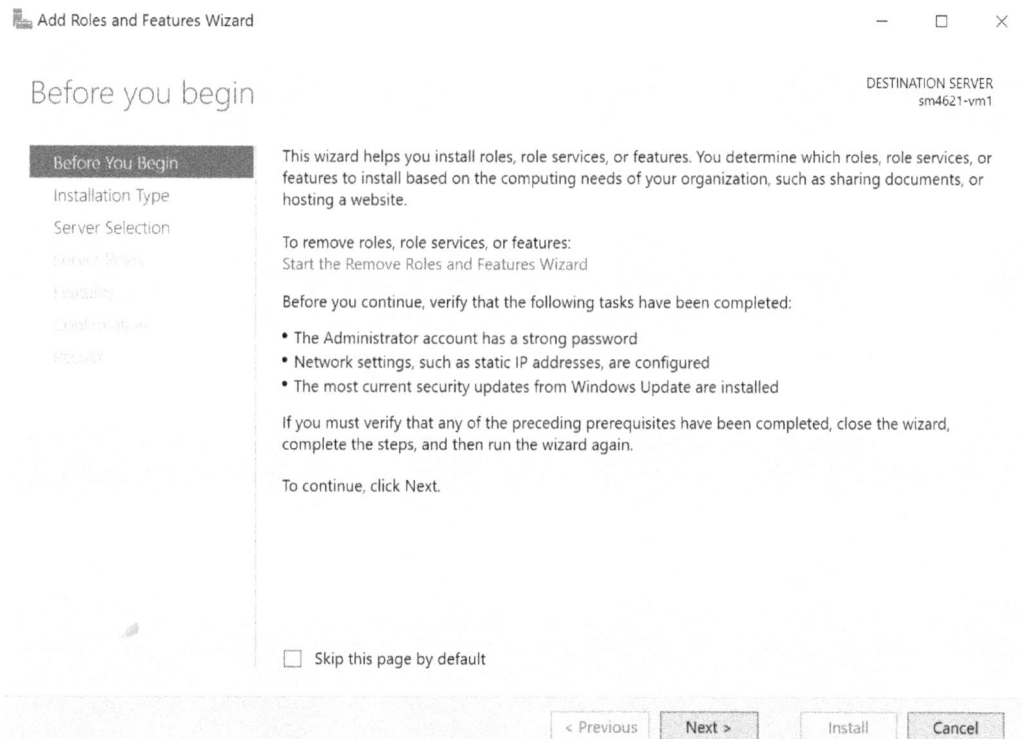

Figure 3.15 – Before you begin

3. On the **Select installation type** page, leave it set to **Role-based or feature-based installation** and click **Next**.

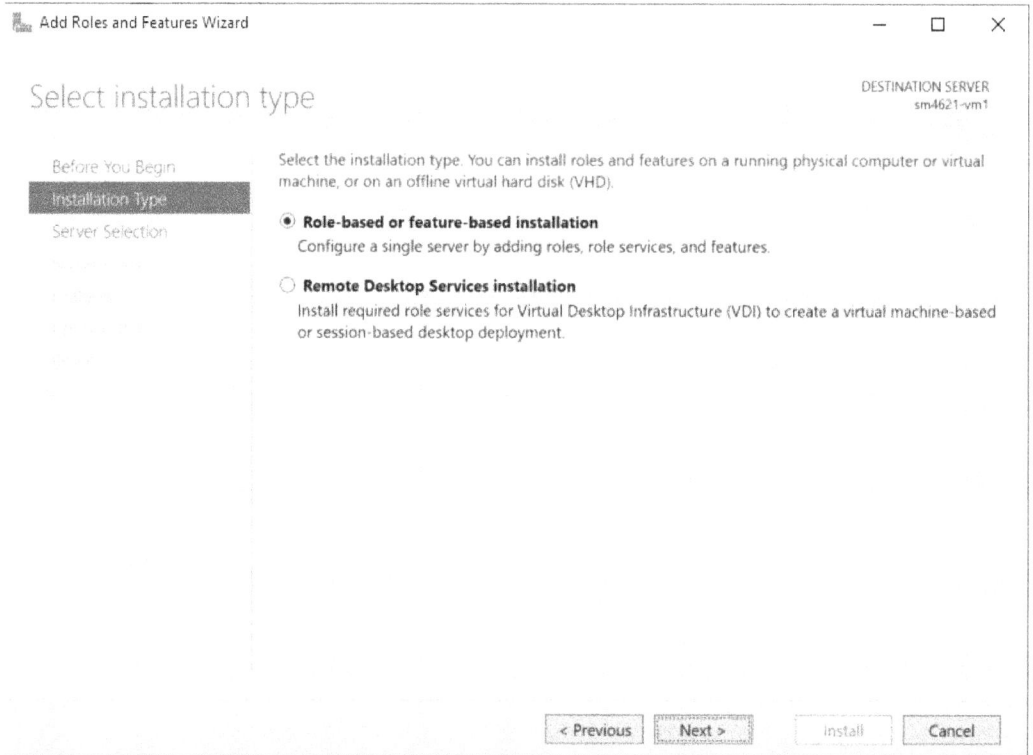

Figure 3.16 – Select installation type

4. On the **Select destination server** page, leave it set to **Select a server from the server pool** and ensure the server where you want to install the Group Policy Management tools is selected; click **Next**.

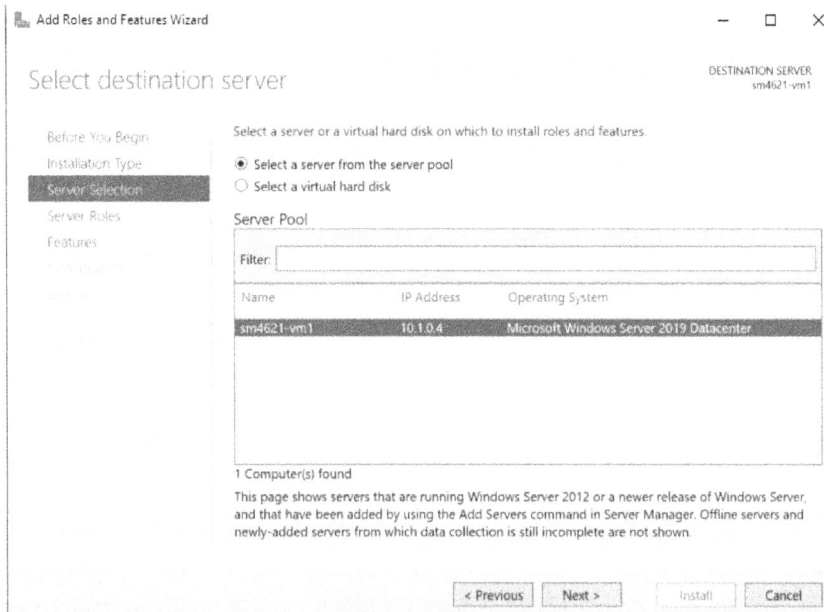

Figure 3.17 – Select destination server

5. On the **Select server roles** page, click **Next**.

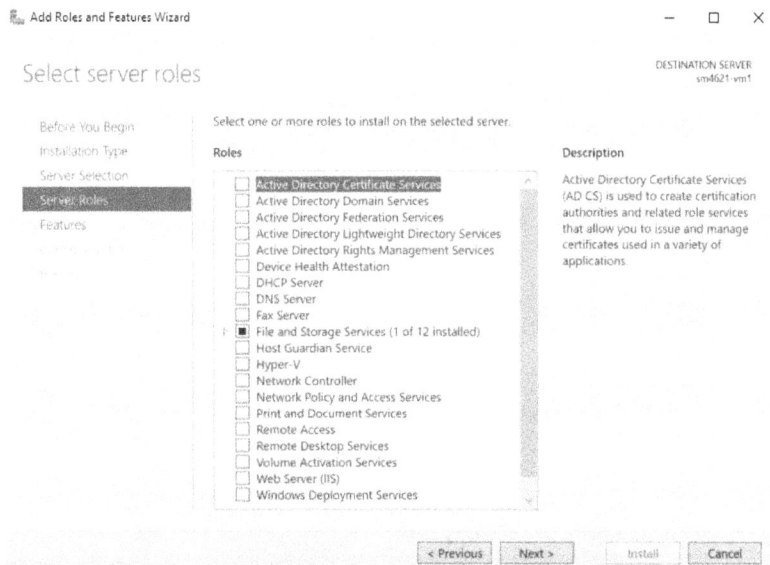

Figure 3.18 – Select server roles

6. On the **Select features** page, select the **Group Policy Management** feature.

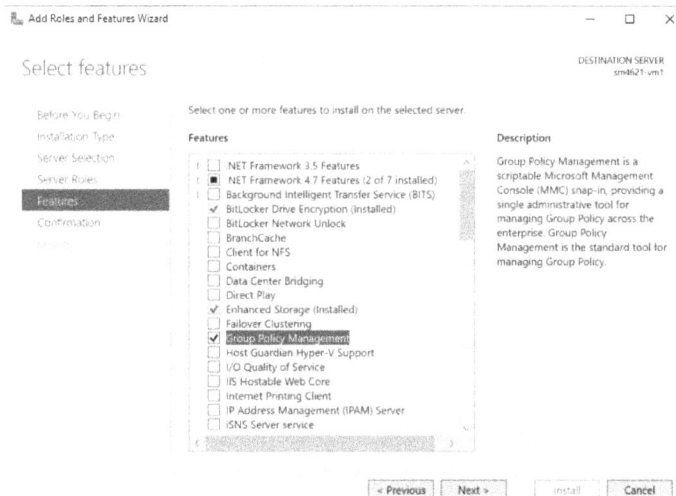

Figure 3.19 – Select features

7. On the **Confirm installation** page, click **Install**. You will see the **Installation progress** page; you can click **Close**.

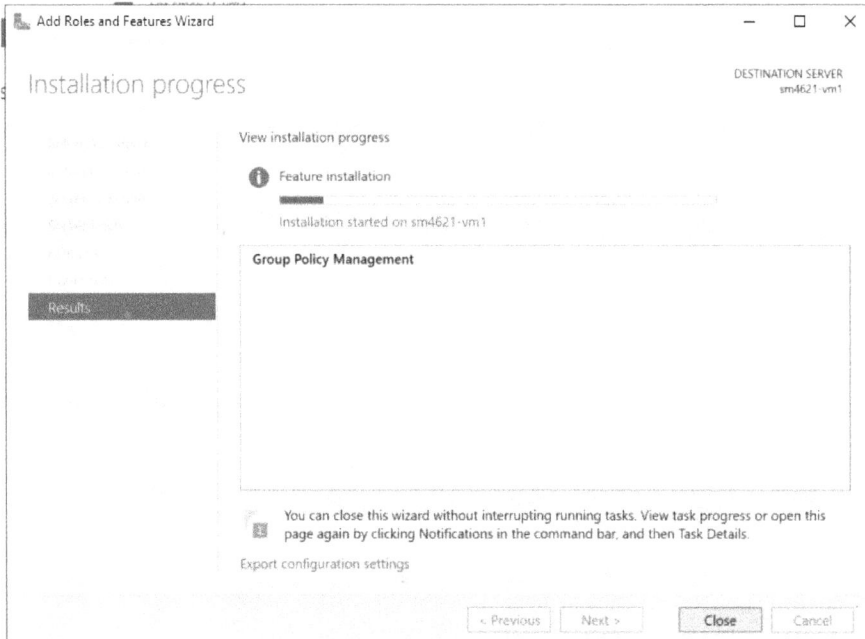

Figure 3.20 – Installation progress

8. You will see a notification that the installation succeeded.

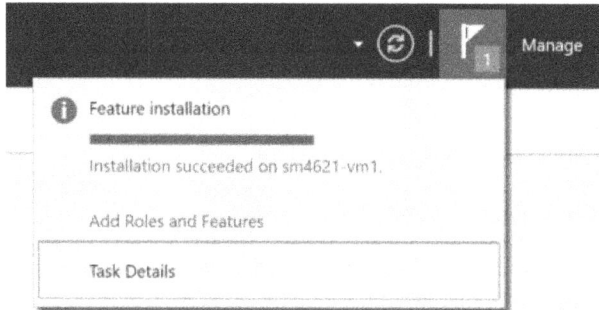

Figure 3.21 – Installation succeeded

9. From **Server Manager**, click **Tools** and **Group Policy Management** to launch the GPMC.

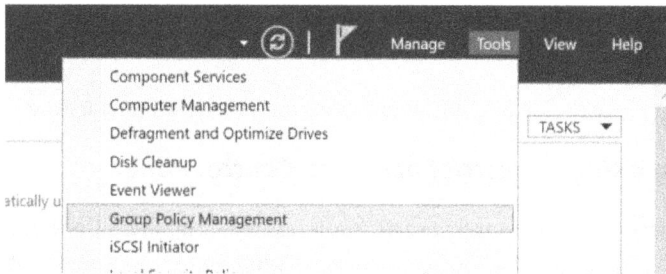

Figure 3.22 – Launch Group Policy Management

10. The GPMC is now launched.

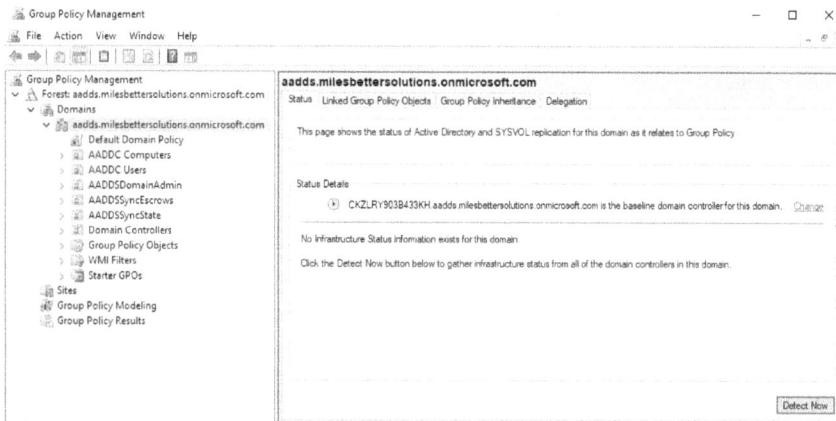

Figure 3.23 – GPMC

With this, we have installed the Group Policy Management tools that allow you to use the GPMC to create and configure GPOs and custom OUs in an Azure AD DS managed domain.

This exercise taught us the skills needed to install Group Policy Management tools. The next section teaches us the skills for implementing Group Policy.

Exercise – Implementing Group Policy

This section will look at implementing Group Policy.

For this exercise, we will continue to use the Azure AD DS managed domain we created in the *Azure AD tenant* exercise in the previous chapter. If needed, please complete this exercise and then return.

You will need the following requirements in place to complete this exercise:

- A Windows Server management VM joined to the Azure AD DS managed domain with the Group Policy Management tools installed. If needed, please complete the previous exercise in this chapter and then return. Alternatively, please use your own environment so that you can follow this exercise.

We have broken down the exercise into individual tasks for a better understanding.

Task 1 – Creating a GPO to restrict access to Control Panel

This task focuses on security operations by restricting access to a computer's Control Panel. You apply the same approach to restrict access to Server Manager, Command Prompt, PowerShell, and so on.

This policy will be configured as a *User Configuration* setting and will apply to all computers a user accesses that this policy applies to; by default, a GPO applies to the *Authenticated Users* group. You could adjust this behavior by applying security filtering.

We will link the created GPO at the domain level. This means it will be applied to all users and computers in the domain and all child OUs through inheritance. For simplicity, we will not demonstrate override, blocking, or enforcement:

1. From the server, once the Group Policy Management tools are installed, launch **Group Policy Management** from **Tools** in **Server Manager**.

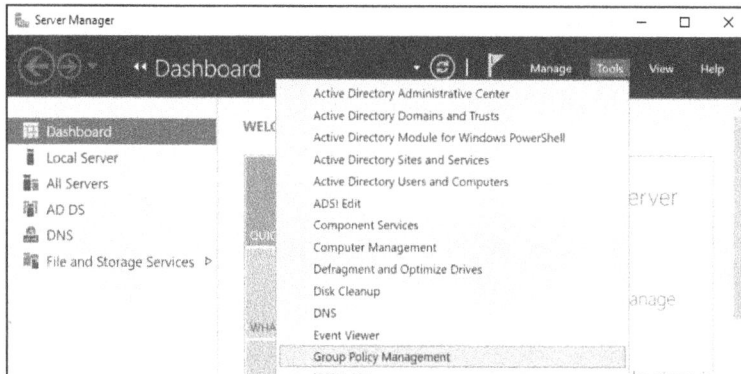

Figure 3.24 – Launch the GPMC

2.   From the GPMC, right-click the domain, click **Create a GPO in this domain, and Link it here…**.

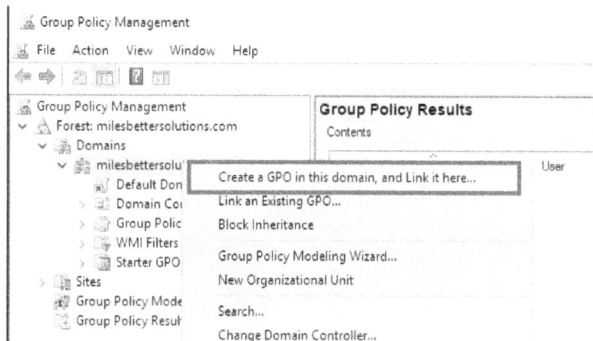

Figure 3.25 – Create the GPO

3.   Provide a name for the GPO and click **OK**.

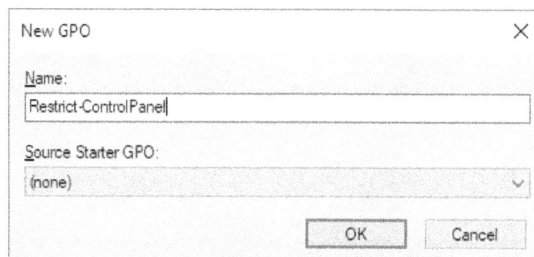

Figure 3.26 – Name the GPO

4.   Right-click the GPO that we just created and click **Edit…**.

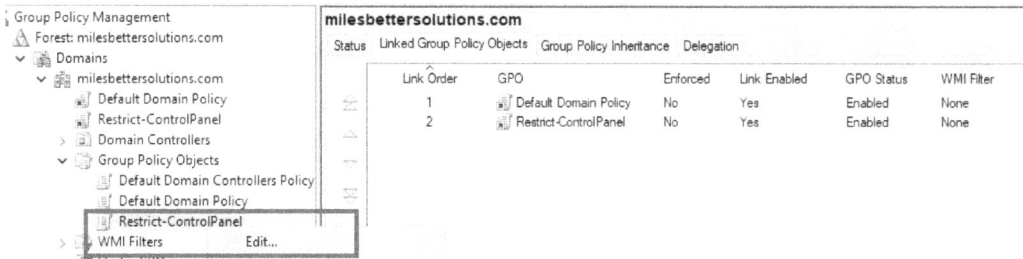

Figure 3.27 – Edit the GPO

5.   From **Group Policy Management Editor**, navigate to **User Configuration | Policies | Administrative Templates: Policy definitions (ADMX files) retrieved from the central store | Control Panel**.

Then, double-click **Prohibit access to Control Panel and PC settings**.

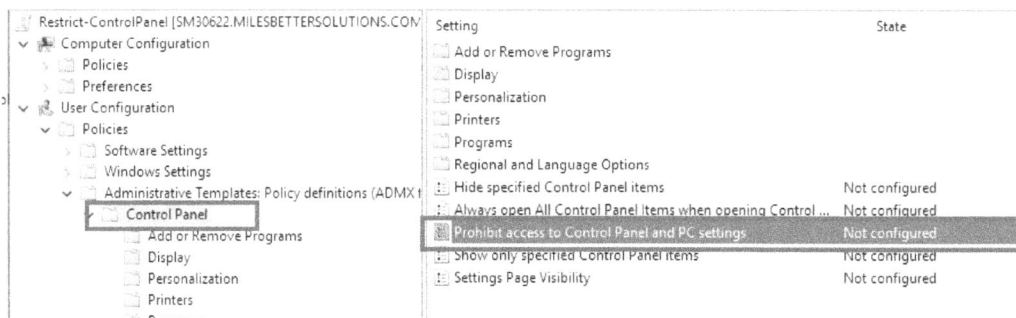

Figure 3.28 – Control Panel admin template

6. Select **Enabled** in the policy settings and then click **OK**.

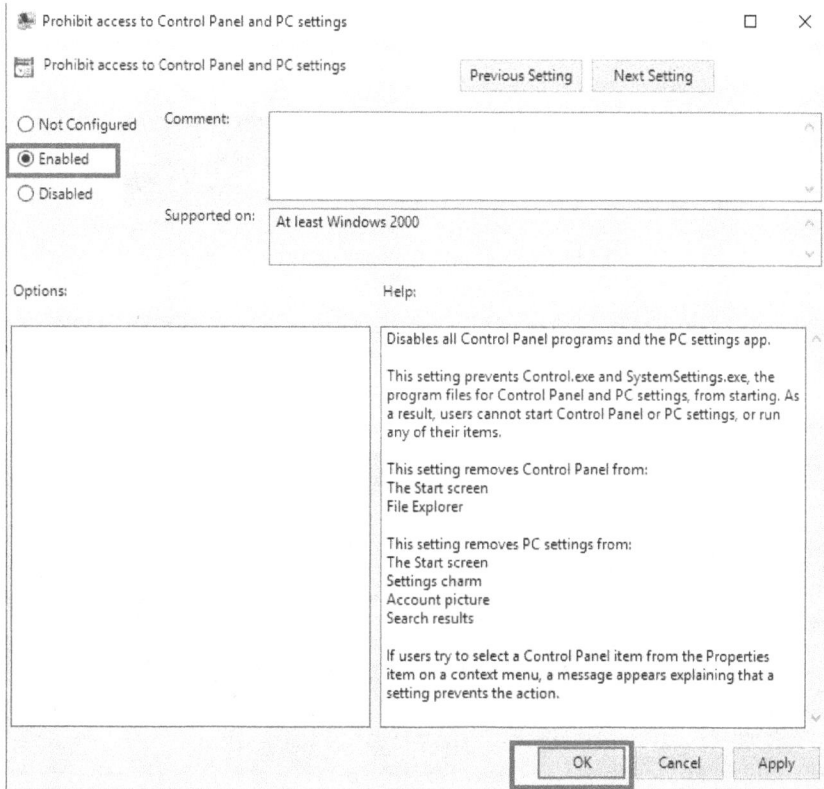

Figure 3.29 – Policy settings

Task 2 – Forcing an update and verifying that the policy is applied

In this task, we will force the policy to be applied and verify it:

1. From Command Prompt, run the `gpupdate /force` command:

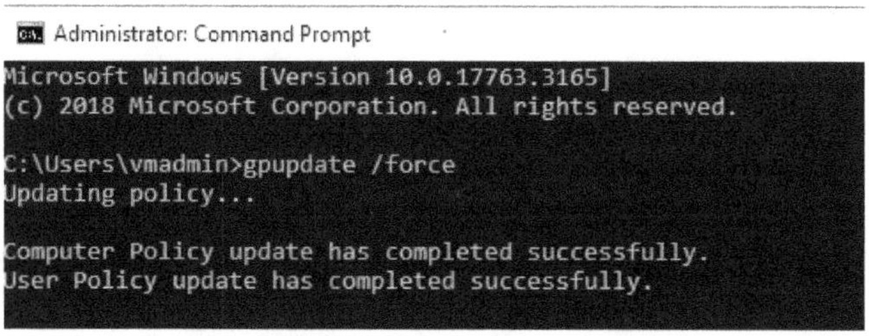

Figure 3.30 – Force the policy update

Confirm that the command was completed successfully.

2. From Command Prompt, run the `gpresult /R` command:

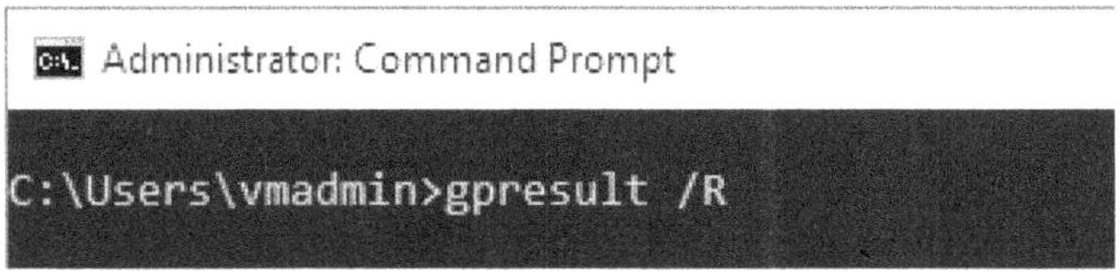

Figure 3.31 – Run the policy report

3. From the report, verify that the created GPO appears listed under **USER SETTINGS | Applied Group Policy Objects**:

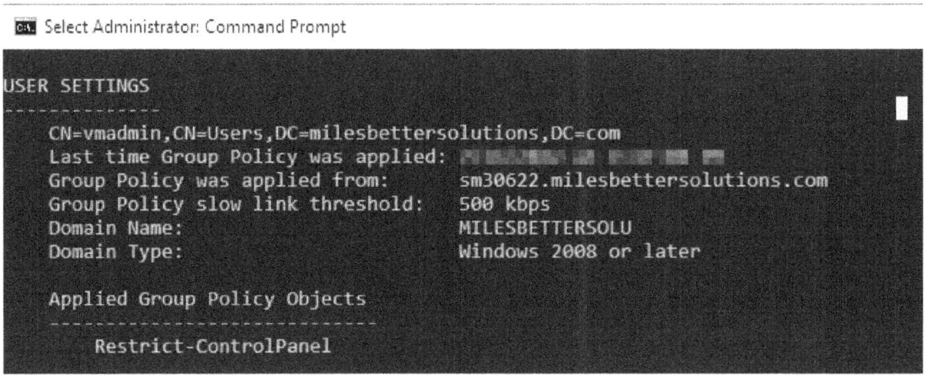

Figure 3.32 – Policy applied verified

Task 3 – Testing that the policy works

In this task, we will confirm that the policy setting works and that access to the computer's Control Panel is restricted.

Attempt to access the computer's Control Panel and ensure you receive a restriction message preventing access:

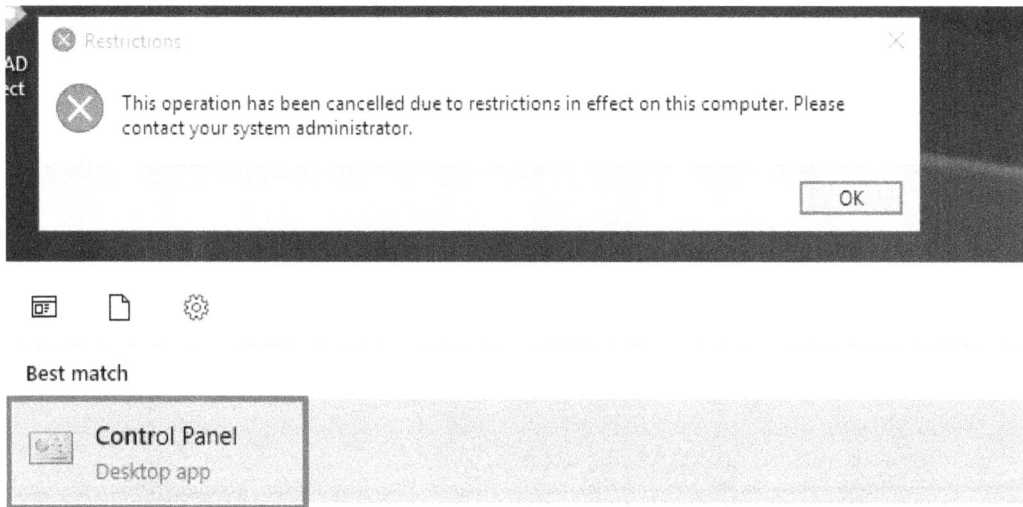

Figure 3.33 – Access to Control Panel restricted

With this, we have implemented a group policy that allowed us to restrict access to a computer's Control Panel.

This exercise taught us the skills needed to implement group policies.

Now, let's summarize this chapter.

Summary

This chapter has provided coverage for the AZ-800 Administering Windows Server Hybrid Core Infrastructure exam learning objective: *Manage Windows Server by using domain-based Group Policies*.

We learned about the concepts of Group Policy and its components and how to configure, manage, and troubleshoot. Two hands-on exercises then finished the chapter to provide you with additional practical skills.

This chapter aimed to take your knowledge beyond the exam objectives; we added new skills and learning with the content provided. This further develops your knowledge and skills for on-premises network infrastructure services and enables you to be prepared for a real-world, day-to-day hybrid environment-focused role. In the next chapter, we will learn about hybrid identities.

Further reading

This section provides links to additional exam information and study references:

- *Microsoft Certified: Windows Server Hybrid Administrator Associate*: `https://docs.microsoft.com/en-us/learn/certifications/windows-server-hybrid-administrator/`

- *Exam AZ-800: Administering Windows Server Hybrid Core Infrastructure*: `https://docs.microsoft.com/en-us/learn/certifications/exams/az-800`

- *Exam AZ-800: skills outline*: `https://query.prod.cms.rt.microsoft.com/cms/api/am/binary/RWKI0r`

- *Microsoft Learn*: `https://docs.microsoft.com/en-us/learn/modules/implement-group-policy-objects/`

Skills check

Challenge yourself with what you have learned in this chapter:

1. Explain what Group Policy is.
2. Explain what a GPO is.
3. What is the difference between a local GPO and a domain GPO?
4. How do group policies in Azure AD DS differ from those in AD DS?
5. What are the two built-in GPOs?
6. Can you create custom GPOs and OUs in Azure AD DS?
7. For GPO administration, what privileges are required in the Azure AD DS managed domain?
8. What is essential regarding GPOs for AD DS and Azure AD DS in a hybrid identity environment?
9. What are the two components of a GPO?
10. What AD objects can a GPO be linked to?
11. What AD objects can't a GPO be linked to?
12. What are the two top-level configuration node sections called?
13. What are GPO administrative templates?
14. What is the difference between the `.admx` file and the `.adml` file?
15. What is meant by GPO scope?

16. What is the order of GPO processing?

17. Explain inheritance, blocking, and enforcement.

18. What is the primary GUI management tool?

19. List the CLI tools that can be used.

20. What attributes does a GPO have?

4

Implementing and Managing Hybrid Identities

This chapter covers the AZ-800 Administering Windows Server Hybrid Core Infrastructure exam learning objective: *Implement and manage hybrid identities.*

In the previous chapter, we added skills to understand the concepts, creation, and management of Group Policy.

We will start this chapter with some concepts and an understanding of hybrid identity. We will then look at the AD Connect directory synchronization tool that is the enabler for a hybrid identity and the hybrid identity authentication methods available. We will then conclude with a hands-on exercise to develop your skills further. The following topics will be covered in this chapter:

- What is a hybrid identity?
- Azure AD Connect
- Hybrid identity authentication methods
- Exercise – Implementing Azure AD Connect

In addition to the topics listed in this chapter, this chapter's goal is to take your knowledge beyond the exam objectives to prepare you for a real-world, day-to-day hybrid environment-focused role.

What is a hybrid identity?

In this first section of this chapter, we will recap some definitions and concepts to set a baseline and foundation of knowledge to build from.

Microsoft provides two **identity provider (IDP)** and **Directory Service (DS)** solutions; **Active Directory (AD)** and **Azure Active Directory (Azure AD)**.

In a nutshell, **AD** is an **IDP** *dedicated* to a *single organization*. It runs as an installed service as part of Windows Server.

Azure AD, in a nutshell, is Microsoft's multi-tenant, cloud-based, and fully managed IDP and DS. Azure AD can be considered Microsoft's **Identity as a Service (IDaaS)**.

A **hybrid identity** provides users with a *common identity*. This is simply a means to provide users access to resources and services regardless of whether they are cloud-based or on-premises. This approach means less confusion for users and one identity that can be used independently of location or IDP. This can be visualized in the following illustration:

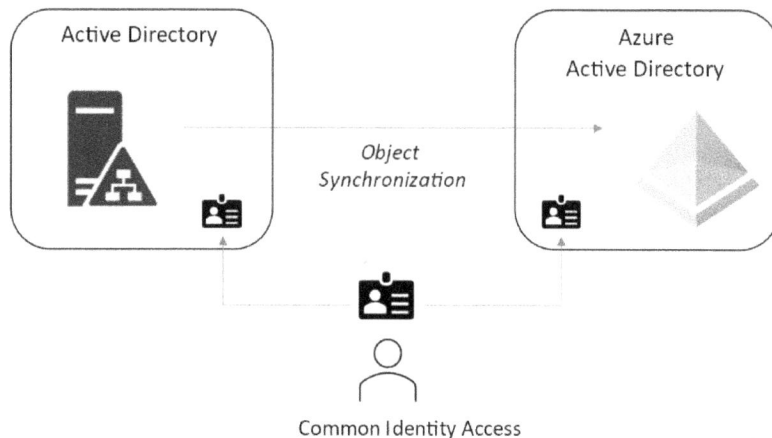

Figure 4.1 – AD and Azure AD integration

In this section, we provided a recap of Microsoft's IDPs, AD, and Azure AD, which we covered in the first two chapters, and defined what a hybrid identity is. The following section looks at the Azure AD Connect tool, the foundation for enabling hybrid identity authentication methods.

Azure AD Connect

Azure AD Connect is a free download tool enabling *hybrid identities* and seamlessly connecting and integrating AD with Azure AD.

Azure AD Connect is implemented by deploying the service to Windows Server, usually to a dedicated server. However, depending on your needs, it could also be run on a domain controller.

The following are the operations of AD Connect when installed on the chosen Windows server:

- Users, groups, and contact objects that exist or are added to AD configured for directory synchronization are copied to Azure AD through a synchronization process using Azure AD Connect. Note that users have no automatic license assignment, such as M365.

- Any changes to an object's attributes in AD will be updated to the synchronized object in Azure AD. The *Synchronization Service Manager* (the `miisclient.exe` process) of Azure AD Connect allows you to be selective about what attributes are synchronized, so not all are copied if required.

- Any deleted users, groups, and contacts from AD are also deleted from Azure AD.

- The single source of authority for objects comes from AD as the *authoritative IDP* and not Azure AD. This means any changes required in the Azure synchronized objects must have their attributes and changes made in AD and allow those changes to be updated in the Azure AD objects by the synchronization process.

- The previous point means that Azure AD objects are *read-only* and cannot be changed directly in Azure AD unless a writeback solution is implemented for password resets in Azure AD. **Self-service password reset (SSPR) writeback** is an optional feature that allows password changes to Azure AD, leading to the same identity having different passwords between IDPs. This can be solved with password writeback, so password changes made in Azure AD can be sent back to AD to update the objects attribute so that the same password is maintained in both AD and Azure AD directories.

Azure AD Connect cloud sync

Similar capabilities of Azure AD Connect are also available in the **Azure AD Connect cloud sync** service.

This is based on a *provisioning agent* approach rather than a fully installable application. The configuration and orchestration are handled from Azure AD rather than from AD itself. All information is stored within Azure AD rather than locally, where the AD Connect tool is installed and running. The tool has a use case for mergers and acquisitions where many multiple disconnected and isolated forests are required to be integrated with Azure AD. Multiple provisioning agents can be deployed to provide high availability for directory synchronization.

Azure AD Connect planning

The following planning activities must be considered before using the AD Connect tool. Planning is everything when it comes to implementing a hybrid identity:

- Identifying a domain-joined server to download and install the Azure AD Connect tool and components of the service; based on needs, the tool can be installed directly on a DC server in the domain to be synchronized or on a dedicated server. The following are the supported Windows Server OS versions:

  - Window Server 2016 or later; standard or datacenter edition.

  - Windows Server Essentials 2019.

- • The server where AD Connect will be installed must run a full GUI; Windows Server Core is not supported.

- • Identifying schema extensions and custom attributes.

- • Identifying domain functional levels.

- • Identifying the preparedness of AD with regards to *AD object attributes* and **User Principle Names** (**UPNs**). If you have a non-routable domain, such as a *local* one, that is used for user UPNs; then you should use a domain that can be validated with Azure AD – for instance, a *valid internet domain*, such as .com or .net.

You then need to set the UPN suffix for users to use this Azure AD-verified domain. You could change the primary domain of your Azure AD DS domain if you wished to synchronize to Azure AD, but a better approach would be to add a UPN suffix to the verified domain you wished to add to Azure AD. This change is made in the **Active Directory Domains and Trusts** management console from the **UPN Suffixes Properties** tab. The following illustration shows where this configuration is done:

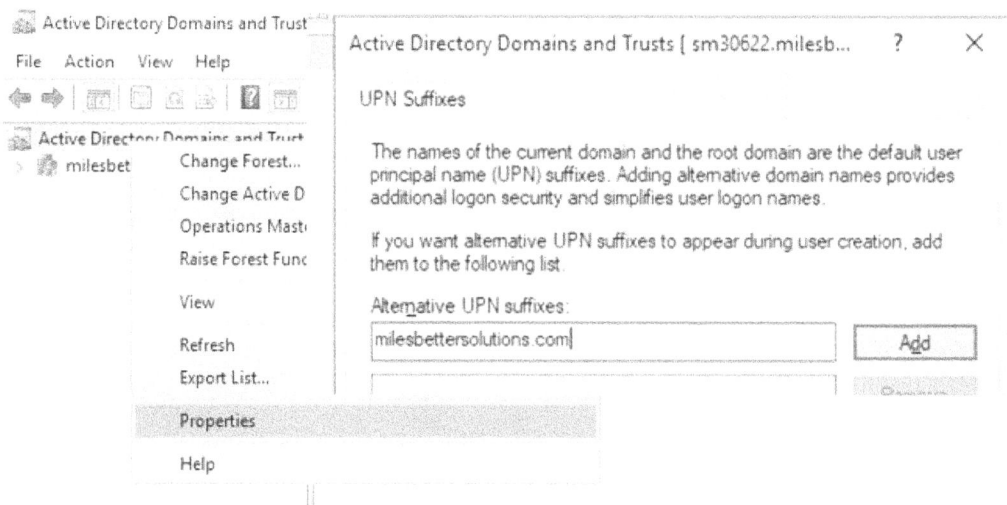

Figure 4.2 – Adding a UPN suffix

This topic will be covered in the hands-on exercise section of this chapter.

You can add multiple UPN suffixes and add and verify them in Azure AD if this meets your needs:

1.   Validate that you own the chosen domain(s) to add to Azure AD for directory synchronization and ensure you have access to DNS records to make a change. This is to be able to verify the domain(s) when adding the domain to Azure AD.

2. Once any non-routable domains for synchronization have been addressed and prepared, you can proceed to add and verify the user's *AD DS UPN suffix* domain(s) to Azure AD for use with directory synchronization and **single sign-on** (**SSO**). Then set this domain in Azure AD as the primary domain for the tenant, that is, you should not use the tenant's default `.onmicrosoft.com` domain.

Staging mode

AD Connect can also be used in staging mode, allowing changes to be made to a configuration and previewed before the server is made active. It also allows pre-production testing by verifying that all changes will be made as expected in production when you run a full import and synchronization through staging mode. Staging mode can be used in the scenarios of high availability, testing, the deployment of new changes to the configuration, and decommissioning or replacing servers.

Staging mode can be selected during the installation of AD Connect but does not run any exports. Staging mode does not run password sync or writeback. When staging mode is disabled, the exports will run, and password sync and writeback will be enabled.

Express and custom installation

The **express installation** method was designed to support a configuration that worked for *most customers most of the time,* and new installations use this about 90% of the time.

The *express installation* assumes the following:

* A single Windows Server AD forest

* An enterprise administrator account that can be used for the installation

* Less than 10,000 objects in the Windows Server AD forest

The express installation provides the following configuration outcomes:

* Password hash synchronization from Windows Server AD to Azure AD

* Synchronization of all eligible objects in all domains and OUs

The **custom installation** method is used when your scenario does not match the preceding assumption and configuration outcomes.

The *custom installation* assumes the following:

* No access to an enterprise administrator account

* More than one Windows Server AD forest to be synchronized

* Forest domains that are unreachable from the Connect server

- Federation or pass-through authentication is planned for user sign-in
- More than 100,000 objects in the Windows Server AD forest need to use the full SQL Server
- Group-based filtering is planned to be used, not just the filtering of domains or OUs

AD health check tools

It is critical to have an operationally healthy AD instance before implementing directory synchronization through Azure AD Connect. Typical issues that will be flagged and can be remediated will be related to UPNs and proxy addresses. The **IdFix** tool will check for errors and allow for any remediation once identified.

In this section, we introduced the Azure AD Connect tool. We also looked at the AD health-check tools. In the following section, we will look at the hybrid authentication methods.

Hybrid identity authentication methods

Hybrid identity requires an authentication method that supports the integration of AD and Azure AD. Using AD Connect, the following sign-on methods can be configured for users to enable hybrid identity in an organization:

- Azure AD **Password Hash Synchronization** (**PHS**)
- Azure AD **Pass-Through Authentication** (**PTA**)
- AD Federation Services

The following screenshot shows the sign-on methods and options that are available:

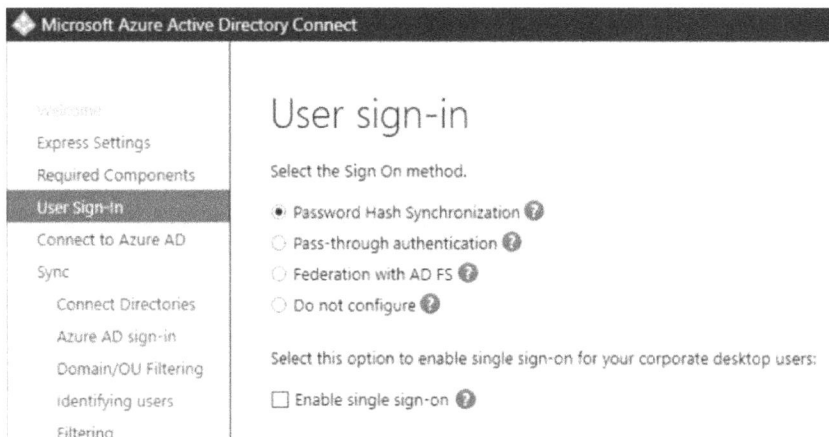

Figure 4.3 – Cloud sign-on methods

The authentication method chosen will depend on the needs of the scenario.

These hybrid identity authentication methods are categorized as *cloud authentication* and *federated authentication*.

In the following sections, we'll discuss each of these authentication methods.

Cloud authentication

With the cloud authentication model, the user's sign-on credentials are processed by *Azure AD* as the *IDP*.

Two sign-on methods are provided for hybrid identity cloud authentication: *PHS* and *PTA*.

PHS

In the PHS hybrid identity authentication model, a hash of the user's AD password hash is synchronized to Azure AD as the cloud-based IDP. This means each IDP has a copy of the password, but it is never stored in plaintext. AD stores a hash of the password, and Azure AD, as the cloud-based IDP, stores a *hash of the hash*.

With a copy of the user's password now stored in the cloud-based IDP, Azure AD, a user can sign in to Azure AD-authenticated services such as Microsoft 365 and Dynamics 365 using the *same password* to sign in to AD authenticated services. This provides a *common identity* for users, meaning many usernames and passwords no longer have to be remembered, just one. This gives a simple method for authentication to resources and services, whether they are cloud-based or on-premises.

PHS is the default cloud authentication method implemented when implementing Azure AD Connect.

The following diagram visualizes this *PHS cloud authentication* model:

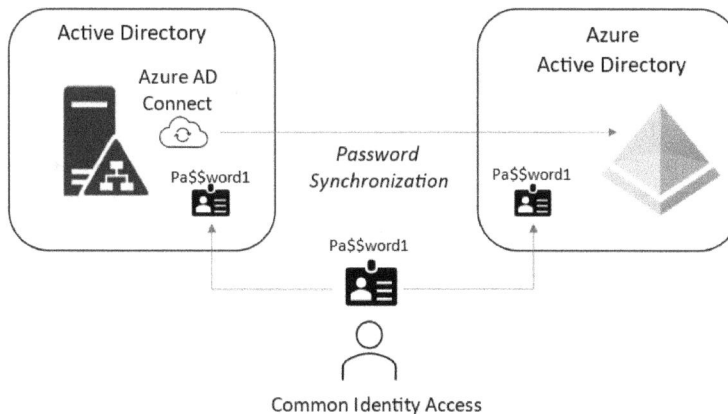

Figure 4.4 – PHS authentication sign-on

Concerns around storing the user's password in a cloud-based IDP (even if the password hash is further hashed) can be alleviated with the *PTA* and *federated authentication* models we will cover in the following sections. PHS can be a backup sign-in method for federated authentication but will require a password to be stored in Azure AD. Your needs and requirements will dictate the appropriate approach to take for your scenario.

PTA

Users can also use a *common identity* in the *PTA* hybrid identity authentication model. This allows a single username and password for signing on to cloud-based resources and services authenticated using Azure AD and signing on to AD-authenticated-based resources and services. As with *PHS*, *PTA* significantly improves user productivity, security, and identity governance.

The difference between this authentication method and PHS is that passwords are *never stored* in Azure AD. All password validations are *always* passed-through to your AD domain controllers to perform the authentication directly and are not validated by Azure AD. This does mean your domain controllers always need to be available and reachable by Azure AD.

It is, however, not a good strategy to have your cloud-based authentication solely reliant on accessing on-premises domain controllers to validate credentials – you can't fail over to the cloud or access cloud-based resources and services if you need to authenticate to an on-premises location you just lost due to flooding, fire, loss of connectivity, or even global pandemic.

One way to achieve this availability of access to resources, and business continuity in the face of losing connectivity to an on-premises location, is to create a *replica domain controller* in Azure as an IaaS VM. In case of a connectivity failure on-premises, the Azure-hosted domain controllers can be used to validate the credentials, ensuring business continuity and continued access to cloud-based resources and services. This approach requires cross-premises hybrid networking, such as a site-to-site VPN or ExpressRoute; ADDS cannot be replicated directly over a public internet endpoint.

Another solution is to use it in conjunction with *PHS directory synchronization*. This provides a backup solution for authenticating users when the on-premises domain controllers can no longer be contacted. However, this does require the password hashes to be stored in Azure AD, which may not be possible for some. There will be a trade-off then for determining which approach is most appropriate for your scenario. The same on-premises access dependency scenario exists for federated authentication, which we will address again in the following section.

The following diagram visualizes the *PTA cloud authentication* model:

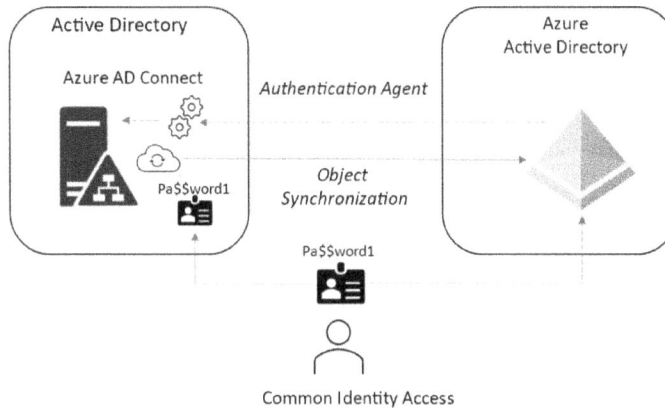

Figure 4.5 – PTA sign-on

A benefit of this authentication model can be that your own password policies and security policies from your AD can be enforced. From a network security perspective, only *outbound traffic connections* from AD to Azure AD are made; *no inbound connections* are made to your AD domain controllers. All communication is *certificate-based*, issued, and managed by Azure AD as a fully managed service.

This section looked at the cloud authentication sign-in methods. In the next section, we look at the federated authentication method.

Federated authentication

Federation simply means a *trust* shared between different IDPs for access to resources and services.

Azure AD is the *trusted IDP* to validate and authenticate user sign-on requests in this scenario. We will look at federation in the following section.

AD FS

Active Directory Federation Services (**AD FS**) is the Microsoft service that provides a sign-on method for this federated authentication.

In the federated authentication hybrid identity authentication model, the user's sign-on credentials are passed to an IDP other than Azure AD to validate and authenticate the user's credentials. This other IDP used to trust Azure AD is *Windows Server AD* and can be hosted on a server on-premises or as an Azure IaaS VM. An *AD FS server infrastructure* is also required to be implemented, which will require two servers for the ADFS proxy in the perimeter network, two servers for the federation servers, and, depending upon the scenario, an optional two servers for SQL Server. In addition, you will need to consider networking, firewall ports, and certificates; this is by no means trivial and is complex at best.

This works similarly to the PTA method; however, PTA is much more straightforward and less complex than federation and does not require any additional infrastructure servers as AD FS does. You should consider where use cases allow evaluating PTA over AD FS; we recognize this is not possible in all scenarios where AD FS-only capabilities exist.

The following diagram visualizes this *federated authentication* model:

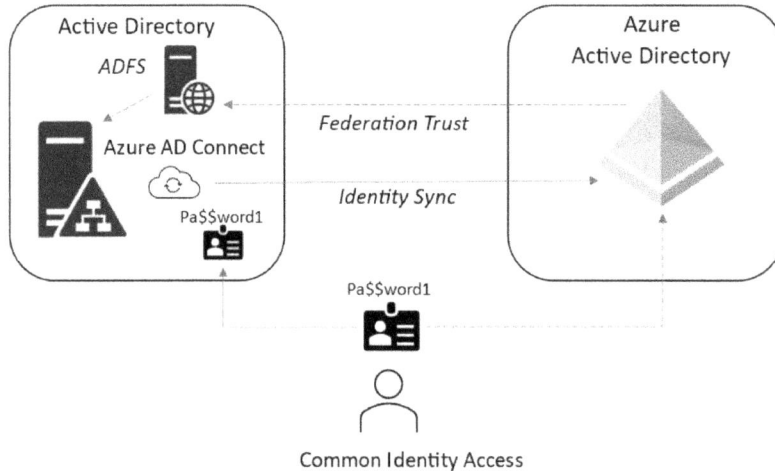

Figure 4.6 – Federated authentication sign-on

As we learned in the previous section on the PTA cloud authentication model, this approach does mean your domain controllers always need to be available and reachable by Azure AD. The same as we saw with the PTA authentication model is that *PHS directory synchronization* can be used as a backup sign-in method in the case of failure of the AD FS infrastructure to provide a business continuity strategy.

In this section, we introduced the hybrid identity authentication methods. In the following section, we look at some use case scenarios based on our needs.

Choosing a suitable hybrid authentication method

With identity being the control plane for security, it is vital to ensure that each authentication method is understood and chosen based on needs.

Cloud and federated authentication methods each have some limitations; not all authentication models support all use case scenarios.

Only PTA and AD FS methods can be utilized when you want to use the following hybrid scenarios:

- Prevent password hashes from being stored in the cloud (Azure AD)

Only the AD FS method can be utilized when you want to use the following hybrid scenarios:

- Implement on-premises **multi-factor authentication (MFA)**

- Provide smartcard authentication

In the following illustration, we look at a decision tree to aid in the process of choosing the most appropriate authentication method to meet your needs:

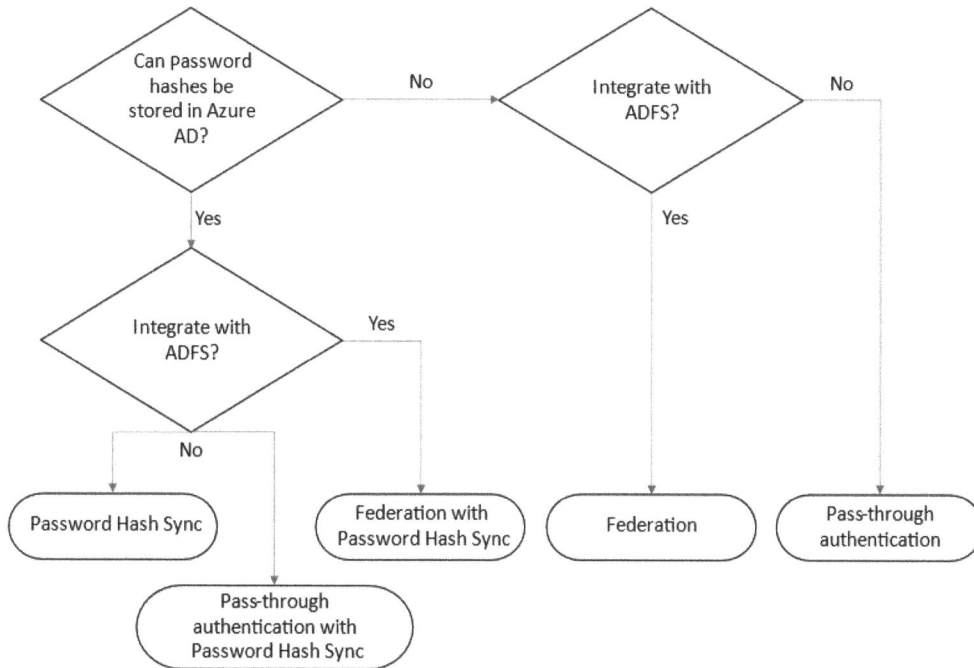

Figure 4.7 – Hybrid identity authentication methods

For clarity and to reduce complexity, the diagram does not intend to represent all decision points for all scenarios. Only core scenarios are shown for simplicity.

This section looked at choosing the most suitable hybrid identity approach to meet your needs. In the next section, we will look at SSO.

SSO

The option for **SSO** functionality can also be provided with a hybrid identity. This allows users already signed in to a device with their AD credentials to be signed in to cloud resources and services automatically.

SSO works with federated authentication, such as *AD FS*, and cloud authentication, such as *PTA*. This ensures that any requests for access to Azure AD-authenticated resources are validated by AD as the IDP and not Azure AD. It is more secure in its network operation as it does not require inbound access; all traffic is initiated as outbound from the sync agents and is secured through Microsoft-managed certificates.

PTA provides a similar capability for *SSO* as *AD FS*, although it is less complex to implement and operate as it requires no AD FS infrastructure.

The term *SSO* should not be confused with the *same sign-on*, which can also confusingly be referred to as *SSO*.

Directory synchronization provided through the cloud authentication method of *PHS* allows users to access cloud resources using the *same* credentials. Still, they will be required to enter them again when accessing Azure AD-authenticated resources when already signed into AD.

This section looked at implementing SSO. In the next section, we will complete a hands-on exercise section to re-enforce some of the concepts covered in this chapter.

Hands-on exercise

To support your learning with some practical skills, we will utilize the concepts and understanding gained from the previous sections of this chapter and practically apply them.

We will look at the following exercise:

- Exercise – Implementing Azure AD Connect

Getting started

To get started with this section, if you do not already have an Azure subscription, you can create a free Azure account at `https://Azure.microsoft.com/free`. This free Azure account provides the following:

- 12 months of free services

- $200 credit to explore Azure for 30 days

- Over 25 services that are always free

Let's move on to the exercise for this chapter.

Exercise – Implementing Azure AD Connect

This section will look at implementing Azure AD Connect. This will allow you to synchronize your AD objects with Azure AD and provide users with a common identity. This means the same AD username and password can be used to access Azure AD-authenticated resources and services.

To begin this exercise, we will need an *AD DS domain* and an *Azure AD tenant* that currently does not have Azure AD Connect directory sync implemented. Your Azure AD tenant should look the same as the following illustration to proceed, showing that it is not installed:

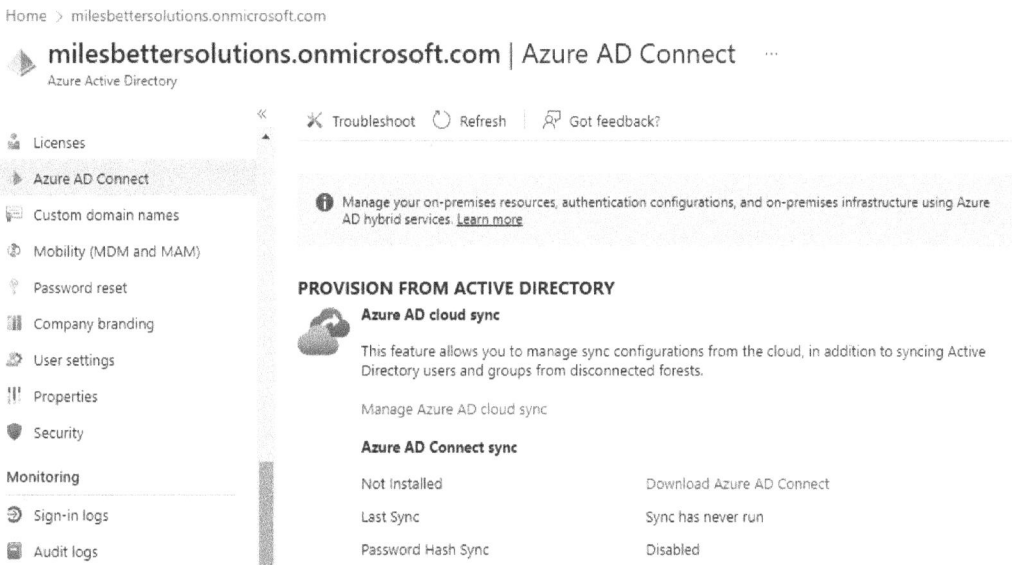

Figure 4.8 – Azure AD Connect in the Azure portal

For this exercise, a *custom domain name* must be added to Azure AD and verified. This domain name should be the *user's UPN suffix* from the AD DS domain that will be synchronized to Azure AD. This must be a domain you own and for which you can update DNS records.

The following illustration shows this represented in the Azure portal, and we will cover more in the task instructions to follow:

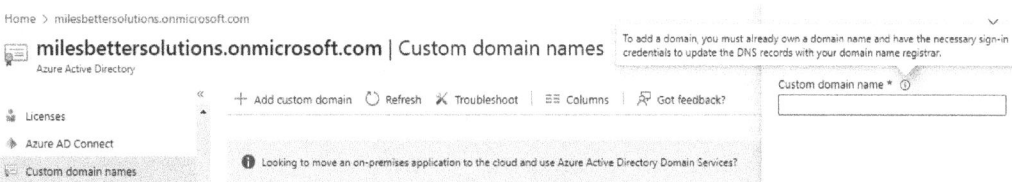

Figure 4.9 – Adding custom domain names in Azure AD

For this exercise, you could use the AD instance we created in the *Install AD DS on Windows server exercise* in *Chapter 1*. You should adjust the steps according to the environment you will use.

You will need the following accounts and permissions to complete the exercise:

- An AD account with *Enterprise Administrator* permissions and *Local Administrator* rights to the server where the Azure AD Connect tool is installed

- An Azure AD account with *Global Administrator* permissions

- A public domain that you own that has access to DNS records you can manage

We have broken down the exercise into individual tasks for better understanding:

Task 1 – Preparing a non-routable domain for AD Connect directory synchronization

> **Note**
> This task is only applicable to readers who have a non-routable domain. You may skip this task if this is not relevant to your scenario.

In this task, we prepare a non-routable AD DS domain (such as a .local namespace) for directory synchronization by adding a UPN suffix that can be added to Azure AD and verified for use with Azure AD Connect. We then change the user's UPN suffix to this domain:

1. Log in to an *AD DS DC* and select **Tools | Active Directory Domains and Trusts** from **Server Manager**:

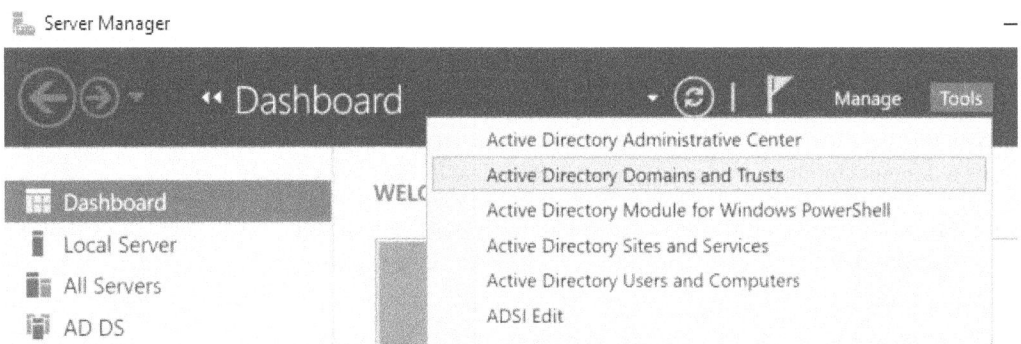

Figure 4.10 – Server Manager tools

2. Right-click the root level tree in the **Active Directory Domains and Trusts** console and select **Properties**:

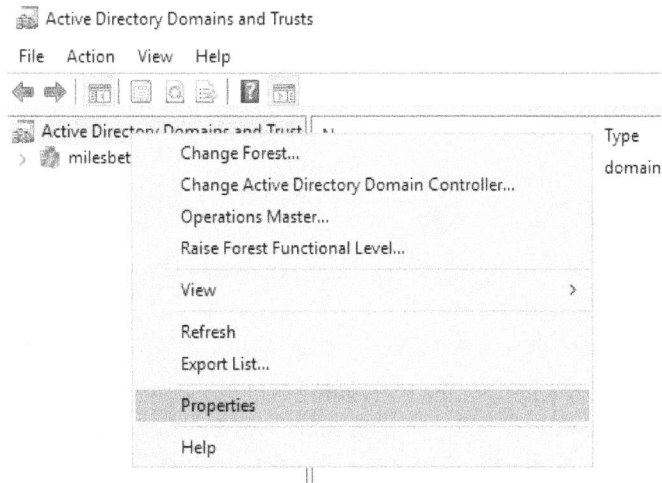

Figure 4.11 – The Active Directory Domains and Trusts console

3. From the **UPN Suffixes** tab, enter the UPN suffixes into the **Alternative UPN suffixes** field, and then click **Add**. Then, close the **Active Directory Domains and Trusts** console:

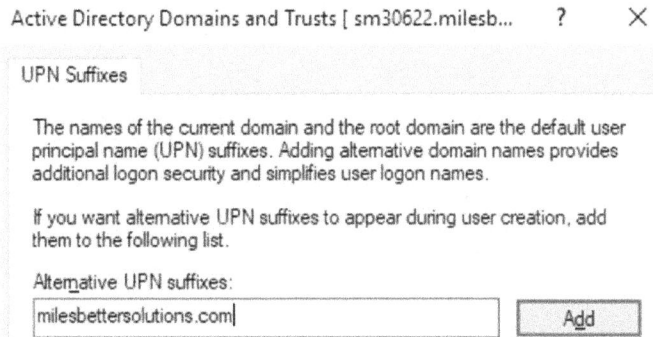

Figure 4.12 – Adding UPN suffixes

4. From **Server Manager**, select **Tools | Active Directory Users, and Computers**:

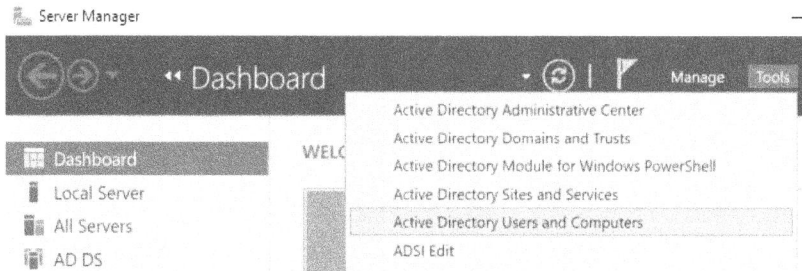

Figure 4.13 – Server Manager tools

5. Please navigate to a user whose *UPN suffix* you wish to change. Right-click and select **Properties**.

6. Select the **Account** tab, and from the **UPN suffix** field dropdown, select the UPN suffix to be used. In this case, we will select the domain suffix, @milesbettersolututions.com, which we added in a previous step, using the **Azure Directory Domains and Trusts** console to replace the non-routable domain of @milesbettersolutions.local with the user's new UPN and their sign-in address:

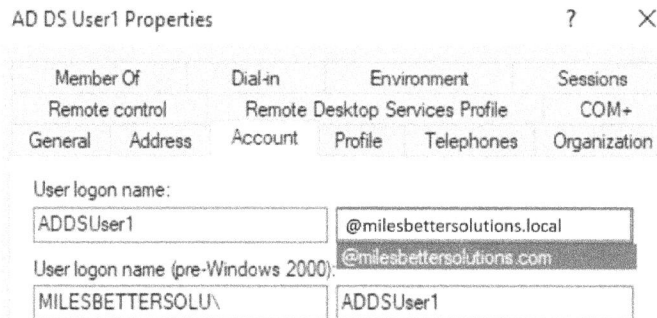

Figure 4.14 – UPN suffix selection

7. Repeat for *each user* (use a multi-select), close the **Active Directory Users and Computers** console, and prepare for the next task.

Task 2 – Adding the custom domain name to Azure AD

In this task, we will add a custom domain to Azure AD and verify it to be used with Azure AD for directory synchronization. You will require a public domain that you own, and that has access to DNS records you can manage for this exercise:

1. Log in to the *Azure portal* with an account with a **Global Administrator** role for the tenant.

2.   Navigate to **Azure Active Directory** and select **Custom domain names**:

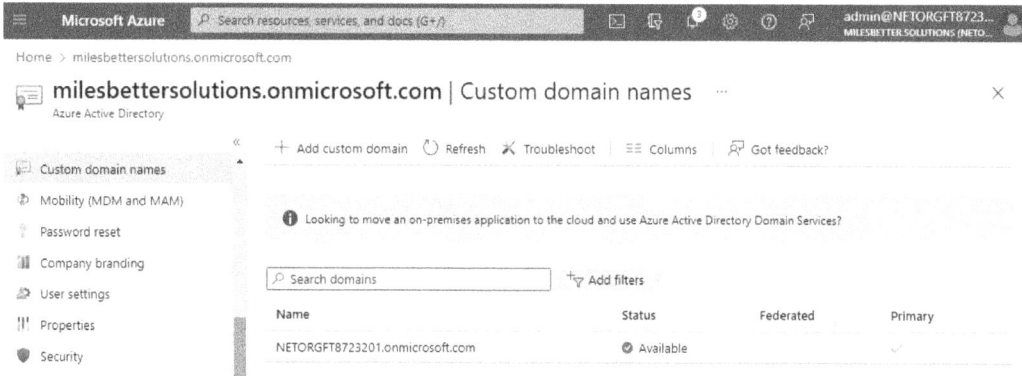

Figure 4.15 – Custom domain names in Azure

3.   Click **Add custom domain**, enter the required name, and then click **Add domain**:

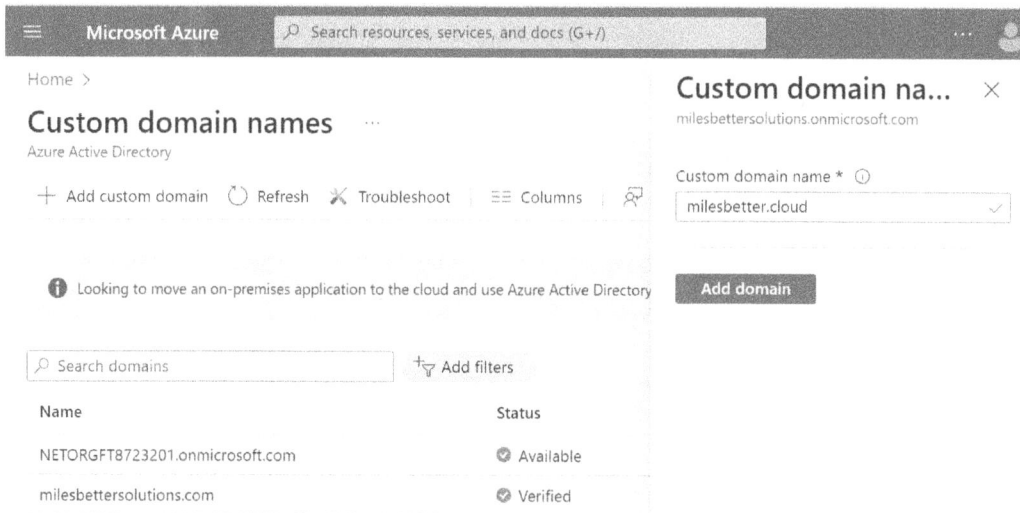

Figure 4.16 – Adding a custom domain name in Azure

4. You will now be required to *verify* your domain. As the information box shows, you should now create the DNS records with your registrar:

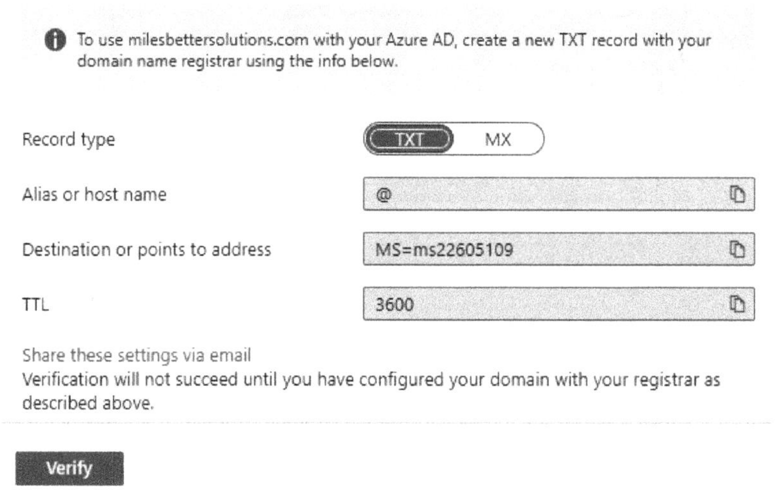

Figure 4.17 – Domain verification

5. For this exercise, my domain name and DNS are managed via GoDaddy as the registrar. You should follow the individual process for your domain name registrar and then return to the next step:

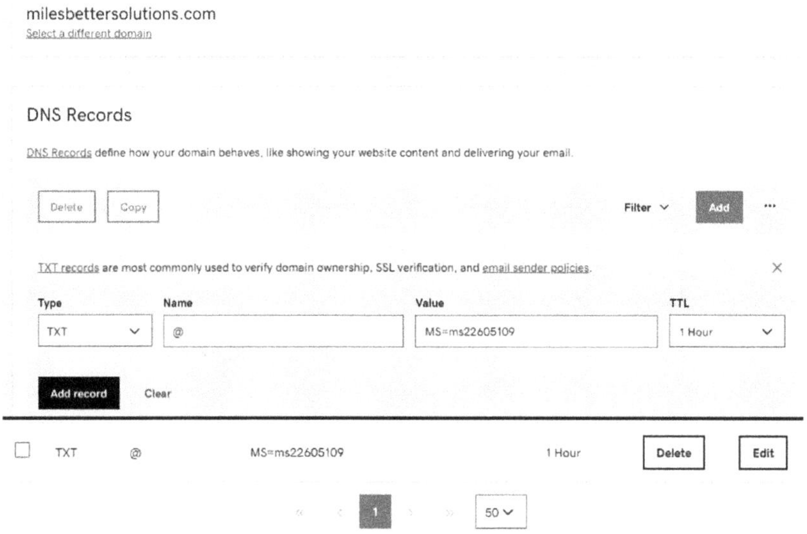

Figure 4.18 – Adding verification DNS records using your registrar

6. After the required DNS record has been set using your registrar, you may have to wait for the DNS records to update. You can then click **Verify** in the Azure portal:

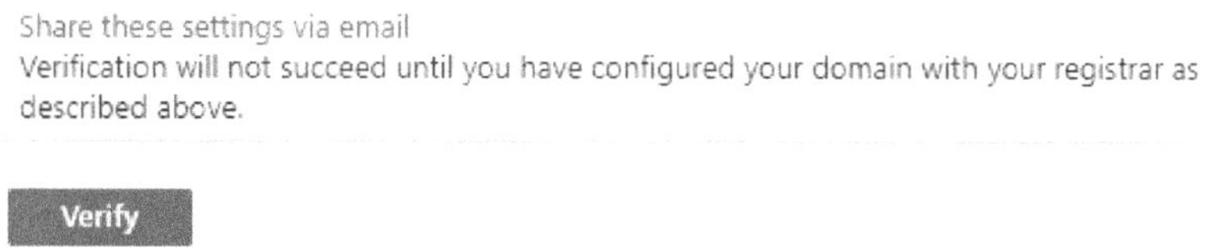

Share these settings via email
Verification will not succeed until you have configured your domain with your registrar as described above.

Verify

Figure 4.19 – Domain verification in the Azure portal

7. You will receive a notification that the verification of the domain was successful:

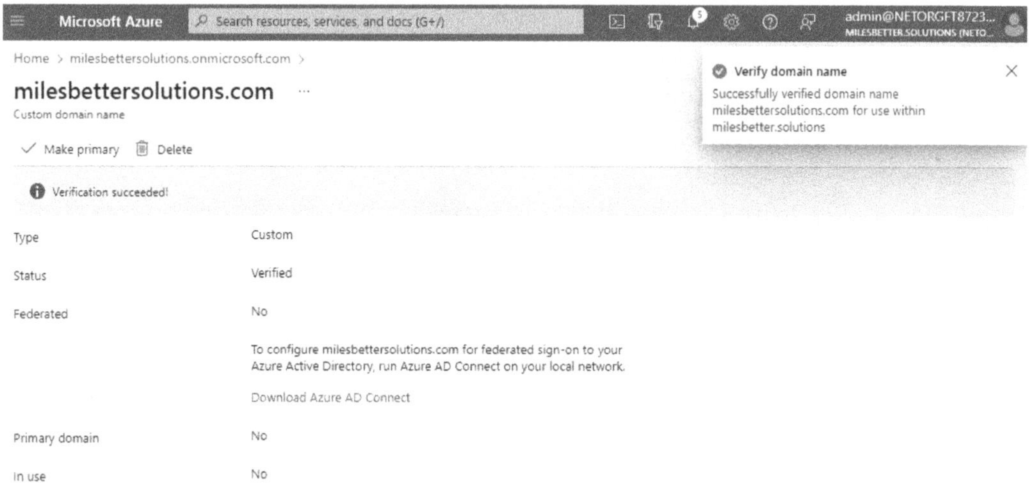

Figure 4.20 – Successful domain name verification

8. Now that you have successfully verified the domain, you should make it the primary domain for the Azure tenant by clicking **Make primary**:

Figure 4.21 – Setting a primary domain for a tenant

9. You will receive a notification that the domain has been made *primary* for the tenant:

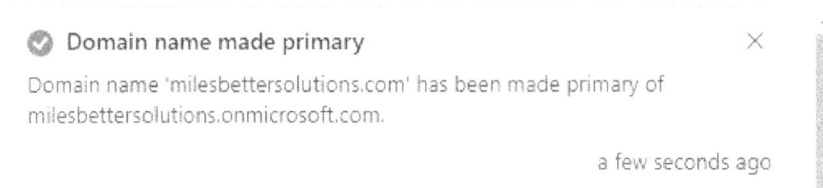

Figure 4.22 – Successful primary domain name change for the tenant

10. In the **Custom domain names** blade, you can now see that the domain is shown as the *primary domain*:

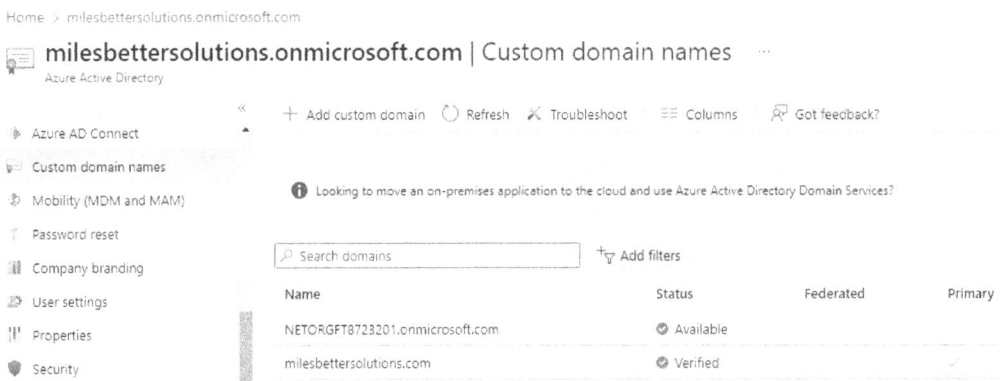

Figure 4.23 – Updated primary domain in the Azure portal

11. You can now close the **Custom domain names** blade, ready for the next task.

Task 3 – Downloading and installing Azure AD Connect

We will download and configure the Azure AD Connect tool in this task. For this exercise, we will follow a simple use case of PHS directory synchronization:

1. From a server where you will download and install the tool, log in to the *Azure portal* with an account with a **Global Administrator** role for the tenant.

2. Navigate to **Azure Active Directory** and select **Azure AD Connect**. Then, click **Download Azure AD Connect**:

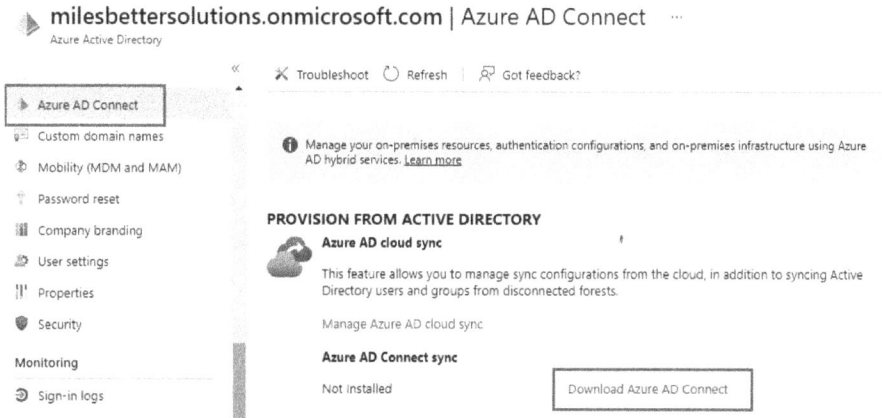

Figure 4.24 – Provisioning Azure AD Connect

3. Review the information under **Details | System Requirements | Install Instructions** and then click **Download**:

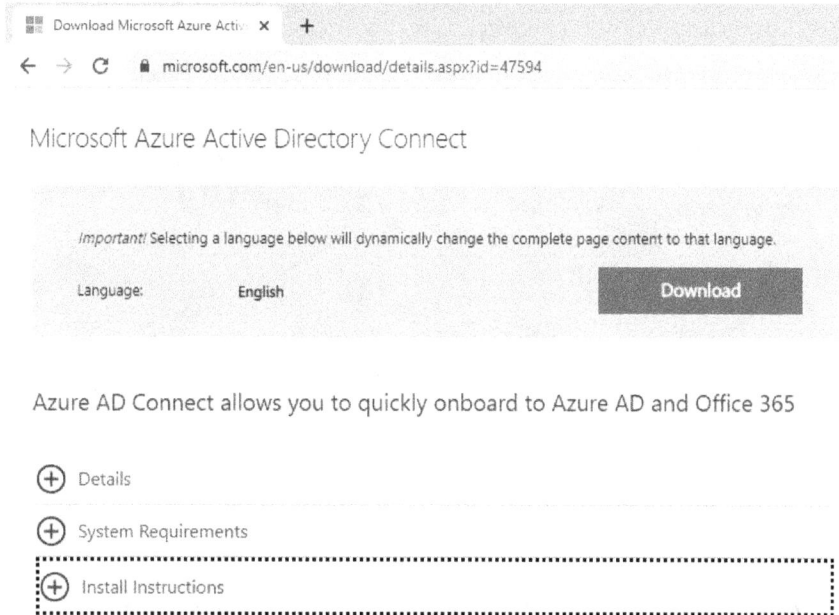

Figure 4.25 – Downloading Azure AD Connect

4.   Navigate to the download location and click the **AzureADConnect** installer to start the installation:

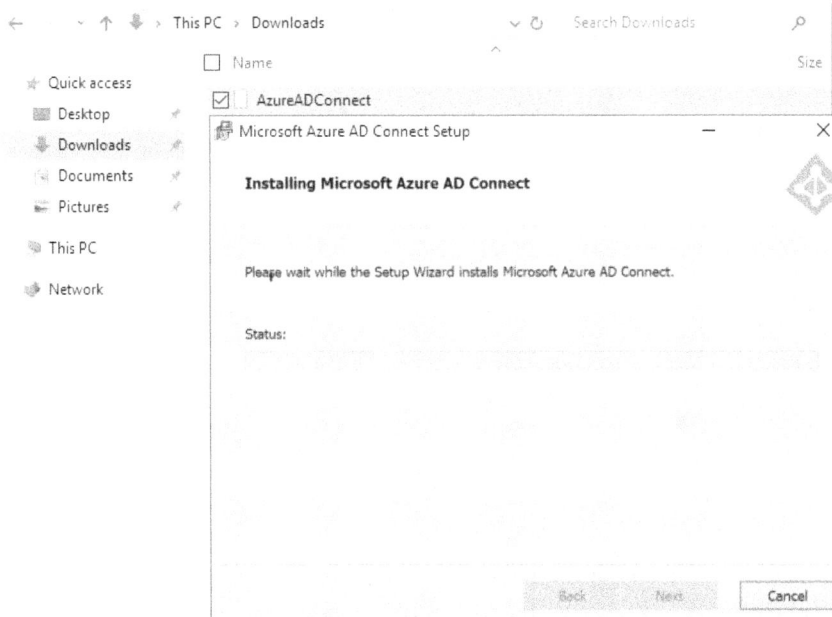

Figure 4.26 – The AzureADConnect installer

5.   Review the **Welcome to Azure AD Connect** page information, and check that **I agree to the license terms and privacy notice.** checkbox, and click **Continue**:

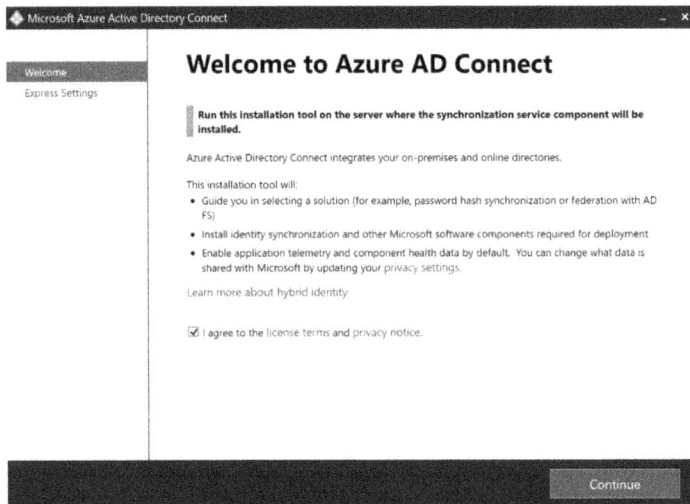

Figure 4.27 – Azure AD Connect welcome screen

6. For this exercise, click **Use express settings** on the **Express Settings** screen:

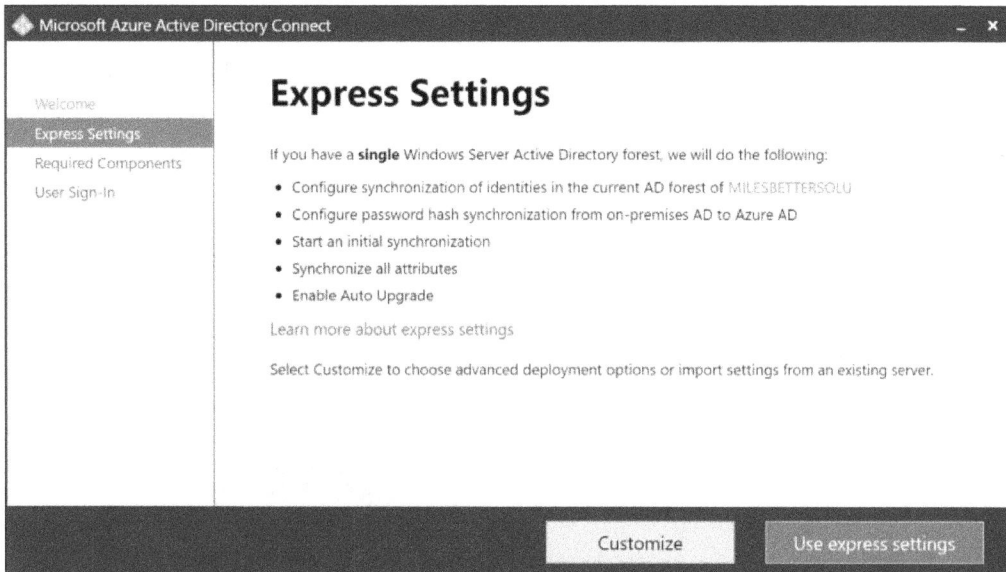

Figure 4.28 – Azure AD Connect express settings

7. Review the *helper information* on the **Connect to Azure AD** screen and enter the *username* and *password*. Then, click **Next**:

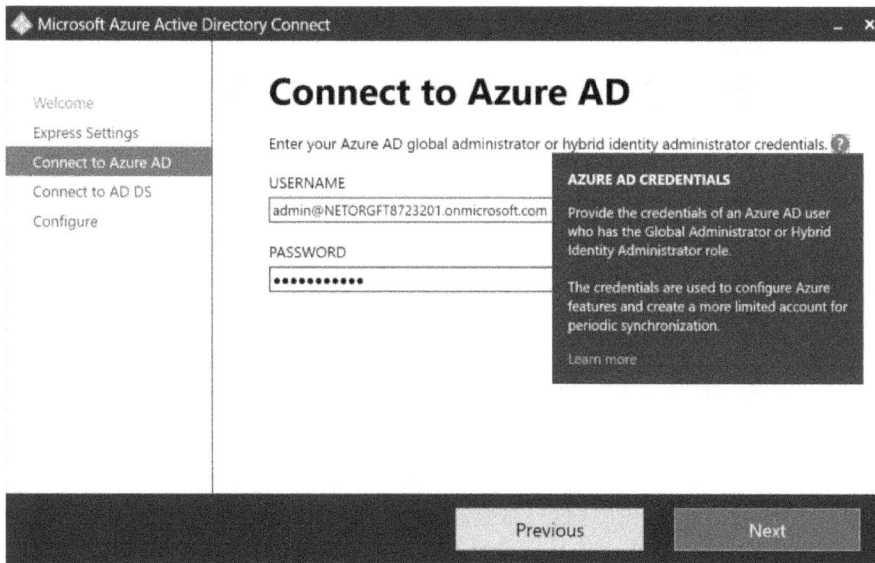

Figure 4.29 – Connecting to Azure AD

8. Review the *helper information* on the **Connect to AD DS** screen and enter the *username* and *password*. Then, click **Next**:

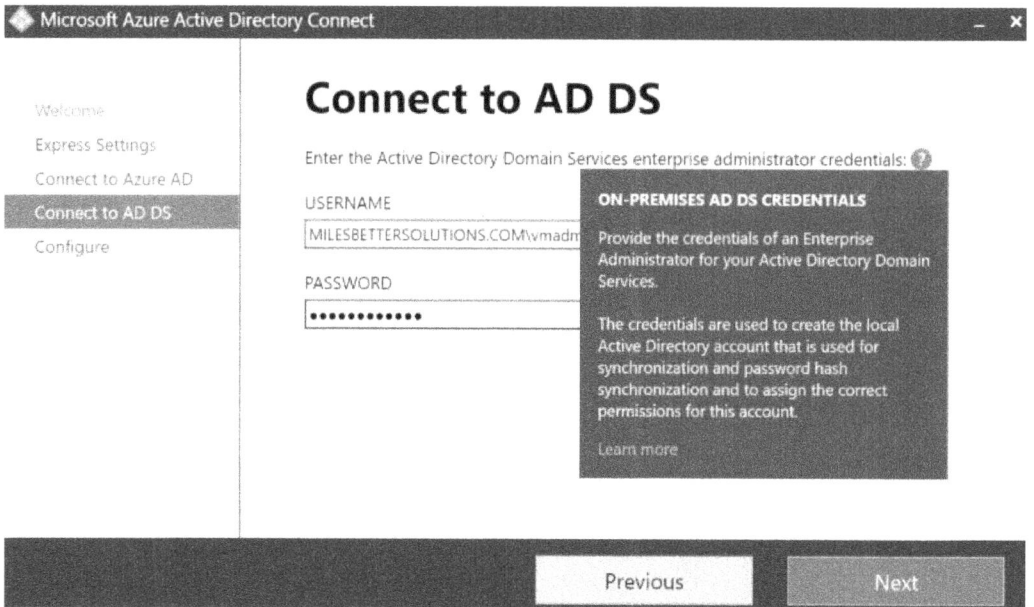

Figure 4.30 – Connecting to AD DS

9. Review the information on the **Ready to configure** screen and then click **Install**. The configuration will begin and progress:

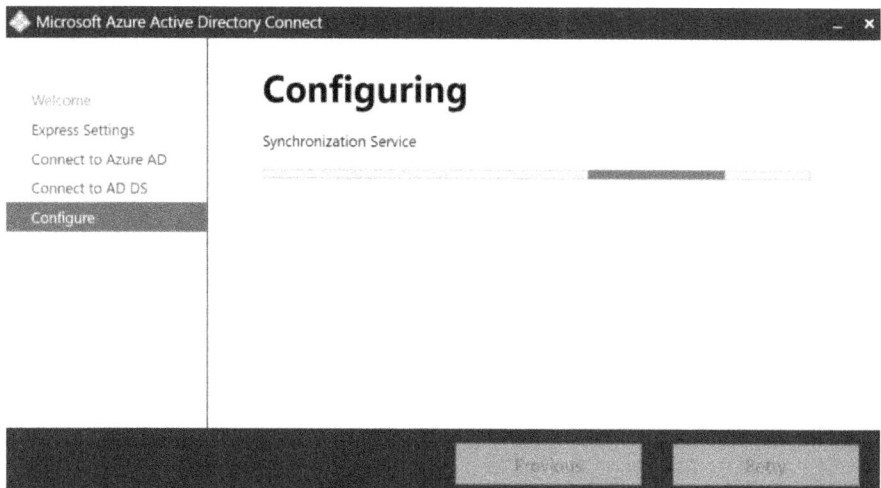

Figure 4.31 – Configuration in progress

10. Review the information on the **Configuration complete** screen and click **Exit**:

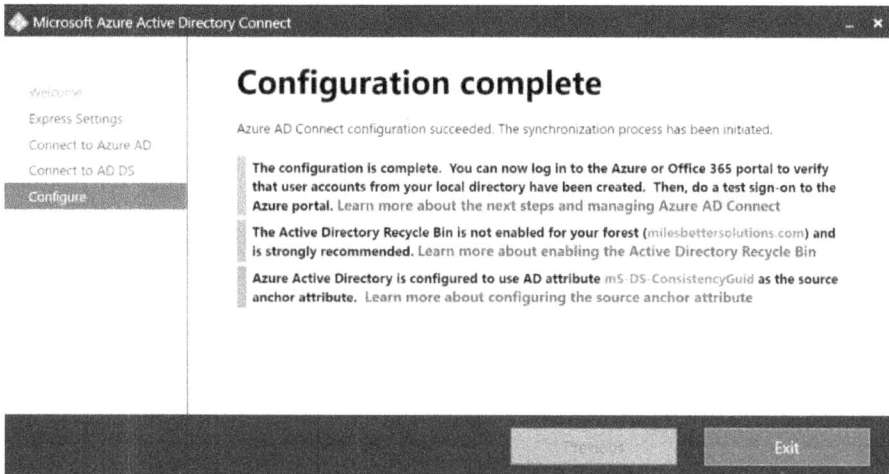

Figure 4.32 – Azure AD Connect configuration complete

11. Log in to the *Azure portal*, navigate **Azure Active Directory**, and select **Azure AD Connect**. You will see that the **Sync Status** and **Password Hash Sync** options now show as **Enabled,** and a time that the **Last Sync** occurred is provided. Synchronizing AD objects in the Azure AD portal can take up to 15 minutes to appear:

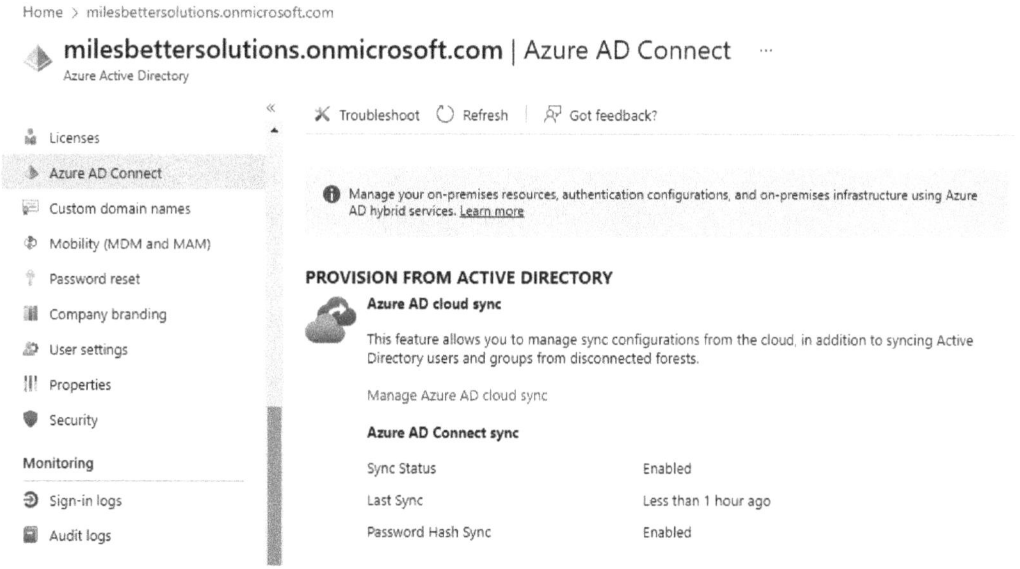

Figure 4.33 – The provision status of Azure AD Connect

In the following task, we look at verifying synchronization.

Task 4 – Verifying synchronization

This task looks at verifying directory synchronization using the Synchronization Service Manager:

1. From the server where *AD Connect* was installed, click the **Start** menu and select **Synchronization Service Manager** from the **Azure AD Connect** menu item.

2. On **Synchronization Service Manager**, from the **Operations** tab, review the tasks for **Connector Operations** and **Synchronization Statistics**. This is represented in the following figure:

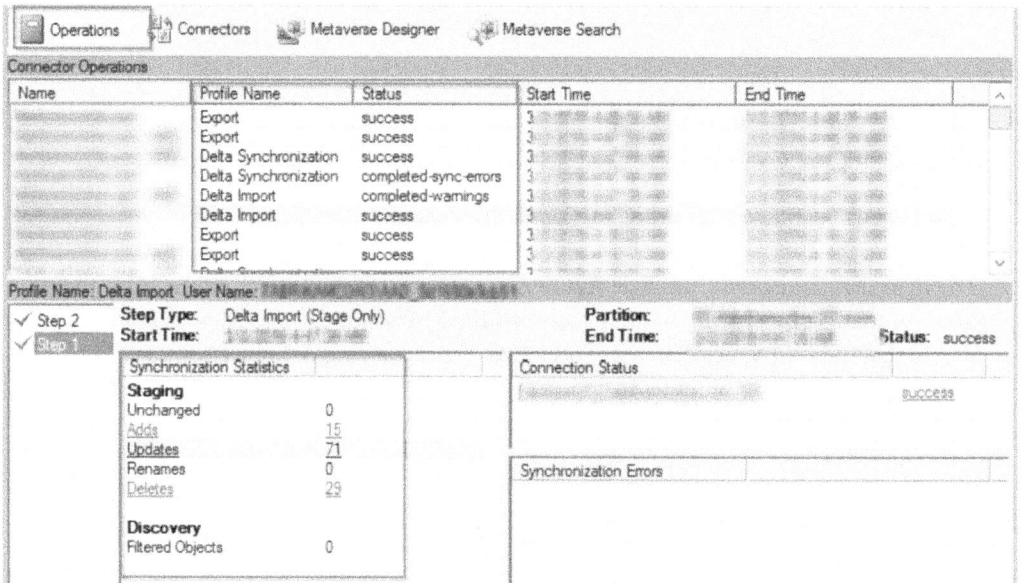

Figure 4.34 – Synchronization Service Manager operations

3. Click the **Connectors** tab and review the two connector types and the **Actions** that can be taken. One connector is for **AD DS focus**, and one is for **Azure AD focus**. This is represented in the following illustration:

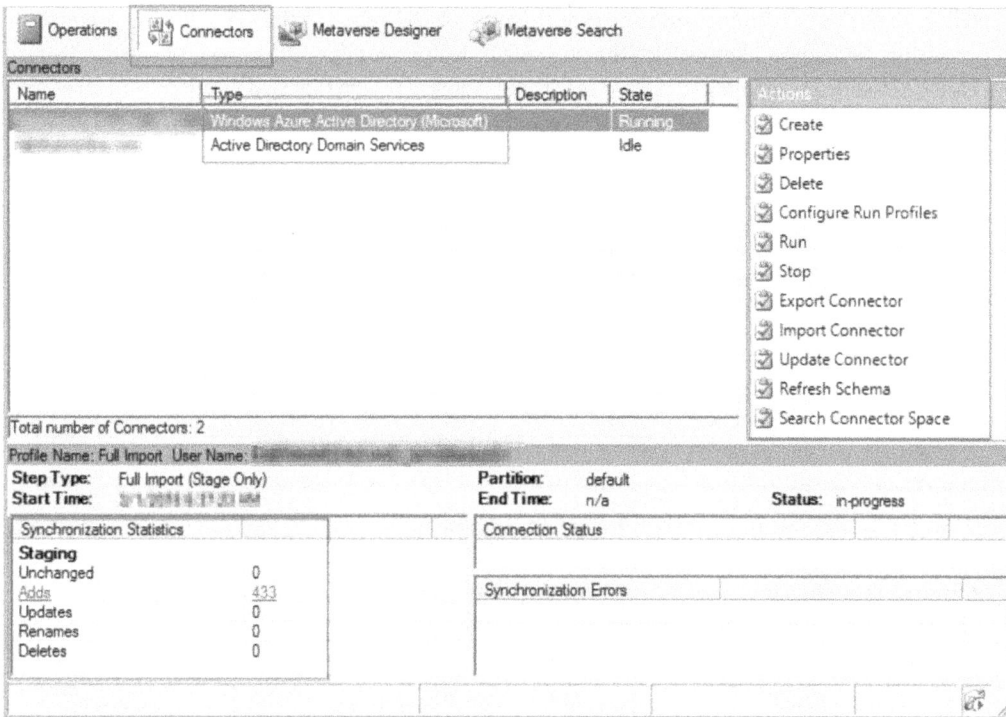

Figure 4.35 – Synchronization Service Manager connectors

4. You may now close **Synchronization Service Manager**.

5. You synchronize changes also by using the following PowerShell `cmdlet` command:

```
Start-ADSyncSyncCycle
```

In the following task, we will look at tasks for managing users in Azure AD that have been directory synchronized.

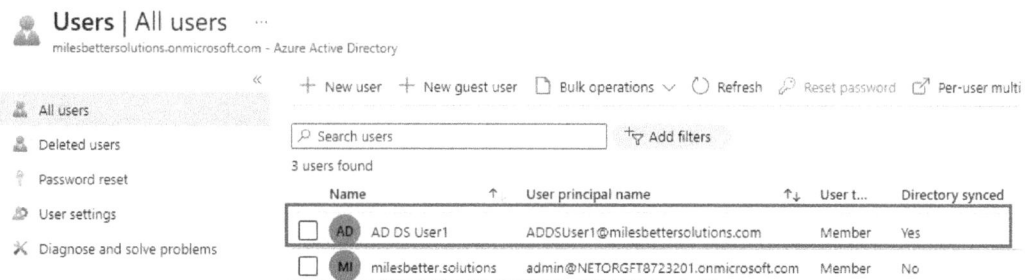

Figure 4.36 – A directory synchronized user

7.   Click on the user account and note that the user's UPN remains the same for on-premises, providing a common identity with the same SSO capabilities to use the same UPN for accessing either AD-authenticated resources and services or Azure AD-authenticated resources and services:

Figure 4.37 – A user account UPN

8.   Observe that you *cannot edit* any of the attributes for this directory synchronized user. All changes must be made to the object in AD and allow synchronization to update the attributes in Azure AD:

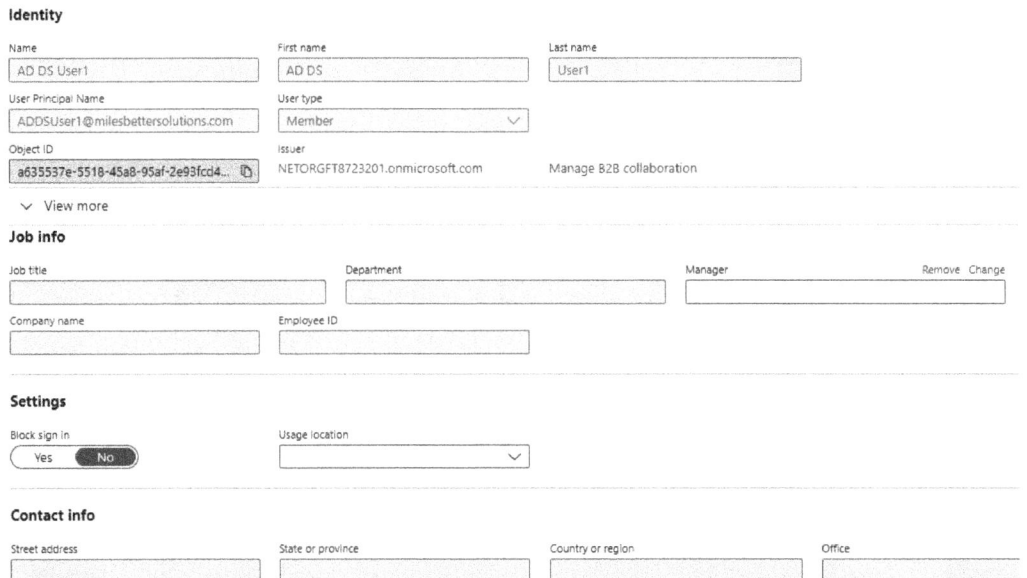

Figure 4.38 – Editing user attributes

9.   Observe that to use the **Reset Password** option, you must have enabled SSPR writeback:

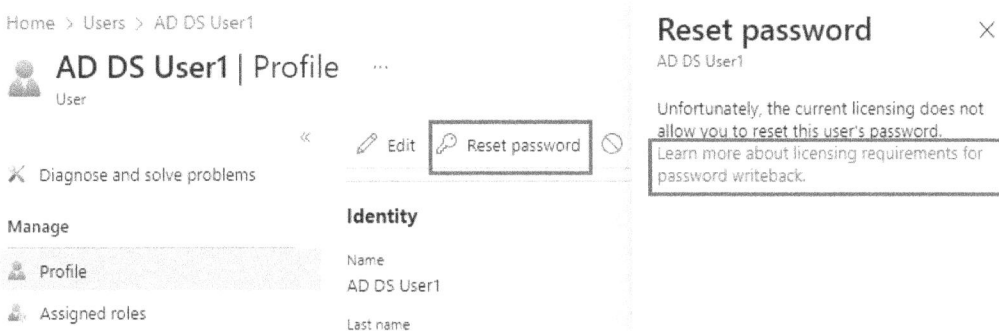

Figure 4.39 – Resetting a password

10.   Once the user accounts are synchronized into Azure AD, you can *assign licenses* as required:

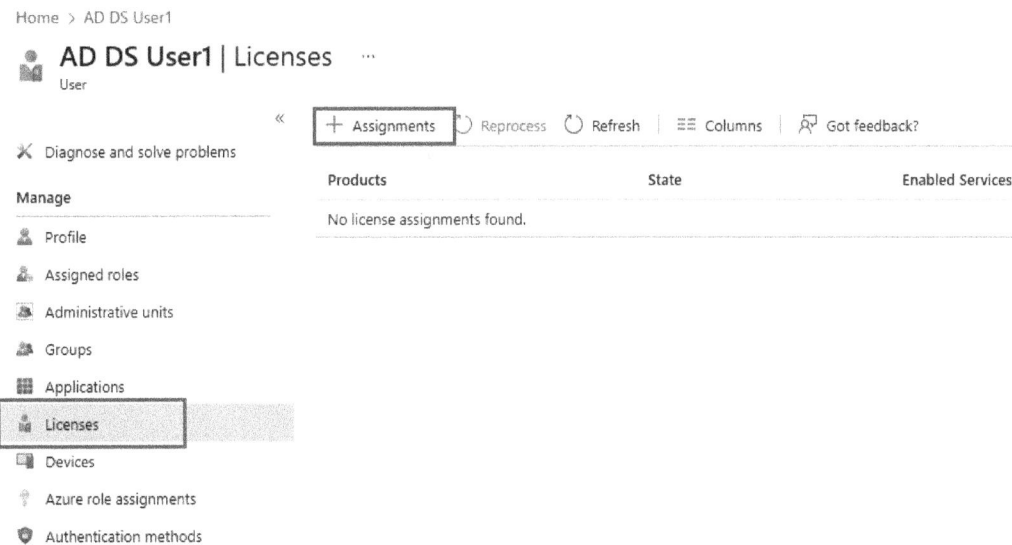

Figure 4.40 – Assigning user licenses

With this, we have installed Azure AD Connect and synchronized our AD objects with Azure AD.

This exercise taught us the skills needed to implement AD Connect. Now, let's summarize this chapter.

Summary

This chapter provided coverage for Exam AZ-800 Administering Windows Server Hybrid Core Infrastructure: *Implement and manage hybrid identities*.

The key lessons to take away from this chapter are understanding the concepts of hybrid identities, how to implement and manage Azure AD Connect, the differences between cloud sync, the different hybrid identity authentication methods, and when to choose between them based on your needs.

We then finished the chapter with a hands-on exercise to simulate a real-world scenario and develop our skills further.

You will have gained new skills through the information provided, and the chapter's goal has been for you to take your knowledge beyond the exam objectives so that you are prepared for a real-world, day-to-day hybrid environment-focused role.

The next chapter will teach us about implementing and managing on-premises network infrastructure.

Further reading

This section provides links to additional exam information and study references:

- *Microsoft Certified: Windows Server Hybrid Administrator Associate*: `https://docs.microsoft.com/en-us/learn/certifications/windows-server-hybrid-administrator/`

- *Exam AZ-800: Administering Windows Server Hybrid Core Infrastructure*: `https://docs.microsoft.com/en-us/learn/certifications/exams/az-800`

- *Exam AZ-800: study guide*: `https://query.prod.cms.rt.microsoft.com/cms/api/am/binary/RWKI0r`

- *Microsoft Learn*: `https://docs.microsoft.com/en-gb/learn/modules/implement-hybrid-identity-windows-server/`

Skills check

Challenge yourself with what you have learned in this chapter:

1. What are the two Microsoft IDPs?
2. What is a hybrid identity?
3. What is Azure AD Connect?
4. What is Azure AD Connect cloud sync?
5. What are the operations of Azure AD Connect when installed?

6. Who is the authoritative IDP in hybrid identity?

7. What solution can be implemented to prevent an identity from having passwords that are out of sync between directories?

8. What planning activities should be carried out before implementing the Azure AD Connect tool?

9. What is the significance of preparing AD object attributes such as UPNs?

10. What account permissions are required for implementing Azure AD Connect?

11. What two tools can be used to identify and remediate issues in AD?

12. What are the two categorizations of hybrid authentication methods and their crucial differences?

13. In what scenarios can only the AD FS method be used?

14. In what directory scenario can PTA not be implemented?

15. What is the difference between single sign-on and same sign-on?

Part 2:
Hybrid Networking

This part will provide complete coverage of the knowledge and skills required for the skills measured in the *Implement and manage an on-premises and hybrid networking infrastructure* section of the exam.

This part of the book comprises the following chapters:

- *Chapter 5, Implementing and Managing On-Premises Network Infrastructure*
- *Chapter 6, Implementing and Managing Azure Network Infrastructure*

Implementing and Managing On-Premises Network Infrastructure

This chapter covers the AZ-800 Administering Windows Server Hybrid Core Infrastructure exam learning objective: *Implement and manage on-premises network infrastructure.*

In the previous chapter, we added skills to understand Hybrid Identities.

We will start this chapter by understanding **Domain Name System** (**DNS**) for Windows Server; we will cover its implementation and management. We will next look at the **Dynamic Host Configuration Protocol** (**DHCP**) and **IP Address Management** (**IPAM**) network services and finally cover implementing and managing remote access services. We will then conclude with a hands-on exercise to develop your skills further. The following topics are included in this chapter:

- Implementing and managing Windows Server DNS
- Implementing and managing DHCP
- Implementing and managing IPAM
- Implementing and managing remote access
- Hands-on exercises

In addition to these topics, this chapter aims to take your knowledge beyond the exam objectives to prepare you for a real-world, day-to-day hybrid environment-focused role.

Implementing and managing Windows Server DNS

This section will look at implementing on-premises and hybrid name resolution.

The first skills area we will look at in this section is *what is DNS?*

What is DNS, and how does it work?

Name resolution is translating or resolving a *name* to an *IP address*. IP addresses can also be resolved to a name through a reverse lookup.

DNS is an industry-standard name resolution service and the primary name resolution method used by Windows. Other methods include a **host file**, an **LMHOSTS file**, and **Windows Internet Name Service** (**WINS**), which is a **Network Basic Input/Output System** (**NetBIOS**) name resolution service.

DNS uses **resource records** to hold the names and information, such as IP addresses and arbitrary text information it holds for that domain. As with phone contacts, people can remember names more easily than they can numbers, especially in the case of multiple numbers (home, work, mobile, and so on) for the same name.

DNS resource records can be created manually, or clients can automatically register records with a DNS server. These records are then stored in *zones*, which are stored in *zone files*.

The DNS service can be installed on Windows Server as a server role. DNS clients will then contact the server to look up the name to be resolved to an IP or look up arbitrary text in a text field record.

A typical scenario of name resolution with DNS would be a client wishing to resolve a **fully qualified domain name** (**FQDN**) such as `https://server1.milesbetter.solutions` to an IP address.

In this scenario, the DNS server looks in its local database to see whether it has a record of this stored. If it does, the IP address of `10.5.1.4` would be returned to the client by the DNS server.

If the DNS server does not have the information to return to the client for this name in its database, it will make a request to another DNS server either through *root hints* or *forwarding* to provide the requested information.

DNS servers also cache this request and look up information, so the next time the DNS server is asked for the same request, it now has a copy of this information to respond with the next time it is requested; all the while, the cached content remains valid. A client needing to resolve DNS records is a *DNS resolver*, and it will use a *DNS resolver cache* to hold the DNS server response requests for their lifetime.

This process of name resolution is represented in the following diagram:

What is the IP address for:
htttps://server1.milesbetter.solutions   DNS Server

DNS Client   The IP address is:
                    10.5.1.4

Figure 5.1 – Name resolution

A hierarchical structure is used for a DNS, and a DNS namespace is used for a naming structure. The *root domain* starts the hierarchy with subdomains at the next level, which can then have any number of subdomains. This DNS domain naming scheme is represented in the following diagram:

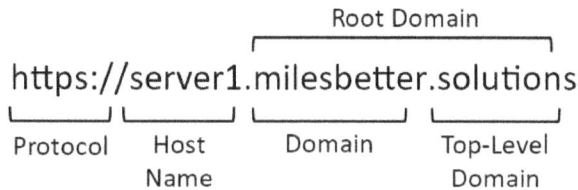

Figure 5.2 – DNS domain names

Namespaces can be *private* or *public*. A public namespace is where a domain you own is registered for public use on the internet, such as milesbetter.solutions or stevemiles.cloud.

A subdomain can often be used to segment a namespace to represent an **Active Directory Domain Services** (**AD DS**) domain such as adds.milesbetter.solutions or a geographical basis such as eu.milesbetter.solutions or us.milesbetter.solutions.

All these approaches allow aspects such as digital certificates to be issued and used for these public internet domains.

However, in non-internet-facing or private organization networks such as lab, test, or that of a highly secure environment, then a private namespace that is not internet-published, accessible, or internet-registered may be appropriate. Examples of a private namespace would be milesbetter.local or milesbettter.lab.

A specific part of a namespace hierarchy is represented on a DNS server by a *zone*—for example, a milesbetter.Solutions zone would be created on the DNS server responsible for responding to clients who request lookups for this domain. These servers that hold the copy of the zones are known as **authoritative** name servers for the domain.

The zones hold the information the DNS server uses to respond to client lookup requests for this namespace. The information that DNS servers hold and send responses from is stored in DNS resource records in the zones.

There are *forward lookup zones* and *reverse lookup zones*; the following list provides examples of *resource records* in forward lookup zones:

- **Host (A) record**: Resolves a name to an IPv4 address
- **Host (AAAA) record**: Resolves a name to an IPv6 address
- **Alias (CNAME) record**: Resolves a name to another name

- **Service location (SRV) record**: Identifies servers hosting a service, such as identifying the location of domain controllers

- **Mail exchanger (MX) record**: Identifies email servers

- **Text (TXT) record**: Stores arbitrary strings of information

Forward lookup zones are represented in the following screenshot:

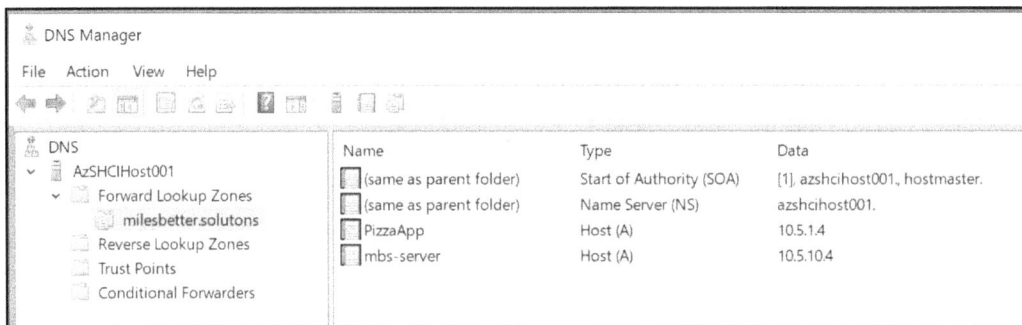

Figure 5.3 – Forward lookup zones

Reverse lookup zones contain **pointer (PTR)** records used to resolve an IP to a name. There is a specific format of [IP-Range}.in-addr.arpa for a reverse lookup zone. An example of a zone name for a reverse lookup zone for 10.5.1.0/24 would be 1.5.10.in-addr.arpa. This is represented in the following screenshot:

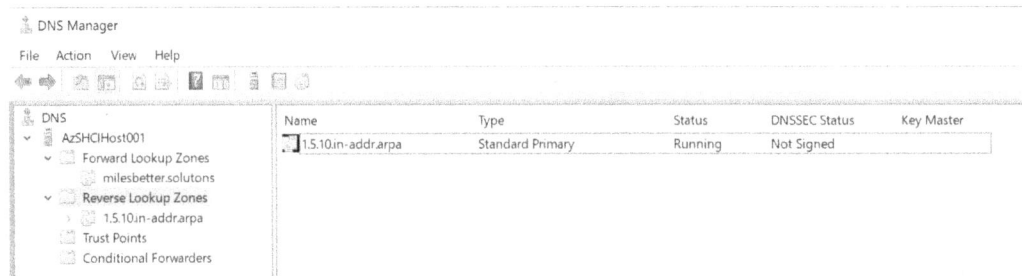

Figure 5.4 – Reverse lookup zone

All forward and reverse lookup DNS zones contain a **start of authority (SOA)** and a **name server (NS)** record. The SOA record contains zone configuration information, with one SOA record per zone. The NS record shows the DNS server for the domain, with one NS record for each DNS server with a zone copy.

You can read more on reverse lookup at the following URL:

`https://en.wikipedia.org/wiki/Reverse_DNS_lookup`

Implementing DNS

The DNS service is installed as a role on Windows Server using Server Manager, Windows Admin Center, or PowerShell with the following command:

```
Install-WindowsFeature -Name DNS -IncludeManagementTools
```

The installation via Windows Server **Server Manager** is represented in the following screenshot:

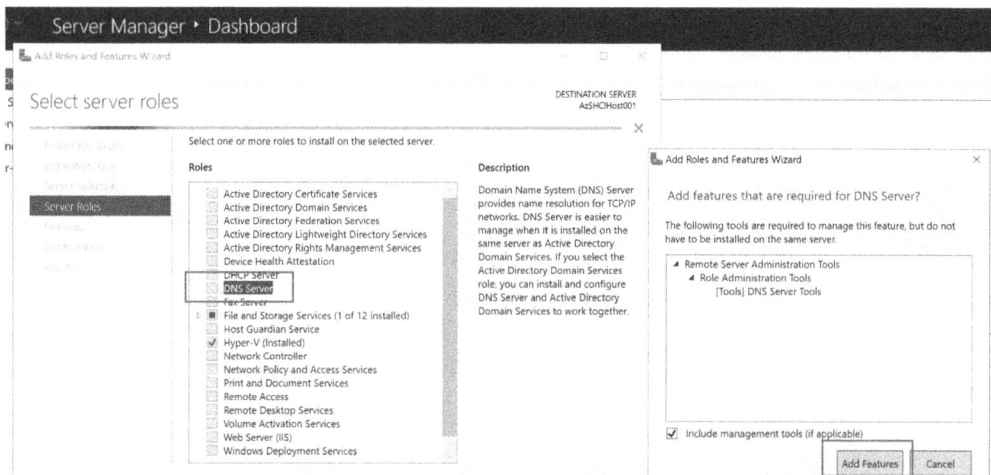

Figure 5.5 – Adding a DNS role

Once the DNS server role is installed, it can be managed from Windows Admin Center or the DNS Manager console. The **DNS Manager** console is represented in the following screenshot:

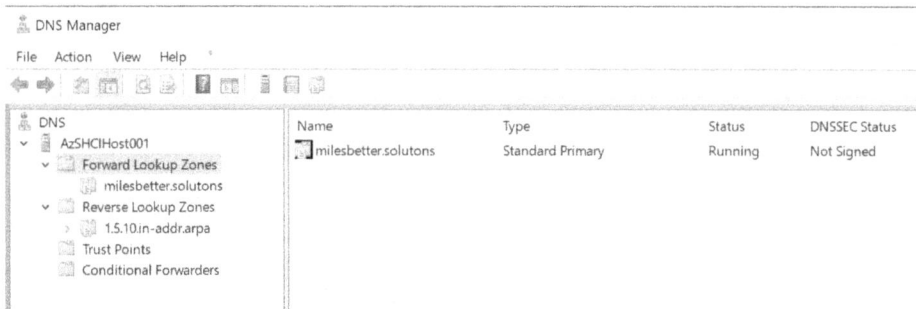

Figure 5.6 – DNS Manager console

The DNS role is commonly installed on *AD DS domain controllers* and, instead of zone files, can use *AD-integrated DNS* to store the DNS information in the AD database. This means the DNS information is made secure and highly available as it is replicated by native AD replication and availability mechanisms.

The following sections look at the topics of zones, records, forwarders, and security.

Zone

When a zone file is created, it must be set as a *primary* or *secondary* zone. The primary zone is used to create, edit, and delete records. The primary zone records can be stored in a zone file or configured to be AD-integrated, which means the AD DS database is used to store the zone information. The secondary zone is a read-only copy of the primary zone and is always stored in a zone file. PowerShell can also be used to manage the DNS zones.

Zone records can be transferred between primary and secondary zones with a *zone transfer*. The secondary zone servers requesting a zone transfer can be controlled by setting the following options:

- **To any server**
- **Only to servers listed on the Name Servers tab**
- **Only to the following servers**

These zone transfer options are represented in the following screenshot:

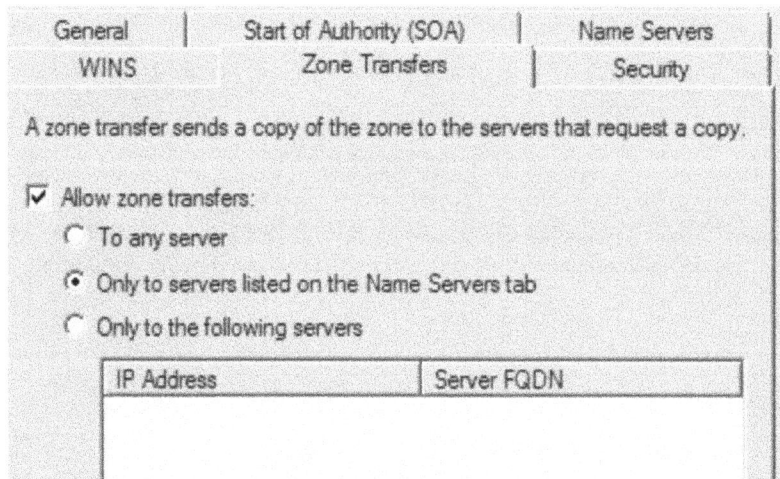

Figure 5.7 – DNS zone transfer options

Incremental zone transfers happen after the initial secondary zone synchronization from the primary zone is complete.

Records

Records can be manually managed using the DNS Manager console, Windows Admin Center, or PowerShell cmdlets.

The zone files are represented in the following screenshot:

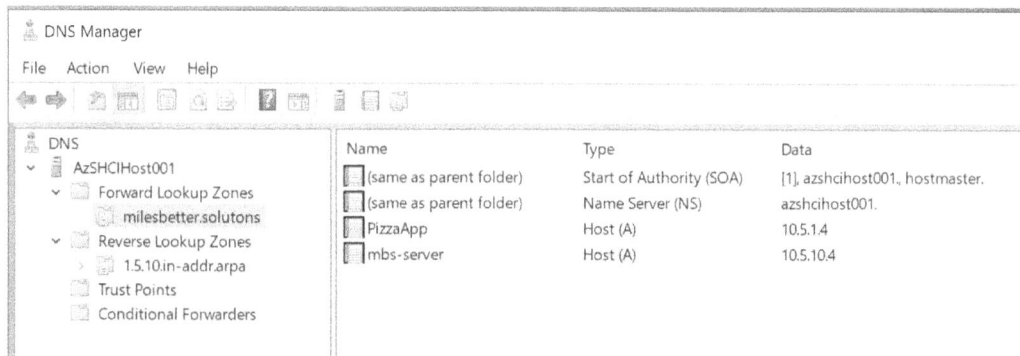

Figure 5.8 – DNS zone files

All records have a **time-to-live (TTL)** value configured. This defines how long the information in the record is valid and how long it is allowed to be cached by clients and servers. If a TTL is set for 36000 secs (1 hour), then the response to the DNS query can only be cached for that period.

During that time, the client will only use the information in the cache and not make a new request. When the TTL expires, the client will make a new request to the DNS server and may receive any updates made to the record since it was requested and cached. The cache should always be cleared for troubleshooting.

Host and PTR records can also be created dynamically, making managing DNS easier. The current IP address for a device is automatically registered after its IP is changed when dynamic DNS is enabled.

Forwarders

DNS servers can be configured with *forwarders*, typically for internet name resolution.

Where the DNS server is requested to resolve a name that is not authoritative for that zone—that is, it does not hold a primary or secondary zone for that domain—it must ask another DNS server for this information. The default is that *root hints* will be configured on each server to find the authoritative servers on the internet; however, that is only possible if the DNS server has access to the internet and the record requested is published on the internet.

Due to network segmentation and security, the DNS servers will not always be directly connected to the internet. They will have to proxy the request through forwarders if the records are not stored locally in a cache.

The forwarders are DNS servers connected to the internet, usually in a perimeter network segment or an **external service provider** (**ESP**) DNS server. These DNS servers can resolve the internet names for these requests outside the local network.

Conditional forwarding is another option and is intended to only forward any request lookups for a *single specific DNS domain* instead of forwarding all requests that cannot be resolved locally.

Security

DNS is secured by validating all responses from a DNS server through the extensions suite of **Domain Name System Security Extensions** (**DNSSEC**). This aims to make the DNS protocol less prone to attacks such as *DNS spoofing* by providing authenticated **denial of service** (**DoS**), data integrity, and origin authority.

The DNS zone is secured using a **zone signing** process when supported by an authoritative DNS server; this does not change the DNS query and response mechanism but adds a validation step. Digital certificates are used for validation in the DNS responses. The process generates new DNSSEC resource records containing digital certificates during zone signing.

When a query for a *DNSSSEC-signed zone* is received by a DNSSEC-aware forwarding or recursive DNS server, it will request that the authoritative DNS server also responds with the DNSSEC records. It will then use the digital certificate contained in the DNSSEC records to validate the response. The DNS server recognizes a zone that supports DNSSEC if the zone has a **DNSKEY** record (trust anchor).

Troubleshooting DNS

This section will look at some commands that can troubleshoot DNS.

An adage says that 90% of the time, it's DNS, and so is the other 10%.

Here are some examples of troubleshooting commands:

- NSLOOKUP: Queries for DNS information and looks up resource records and the DNS server status
- DNSCMD: Manages the DNS server role
- DNSlint: Diagnoses common DNS issues
- IPConfig: Lists the IP configuration details used by a DNS client

This section concludes the topic of name resolution using the DNS network service. We looked at DNS, how it works, and how to implement, manage, and secure it. In the next section, we will look at DHCP and its use to manage leasing IP addresses to clients.

Implementing and managing DHCP

This section will look at managing on-premises IP addresses using the DHCP network service.

The first skills area we will look at in this section is *what is DHCP?*

What is DHCP?

DHCP can be considered a network service component of a **DNS-DHCP-IPAM** (**DDI**) solution; we also cover DNS and IPAM in this chapter.

Concisely, DHCP is a network management protocol used to simplify IP address management through auto-assigning IP addresses and network information.

DHCP is an extended version of **Bootstrap Protocol** (**BOOTP**). The difference between the two is that BOOTP provides *static* configuration of IP addresses, and DHCP provides *automatic* configuration of IP addresses.

Without DHCP, there would be a need for manual network IP address configuration on the network interface for the required network settings and manual assignment of IP addresses; this is highly inefficient and error-prone.

With regards to *DHCP version 6*, this provides IPAM in IPv6 networks. Stateful and stateless configurations are supported.

The DHCP server assigns the IPv6 addresses and network settings in stateful configurations. In stateless configurations, the router assigns the IP addresses, and the DHCP server assigns only the network settings.

How does DHCP work?

DHCP works on the traditional *client/server relationship* model. Windows computers run the DHCP client service, and when set to **Obtain an IP address automatically**, they will communicate with a DHCP server to receive their network information. This setting is found in the **Internet Protocol Version 4 (TCP/IPv4) Properties** dialog, as shown in the following screenshot:

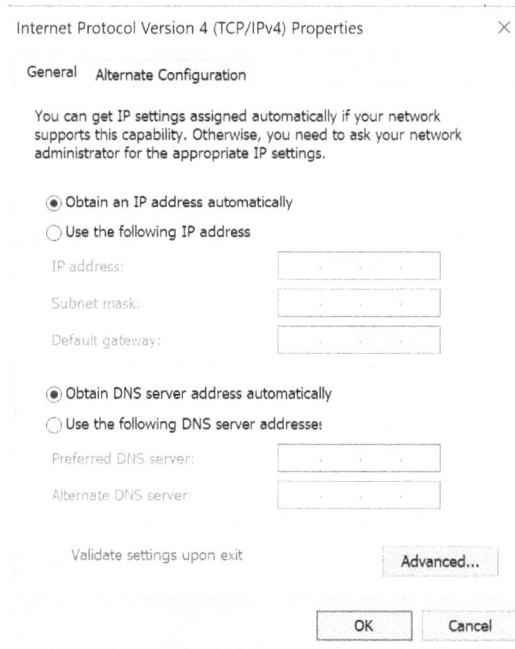

Figure 5.9 – IP properties

The DHCP server has an address pool configured and configuration options. The DHCP server will communicate with the DHCP client, assign IP addresses, and pass the configuration options to the clients requesting.

The DHCP communication flow is given the acronym **DORA**; the four steps are set out here:

- **D**iscover—DHCPDISCOVER

- **O**ffer—DHCPOFFER

- **R**equest—DHCPREQUEST

- **A**cknowledge—DHCPACK

This communication flow is also shown in the following diagram:

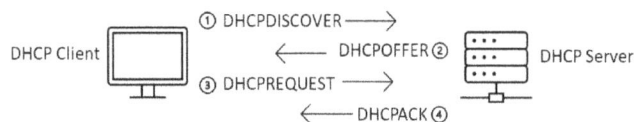

Figure 5.10 – DHCP communication flow

The DHCP address assigned to a client has an amount of time it can be used for before it expires; this period is known as a **lease**, and when 50% of the lease time has been reached, the client automatically attempts the lease renewal.

If the lease length is set to *10 days*, a renewal is attempted on *day 5*.

The request for a renewal goes to the original server that leased the address, which is sent as a unicast DHCPREQUEST command. If this server cannot be contacted, the client waits for *87.5%* of the lease time to be reached and then sends a broadcast (rather than a unicast) DHCPREQUEST command for renewal. This goes to all servers, rather than just the lease originating server; renewals are also attempted during startup. The lease period *resets* when renewals are successful.

DHCP scopes

The DHCP server contains *scopes*, which define an IP address range for clients. Clients will be leased an address assignment from this range; each client will be assigned a unique IP from this range—that is, two clients cannot be assigned the same address from an IP address range.

The following PowerShell cmdlets can also be used to manage the configured scopes:

- `Add-DhcpServerv4Scope`
- `Get-DhcpServerv4Scope`
- `Get-DhcpServerv4ScopeStatistics`
- `Remove-DhcpServerv4Scope`
- `Set-DhcpServerv4Scope`

The scopes can also be added from the DHCP console. Creating scopes using the **New Scope Wizard** dialog is represented in the following screenshot:

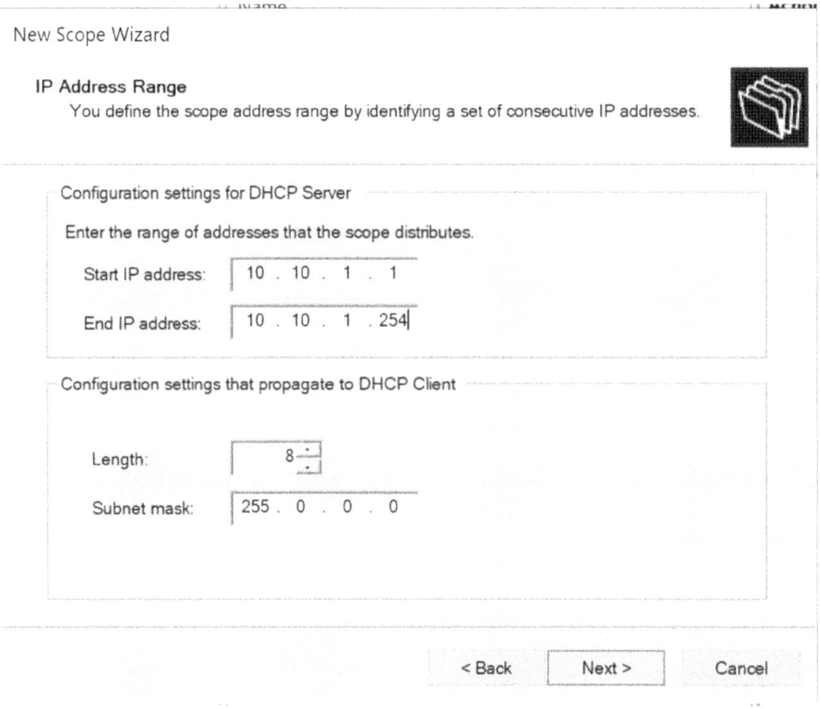

Figure 5.11 – DHCP scope creation

DHCP reservations

You can specify that a client can be permanently assigned a specific address from a range set in the scope; this feature is known as a **reservation**.

Reservations are helpful for clients that are static on the network and are always required to have the same IP; examples of this could be domain controllers, printers, network/firewall appliances, and so on.

The **Reservations** capability is set in the DHCP console, as represented in the following screenshot:

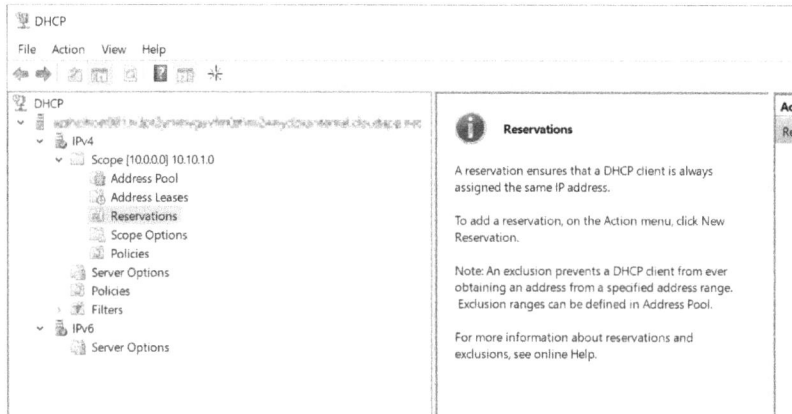

Figure 5.12 – DHCP reservations

The following information will be required when creating a reservation:

- **Reservation name**
- **IP address**
- **MAC address**
- **Description**

These required settings to create an IP address reservation are represented in the following screenshot:

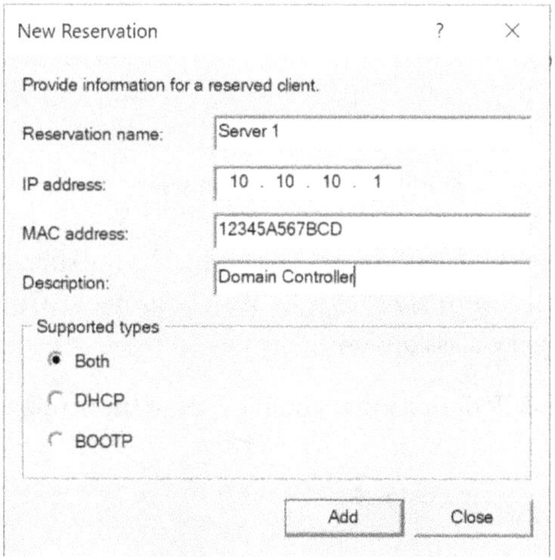

Figure 5.13 – DHCP new reservation

The client with the **assigned Media Access Control [MAC] address** will now always receive the set reserved IP. The client will not automatically receive this IP address. The client will receive the reserved IP address in the following scenarios:

- Renewal of current lease for the client

- A reboot of the client

- Running the `ipconfig` command on the client, like so:

    `ipconfig /release` and then `ipconfig /renew`

Reserved IP addresses for a client can be changed; the device is just assigned a new reservation, and the current one is removed so that IP can be used again for other clients.

DHCP options

Through a capability known as **Options**, DHCP can provide additional network resource information required to be set on a machine's network interface to communicate on the network. These settings are managed via the DHCP console.

The following screenshot represents this:

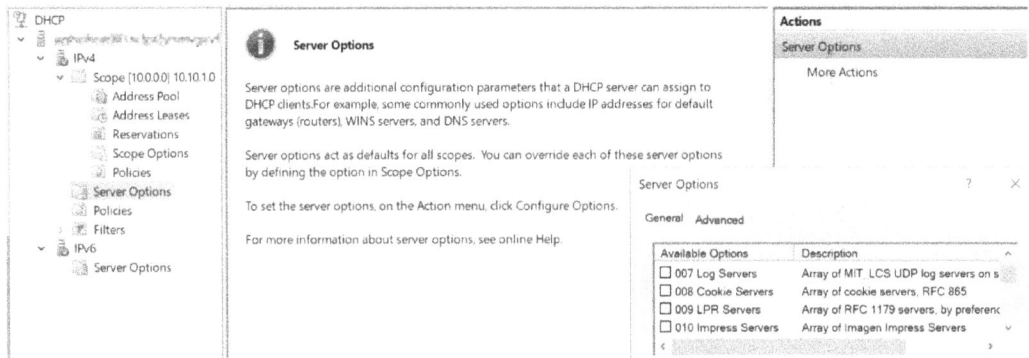

Figure 5.14 – DHCP options

These network settings typically apply to the *scope* level, which is the subnet for clients; they can also be configured to apply at the *server* or *class* level to meet your needs. The scope- and server-level options are both overridden by the class option.

The following are common DHCP options that could be set at the scope level, to be set as network settings on the clients:

- **DHCP option 1: subnet mask**

- **DHCP option 3: router (default gateway)**

- **DHCP option 6: DNS**

- **DHCP option 51: lease time**

More information on all the available DHCP options can be found here:

`https://www.efficientip.com/glossary/dhcp-option`

In this section, we looked at what DHCP is. In the next section, we look at how to implement DHCP.

Implementing DHCP

DHCP is implemented by adding the DHCP service as a server role to Windows Server, which then functions as the DHCP server. You can implement this as a dedicated server for this role or install it on an existing server; the server for the DHCP role must use a static IP, and local admin rights on the server will be required to add the server role.

DHCP authorization

Before a DHCP server can support DHCP clients, it must be *authorized*; this is done to protect the network, as DHCP is based on broadcasts. This is done post-installation and can be authorized in an AD domain by using the DHCP console or the following PowerShell cmdlet:

```
Add-DHCPServerinDC <hostname or IP address of DHCP server>
```

For a single-domain environment, a *Domain Administrator* account is used to authorize a DHCP server; for a DHCP server that assigns IP addresses for subnets for multiple domains, an *Enterprise Administrator* AD account must be used.

A *standalone* DHCP server is not a member of an AD domain. Its behavior is that if it detects an authorized DHCP server, it shuts down so as not to lease addresses and cause a conflict on the network.

Any *unauthorized* DHCP servers detected by clients by running the `ipconfig /all` command should be shut down to prevent clients from getting the wrong networking information.

The DHCP server role can be installed using the GUI or PowerShell.

For the GUI deployment of the role, you can use *Windows Admin Center* or *Server Manager*.

The user account installing the DHCP servers requires local admin rights to the server to add the **DHCP Server** role. The server where the role will be added must have a static IP address assigned.

Adding the **DHCP Server** role via the *Server Manager* method is represented in the following screenshot:

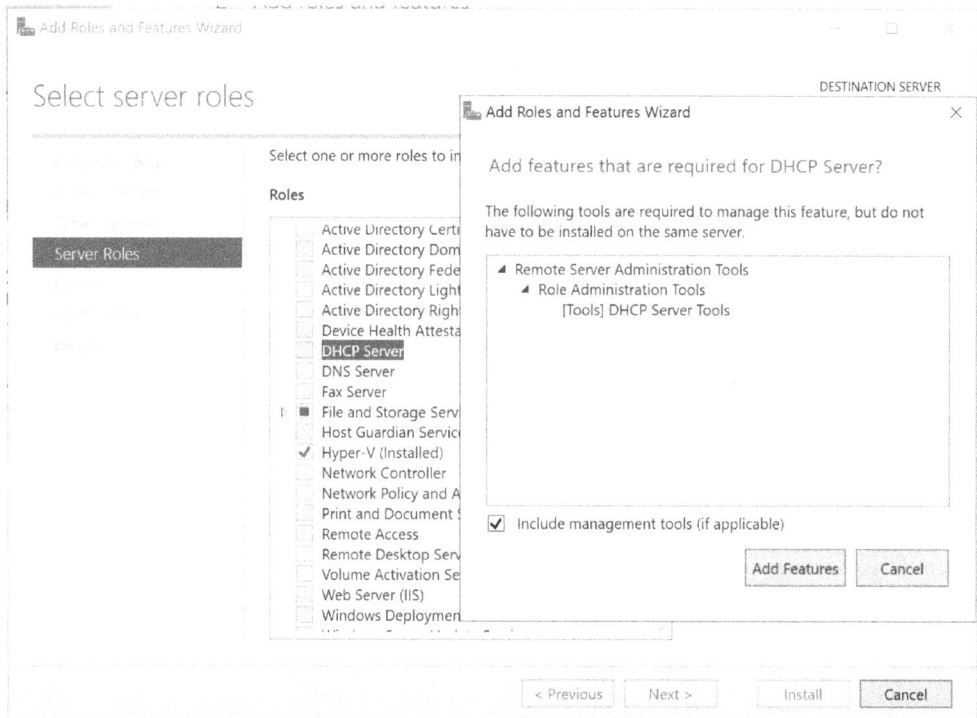

Figure 5.15 – Adding the DHCP role to a server

From PowerShell, the following cmdlets can also be used to install the DHCP Server role:

```
Add-WindowsFeature DHCP -IncludeManagementTools
```

The -IncludeManagementTools parameter is optional.

DHCP availability

A clustering deployment is a **high availability** (**HA**) method in Windows Server 2012 and later OSes for the DHCP service and allows two or more DHCP servers to share information about service availability. This is the preferred HA solution; we will look at DHCP split scopes next.

When configured for *DHCP failover*, the DHCP scope and IP address leases are made available between two servers using *shared storage*. The following diagram represents this:

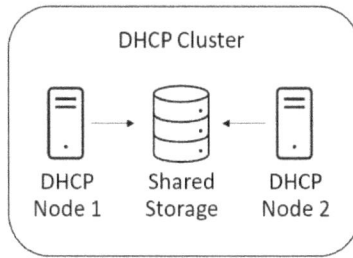

Figure 5.16 – DHCP failover cluster

This allows lease information and IP addresses to be made available if one DHCP server is unavailable; only *two* DHCP servers can be configured in a failover cluster, and the scopes can only be *IPv4*.

In DHCP failover, the failover relationship between the two DHCP servers is configured in the DHCP console through the **Configure Failover Wizard** dialog. The DHCP failover methods that can be configured are as follows:

- Load balance
- Hot standby

Configuring DHCP failover via the DHCP console is represented in the following screenshot:

Figure 5.17 – DHCP failover

With **DHCP split scopes**, this HA solution uses two DHCP servers active on the network. The scopes are *split horizontally* across both servers, so each has control of a portion of the same scope. The following diagram represents this:

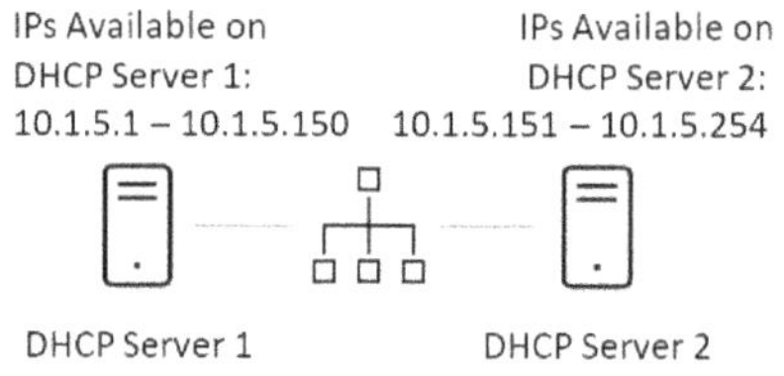

Figure 5.18 – DHCP split scope

The primary server is set with the `Delay configuration` attribute in the **Scope** properties on the secondary server; you control the primary and secondary servers. In this *split scope* scenario, the primary DHCP server responds to client requests, and in the event of becoming unavailable, the secondary server responds to client requests.

In this section, we looked at implementing DHCP, the authorization of DHCP servers, and the HA of IP address information and assignment to clients. In the next section, we will look at how to manage DHCP.

Managing DHCP

DHCP is managed via the DHCP console and DHCP PowerShell cmdlets. When using Server Manager to install the DHCP server role, these tools are installed, or you can use **Remote Server Administration Tools** (**RSAT**).

The following screenshot shows the DHCP console:

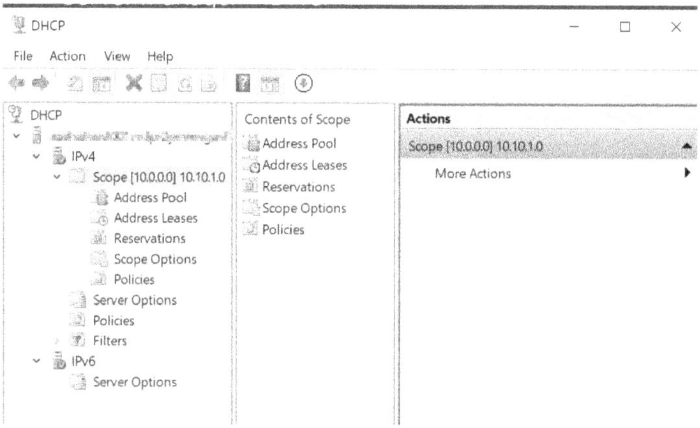

Figure 5.19 – DHCP console

To use Windows Admin Center to manage DHCP, the DHCP PowerShell cmdlets must be installed on the DHCP server. If not installed, a message is displayed, and a button is provided to perform the remote installation.

The following PowerShell cmdlet is used to include the management tools when installing the DHCP server role:

```
Add-WindowsFeature DHCP -IncludeManagementTools
```

DHCP groups

Each DHCP server includes groups used to *delegate management* of DHCP; these groups are set out here:

- **DHCP Administrators local group**: This group is used for local DHCP server management
- **DHCP Users local group**: This group is used to explore configuration and provide the status information of the local DHCP server

When using the Server Manager installation method, these groups are created as part of the installation. When the PowerShell cmdlets or Windows Admin Center is used to install the DHCP role, then these groups must be *manually added* as an extra step using the following PowerShell:

```
Add-DhcpServerSecurityGroup -Computer DhcpServerName
```

This section concluded with the topic of IP address assignment using the DHCP network service. We looked at DHCP and how it works, how to implement it, and how to manage it. In the next section, we look at IPAM and its use to manage on-premises IP addressing.

Implementing and managing IPAM

This section will look at managing on-premises IP addressing using the IPAM functionality.

The first skills area we will look at in this section is *what is IPAM?*

What is IPAM, and how does it work?

IPAM is a framework to provide visibility and control, planning, allocation, and management for IPv4 and IPv6 address spaces, as well as centralized control and management of an organization's DHCP and DNS servers.

IPAM works within the context of AD DS and is a domain service-aware function in that the IPAM servers must be domain-joined and not standalone *workgroup* servers.

IPAM is provided as a built-in set of tools within Windows Server that can be implemented through a client/server model, as follows:

- **IPAM server**: Provides the managed server data collection. It utilizes the **Windows Internal Database** (**WID**) or a **SQL Server database** and provides access control through **role-based access control** (**RBAC**).

- **IPAM client**: Provides the frontend client interface for the IPAM server. Using Windows PowerShell cmdlets, it interacts with the managed servers. It can be installed on a *client* or *server* OS.

When a Windows Server with *Desktop Experience* has the IPAM role installed, the IPAM client is automatically installed locally on the same server. If *Server Core* is used, another server or client must install the client remotely from the IPAM server. The **RSAT: IPAM Client** feature is used to install the client on a server or client OS.

IPAM consists of four modules; these are set out here:

- **IPAM discover module**: This is used for discovery to add to the database all located servers (including domain controllers) with the DNS or DHCP roles installed; servers can also be added manually.

- **IP address space management module**: This is used to examine, manage, and monitor an IP address space. IP addresses can be dynamically or manually assigned to clients. Address space utilization can be tracked, and overlapping DHCP scopes detected.

- **DHCP and DNS multi-server management and monitoring module**: This is used to monitor and manage tasks across multiple deployed DHCP and DNS servers and provide a DNS health status.

- **Operational auditing and IP address tracking module**: This is used to manage and monitor configuration changes. Collects lease information from managed DHCP servers and sign-in information and data from AD DS domain controllers and **Network Policy Server** (**NPS**) servers.

These four modules allow for comprehensive DHCP server, DNS zone, and IP address management tasks to be carried out from a **single-pane-of-glass** (**SPOG**) console for greater efficiency and control.

In this section, we looked at what IPAM is and how it works. In this next section, we will look at how to implement IPAM.

Implementing IPAM

There are two topology choices for data collection: a **centralized** or **decentralized** topology.

Centralized topology

In this topology, a *single* IPAM server is deployed across *all* locations. This provides visibility, control, and management tasks using a single-console view for IP address spaces across a forest and multiple forests when a two-way forest trust is in place.

A centralized topology is represented in the following diagram:

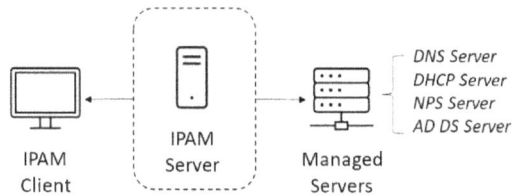

Figure 5.20 – IPAM centralized topology

IPAM can be managed with the client via Server Manager or another computer with the *RSAT: IPAM Client* feature; this can be a Windows Server OS or client OS.

Decentralized topology

In this topology, an IPAM server is deployed in *each* location. This is typical when dealing with large-scale address space management where there is a need to scale out and hand off to distributed local servers. This topology can also be used when you wish to give *autonomy* to a **business unit** (**BU**) or individual locations.

A distributed topology is represented in the following diagram:

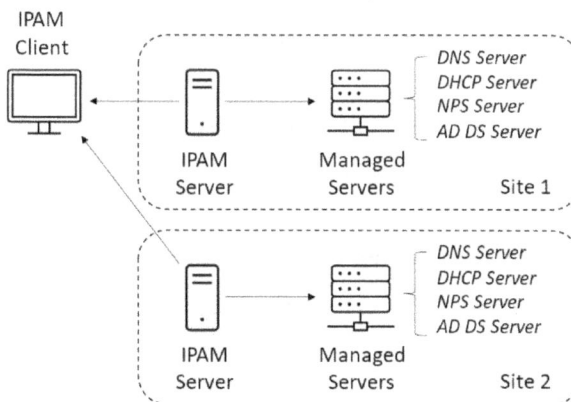

Figure 5.21 – IPAM centralized topology

In addition, a *hybrid topology* can be implemented. In this topology, IPAM servers are deployed at each site as local IPAM servers to monitor the local servers, and centralized IPAM servers can monitor managed servers remotely. You should consider that in this topology, IPAM servers do not replicate information from the data collection of the local servers to the central server; each server is a standalone data point—that is, an **island**.

The following requirements should be considered for deployment:

- The IPAM role should be the only one on the server and perform no other functions
- The IPAM server must be *domain-joined*, and you must be signed in with a *domain account* to install the role
- Installing the IPAM role on an *AD DS domain controller* is not supported
- If the IPAM server role is installed on a DHCP server, no other DHCP servers will be able to be discovered
- The IPAM server requires a WID or SQL Server instance; you cannot run another instance of **Structured Query Language** (**SQL**) on the same server
- An individual IPAM server can support 500 DNS servers, 150 DNS zones, 150 DHCP servers, and 6,000 DHCP scopes
- The required storage space should be planned; approximately 1 GB is required to store the utilization data for 10,000 clients
- There is no checking of routers and switches for IP address consistency

With the previous requirements met, the high-level implementation steps for IPAM servers are outlined as follows:

- Create a server(s) for IPAM based on your chosen topology and domain join.
- Install the IPAM server feature from Server Manager, Windows Admin Center, or PowerShell using the following command:

```
Install-WindowsFeature IPAM -IncludeManagementTools
```

- Provision IPAM servers and required **Group Policy Objects** (**GPOs**) with the following PowerShell command:

```
Invoke-IpamGpoProvisioning
```

- Configure and run DNS and DHCP server discovery.
- Select and manage discovered servers.

Regarding the IPAM client, this can be installed on a Windows Server or client OS using the IPAM Client feature of RSAT.

This section looked at implementing IPAM, including the different deployment topologies. In this next section, we will look at how to manage IPAM.

Managing IPAM

In the previous sections, we learned that IPAM provides a unified view of the three core networking services areas in an IT estate of a *DHCP server*, *DNS zones*, and *IP address management*.

The following are some of the tasks that can be carried out across these three areas:

- Manage DNS server tasks:

  - Examine DNS servers and zones

  - Create DNS zones and records

  - Manage conditional forwarders

  - Access the DNS console for any IPAM-managed server

- Manage DHCP server tasks:

  - Configure DHCP servers

  - Examine, create, and edit scopes

  - Discover assignment of specific IP addresses

  - Create reservations

  - Edit server properties and options

  - Import and replication

  - Configure failover

  - Access the DHCP console for any IPAM-managed server

- Manage IP addressing tasks:

  - Discover DHCP server address spaces and data on utilization

  - Detect overlapping IP address spaces across DHCP servers

  - Import IP address information from CSV files

  - Monitor DHCP and DHCP servers

  - Collect data from NPS servers and AD DS domain controllers, such as sign-in information

RBAC is used to provide granular access control for who can carry out tasks. You define a **role**, **access scope**, and **access policies** to manage these operation tasks. There are several built-in *role-based security groups*. We will not cover these aspects here in this chapter; we have, however, provided links to reference information in the *Further reading* section at the end of this chapter.

This section concluded with the topic of managing on-premises IP addresses using IPAM. We looked at what IPAM is, how it works, how to implement it, and how to manage it. In the next section, we look at using Windows Server to provide remote access to on-premises environments.

Implementing and managing remote access

This section will look at implementing on-premises network connectivity using the options available in Windows Server.

The first skills area we will look at in this section is *what are the remote access options in Windows Server?*

What are the remote access options in Windows Server?

While there are many third-party vendor remote access solutions, in this section, we will look to introduce some of Windows Server's built-in remote access capabilities.

What is DirectAccess?

Microsoft's **DirectAccess** is one of the technology options that can be implemented for secure remote network access, although it should probably now be considered legacy. This remote network access technology was intended as an alternative to the traditional **virtual private network** (**VPN**) approach of secure remote access to an organization's resources on its private networks. It was first introduced with Windows Server 2008 R2 and Windows 7.

What is the Remote Access Service?

The Windows Server **Remote Access Service** (**RAS**) server role can also be used, which implements the following remote network access technology services:

- Routing
- VPN
- **Web Application Proxy** (**WAP**)

The RAS role is installed on Windows Server using Server Manager or PowerShell. The RAS role is managed using the **Remote Access Management** console, the **Routing and Remote Access** (**RRAS**) console, or **PowerShell**.

What is routing?

This comprises routing protocols and routing tables to direct traffic (*network packets*) across a network(s) to their desired destination using the most efficient route. This is represented in the following diagram:

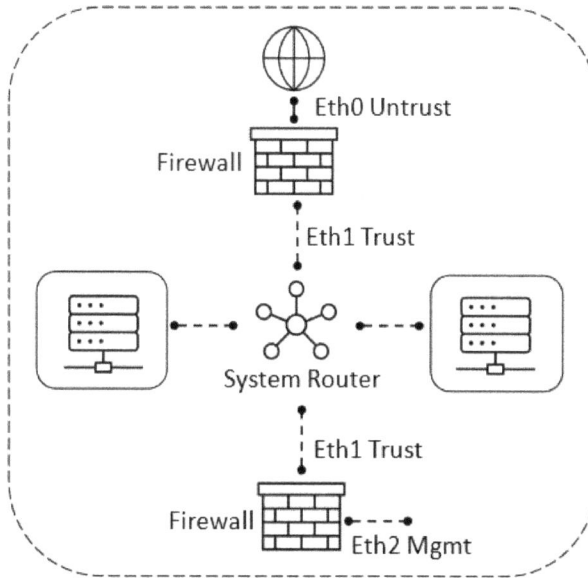

Figure 5.22 – Routing

To use an analogy, we can think of this as satellite navigation for network packets; it requires knowing where the packet is coming from and the destination and will calculate and determine the most appropriate and efficient journey. This is less commonly performed by a server, and often this function is performed by a vendor's specialized hardware appliance.

This section introduced the concept of network routing. In the following section, we look at using VPNs in Windows Server.

What is a VPN?

This can be considered *almost* the de facto way to provide users with a secure way to access an organization's resources and services on its internal private networks when they are not directly connected. An *encrypted tunnel* is set up between the organization's sites and the user's remote location outside the organization's private networks. The following diagram aims to represent this:

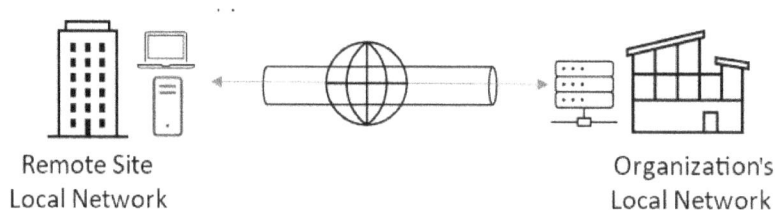

Figure 5.23 – VPN

When selecting a VPN as the remote access solution in Windows Server, you will need to select a *tunneling protocol* and an *authentication option*.

Here are the tunneling protocol options:

- **Secure Socket Tunneling Protocol (SSTP)**

- **Internet Key Exchange version 2 (IKEv2)**

- **Layer 2 Tunneling Protocol with Internet Protocol Security (L2TP/IPsec)**

- **Point-to-Point Tunneling Protocol (PPTP)**

Due to security vulnerabilities in PPTP, IKEv2 should be used, offering advantages over L2TP.

Here are the authentication protocol options:

- **Password Authentication Protocol (PAP)**: The least secure as passwords are in plain text.

- **Challenge-Handshake Authentication Protocol (CHAP)**: Uses challenge-response and a **Message Digest 5 (MD5)** industry standard hashing for response encryption. Due to requiring a reversibly encrypted password, **Microsoft-CHAPv2 (MS-CHAPv2)** should be considered.

- **MS-CHAPv2**: Uses CHAP but with a one-way encrypted password. Improves on CHAP.

- **Extensible Authentication Protocol (EAP)**: More of an authentication framework than a protocol. One of the common authentication methods in this framework is **EAP-Transport Layer Security (EAP-TLS)**. This is a *mutual authentication* method that is certificate-based for its authentication.

This section introduced the VPN remote access option in Windows Server. In the following section, we look at using NPS.

What is NPS?

NPS is a role that can be installed on a Windows Server and is Microsoft's **Remote Authentication Dial-In User Service** (**RADIUS**) implementation. RADIUS is used to provide an authentication exchange between components in a remote access solution and is an industry-standard authentication protocol.

NPS is used to implement authentication, authorization, and accounting for network access and is based on enforcing policies. NPS provides *connection request policies* and *network policies*. These policies have the following four categories: *Overview*, *Conditions*, *Constraints*, and *Settings*.

The authentication methods available are set out here:

- PAP

- **Shiva Password Authentication Protocol** (**SPAP**)

- CHAP

- MS-CHAP

- MS-CHAP v2

- EAP

Due to security concerns with the PAP, CHAP, and MS-CHAP authentication protocols, these should not be considered for a production environment.

For accounting, user authentication requests, access messages, accounting requests, responses, and status updates can be stored in a Microsoft SQL Server database.

NPS cannot be installed on Windows *Server Core* editions.

This section introduced NPS in Windows Server. In the following section, we look at using a **public key infrastructure** (**PKI**) in Windows Server.

What is a PKI?

A PKI is a collective terminology that covers implementing, issuing, and managing digital certificates. These certificates are issued by a **certification authority** (**CA**) and are used to confirm an identity.

Two CA types can provide certificates: a **private CA** and a **public CA**. Each CA type has advantages and disadvantages, as follows:

- A **private CA** provides greater control and lower costs than a public CA and provides customized templates and an auto-enrollment capability. The disadvantage is that private CA-issued certificates are not trusted by external clients and have a higher administrative burden.

- A **public CA** has the advantage that it will be trusted by most external clients and requires lower administration effort. The disadvantage is the higher cost and slower procurement of certificates.

Self-signed certificates can also be created using the following PowerShell cmdlet:

```
New-SelfSignedCertificate
```

This section introduced the PKI in Windows Server. In the next section, we will look at using WAP in Windows Server.

What is WAP?

This provides a way to allow remote users to access internal web applications securely over the internet. It does this using the *reverse proxy* functionality. Users are authenticated by the application using **pass-through** preauthentication or **AD Federation Services (ADFS)** preauthentication.

The following diagram aims to represent this:

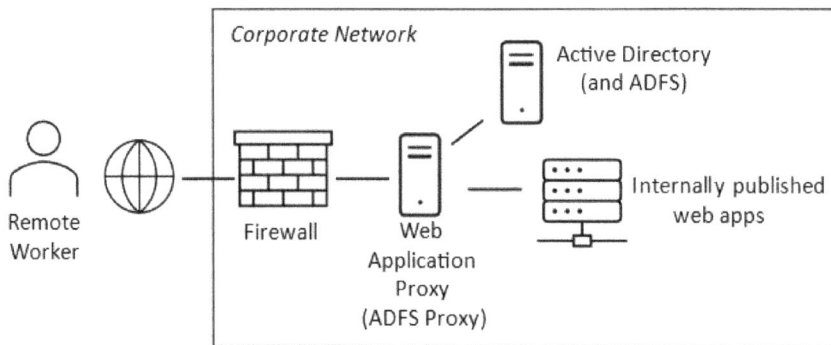

Figure 5.24 – WAP

WAP is installed via Server Manager on a Windows Server or with the following PowerShell cmdlet:

```
Install-WindowsFeature Web-Application-Proxy
-IncludeManagementTools
```

Once the role is configured from the **Remote Access Management** console, your internal web applications can then be published using the **Web Application Proxy** console or the following PowerShell cmdlets:

- `Add-WebApplicationProxyApplication`
- `Get-WebApplicationProxyApplication`
- `Set-WebApplicationProxyApplication`

The following information must be provided to publish a web application through the pass-through preauthentication method:

- The type of preauthentication should be set to *pass-through*

- A friendly name for the web application to be published

- The external URL that will be used to access the application—for example, `https:/pizza.milesbetter.solutions`

- An external certificate with a subject name for the URL—for example, `https:/pizza.milesbetter.solutions`

- The URL of the backend server will automatically be entered, or you edit it—for example, `https:/mbs-server1.milesbetter.solutions`

The alternative is to publish the web application using *ADFS preauthentication*. The benefits of this method are **single sign-on (SSO)** and **multi-factor authentication (MFA)**.

This section introduced WAP in Windows Server. In the next section, we will look at using Azure AD Application Proxy.

What is Azure AD Application Proxy?

Azure AD Application Proxy functions are comparable to WAP, making on-premises web applications available outside corporate networks from a remote client. However, the main difference is that the Application Proxy function (reverse proxy) is now provided as a Microsoft-managed service and uses Azure AD authentication services in place of AD. This replaces the need for a VPN or a reverse proxy with the on-premises-only WAP solution.

Application Proxy works with web applications, web APIs, and applications accessible from a **Remote Desktop (RD)** Gateway.

Azure AD Application Proxy has the components of the *Application Proxy service* that runs as a cloud service hosted by Microsoft and the *Application Proxy Connector*, which runs in the on-premises environment on a server. A user sign-in token from Azure AD is securely passed to the on-premises web application by these two components working together. The following diagram aims to represent this:

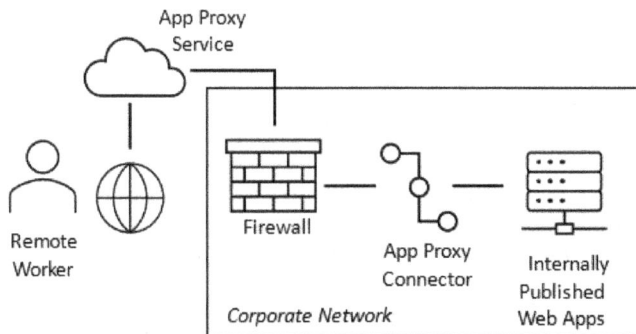

Figure 5.25 – Azure AD Application Proxy

How Application Proxy works

The following steps outline the process for a user accessing an internal web app from a remote client. Users access the application through an endpoint, a URL, or a portal interface and are directed to Azure AD to sign in:

- A token is sent to the user's device by Azure AD.

- The token is sent to the Application Proxy service by the client. The token's **user principal name** (**UPN**) and **security principal name** (**SPN**) are retrieved and sent to the Application Proxy Connector.

- The request is then sent to the on-premises application.

- The response is returned to the user via the Proxy Connector and the Proxy service.

> **Note**
> There are no inbound connections required to be opened on firewalls.

Azure AD Application Proxy is not intended for users on the corporate network, only those remote users wishing to connect to internal apps, such as home workers.

This section introduced Azure AD Application Proxy and concluded with the topic of remote access options in Windows Server. In the next section, we will complete a hands-on exercise section to reinforce some of the concepts covered in this chapter.

Hands-on exercises

To support your learning with some practical skills, we will utilize concepts and understanding gained from the earlier sections of this chapter and put them into practical application.

We will look at the following exercises:

- Exercise—implementing DNS

- Exercise—implementing DHCP

Getting started

To get started with this section, if you do not already have an Azure subscription, you can create a free Azure account at `https://Azure.microsoft.com/free`. This free Azure account provides the following:

- 12 months of free services

- $200 credit to explore Azure for 30 days

- 25+ services that are always free

Let us move on to the exercise for this chapter.

Exercise – implementing DNS

This section will look at implementing the DNS network service using Windows Server.

We will need access to a Windows Server OS computer for this exercise.

Task – installing the DNS role

In this task, we will install the DNS role in Windows Server. Follow the next steps:

1. Log in to your server with an admin account, click **Add Roles and Features** from **Manage** in **Server Manager**, and click **Next** on the **Before You Begin** screen.

2. Click **Role-based or feature-based installation** from the **Installation Type** screen and click **Next**.

3. From the **Server Selection** screen, accept the default selection of **Current Server**, and click **Next**.

4. Select **DNS Server** from the **Roles** list from the **Server Roles** screen and click **Add Features** on the pop-up screen. Then, click **Next**.

5. On the **Features** screen, accept the default selections, and click **Next**.

6. On the **DNS Server** screen, review the information, and click **Next**.

7. On the **Confirmation** screen, click **install**; view installation signs of progress until you receive a message displaying **installation succeeded**, then click **Close** on the **Add Roles and Features Wizard**.

8. You have now installed the *DNS Server role*; you will now find the **DNS** manager in **Server Manager** under the **Tools** menu and the left-hand navigation menu from the **Dashboard**, as illustrated in the following screenshot:

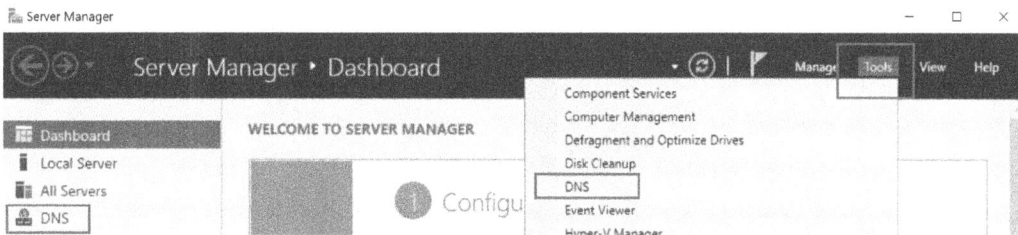

Figure 5.26 – DNS server role installed

In this task, you installed the DNS role. In the next task, you will create a DNS zone.

Task – creating a DNS zone

This task will create a DNS zone for the `milesbetter.solutions` domain; *you should use a zone name that meets your requirements*. Follow the next steps:

1. Launch **DNS Manager** from **Server Manager** under the **Tools** menu.

2. From the **Action** menu at the top of **DNS Manager**, click **New Zone…**, as illustrated in the following screenshot:

Figure 5.27 – Creating a DNS zone

3. On the **New Zone Wizard** screen, click **Next**.

4. On the **Zone Type** screen, leave it as the default of **Primary zone**, then click **Next**.

5. On the **Forward or Reverse Lookup Zone** screen, leave it as the default of **Forward Lookup Zone**, and click **Next**.

6. Review the information on the **Zone Name** screen, enter a **Zone name** value, and click **Next**. For this exercise, we will use an example zone name of `milesbetter.solutions`; *you should use a zone name that meets your requirements*. You can see an illustration of this in the following screenshot:

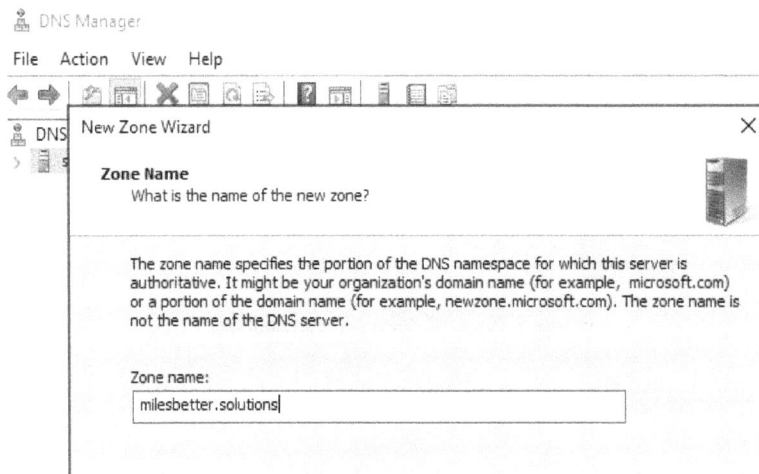

Figure 5.28 – DNS zone name

7. Accept the default filename on the **Zone File** screen and click **Next**. This exercise example will be auto-populated as `milesbetter.solutons.dns`, as illustrated in the following screenshot:

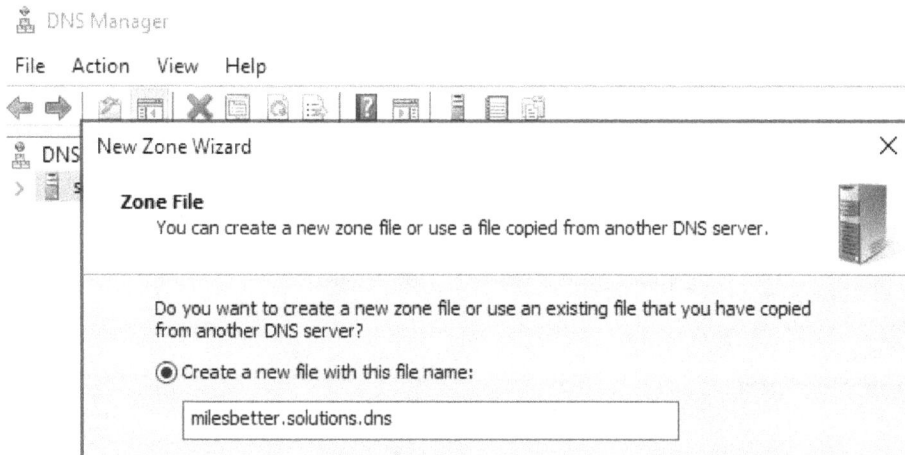

Figure 5.29 – DNS zone file

8. Review the information on the **Dynamic Update** screen, set it to the default of **Do not allow dynamic updates,** and click **Next**.

9. On the **Completing the New Zone Wizard** screen, review the information and click **Finish**.

10. Your *new zone* will appear in **DNS Manager** under **Forward Lookup Zones**, as illustrated in the following screenshot:

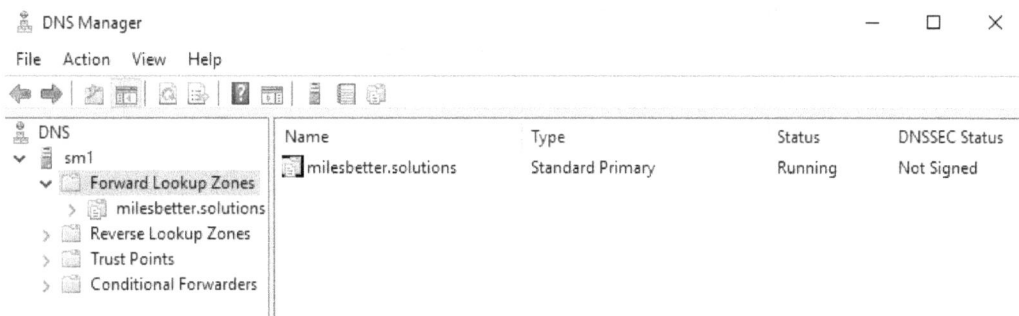

Figure 5.30 – DNS zone created

11. Click into the listed zone created to see the content of the zone file. As shown in the following screenshot, you will see two default SOA and NS records created for this zone automatically. Note the NS for this zone is, at present, only the current host you installed the DNS service on and added the zone to:

Figure 5.31 – DNS zone information

In this task, you created a DNS zone. In the next task, you will create a DNS record.

Task – creating a DNS record

This task will create a DNS record for the `milesbetter.solutions` zone; *you should use a record for the zone you created in the previous task.* Follow the next steps:

1. Launch **DNS Manager** from **Server Manager** under the **Tools** menu.

2. From the **Action** menu at the top of **DNS Manager**, click **New Host (A or AAAA)…**, as illustrated in the following screenshot:

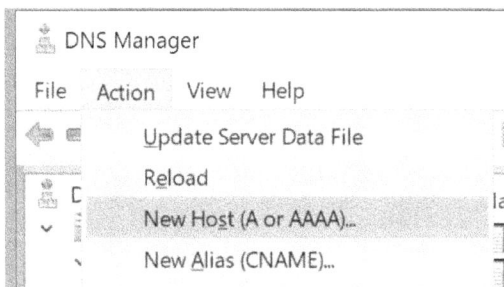

Figure 5.32 – New host record

3. From the **New Host** screen, enter **Name** and **IP address** values. For this exercise, we will use an example record name of `PizzaApp` and an IP of `10.1.0.4`; *you should use a record name and IP address that meets your requirements.* The following screenshot illustrates this stage:

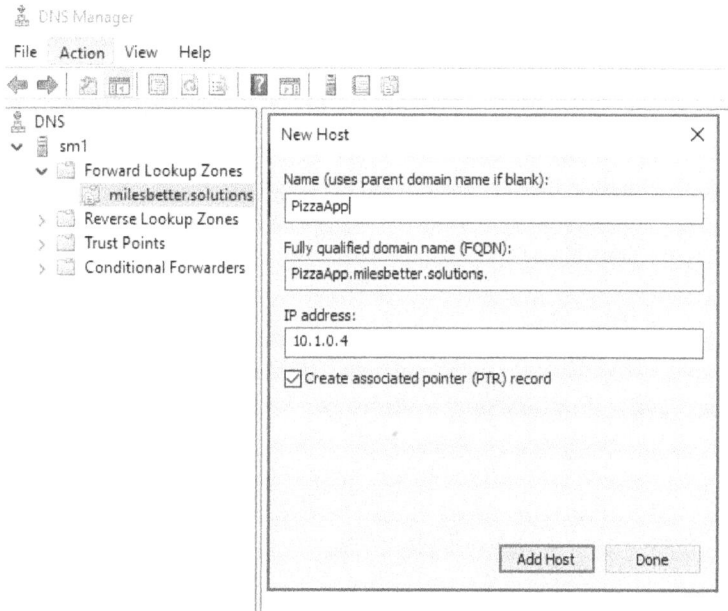

Figure 5.33 – Adding a host record

4. Check the **Create associated (PTR) record** check box and click **Add Host**.

5. On the **DNS** pop-up dialog box where the host record was successfully created, click **OK** and **Done** on the **New Host** screen.

6. From **DNS Manager**, navigate to the created zone, and you will see the newly created host record of `PizzaApp` that shows the record type as **Host (A)** and the data record of the IP address `10.1.0.4`, as illustrated in the following screenshot:

Figure 5.34 – New DNS zone and DNS host record

In this task, you created a DNS record.

With this, we have completed the exercise. This section taught us the skills needed to implement the DNS service in Windows Server and create a DNS zone and a DNS record. The following exercise will look at implementing the DHCP network service.

Exercise – implementing DHCP

This section will look at implementing the DHCP network service using Windows Server.

You will need access to a Windows Server OS computer for this exercise.

Task – installing the DHCP role

In this task, we will install the DNS role in Windows Server. Follow the next steps:

1. Log in to your server with an admin account, click **Add Roles and Features** from **Manage** in **Server Manager**, then click **Next** on the **Before You Begin screen.**

2. Click **Role-based or feature-based installation** from the **Installation Type** screen and click **Next**.

3. From the **Server Selection** screen, accept the default selection of the current server, and click **Next**.

4. Select **DHCP Server** from the **Roles** list from the **Server Roles** screen, click **Add Features** on the pop-up screen, and click **Next**.

5. On the **Features** screen, accept the *default selections*, and click **Next**.

6. On the **DHCP Server** screen, review the information, and click **Next**.

7. On the **Confirmation** screen, click **install**; view installation signs of progress until you receive an **installation succeeded** message, then click **Close** on the **Add Roles and Features Wizard**.

8. You have now installed the *DHCP Server role*; you will now see a notification in Server Manager that you must *complete the DHCP configuration* for the DHCP server. Click **Complete DHCP configuration**, as illustrated in the following screenshot:

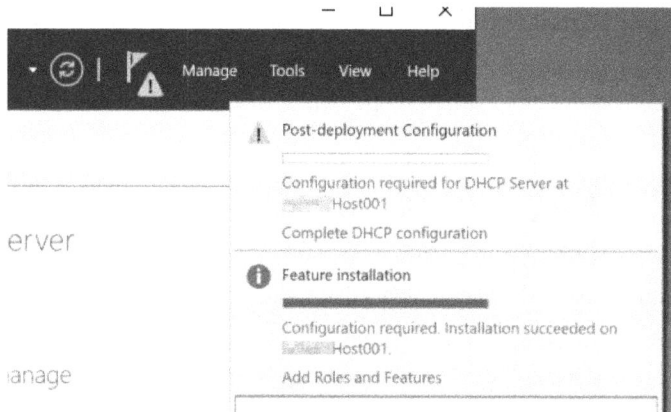

Figure 5.35 – DHCP server role installed

9. From the **DHCP Post-Install** configuration wizard, click **Commit** on the **Description** screen and **Close** on the **Summary** screen.

In this task, you created and installed the DHCP service. In the next task, you will create a DHCP scope.

Task – creating a scope

This task will create a scope. Proceed as follows:

1. Launch **DNS Manager** from **Server Manager** under the **Tools** menu.

2. From the **Action** menu at the top of the **DHCP** console, click **New Scope…**, as illustrated in the following screenshot:

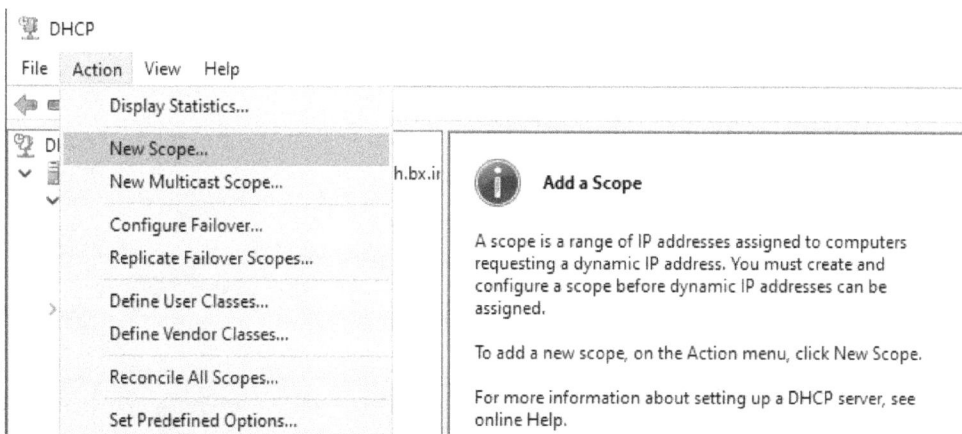

Figure 5.36 – Creating a scope

3. On the **New Scope Wizard** screen, click **Next**.

4. On the **Scope Name** screen, enter a **Name** value and—optionally—a **Description** value. For this exercise, we will use an example scope name; *you should use a scope name that meets your requirements*.

5. On the **IP Address Range** screen, enter a **Start IP address** value, an **End IP address** value, and **Subnet Mask** and **Length** values. For this exercise, we will use an example range; *you should use a range that meets your requirements*.

6. On the **Add Exclusions and Delay** screen, click **Next**.

7. On the **Lease Duration** screen, enter a length, and click **Next**.

8. On the **Configure DHCP Options** screen, select **Yes, I want to configure these options now**, and click **Next**.

9. Enter an **IP address** value on the **Router (Default Gateway)** screen and click **Next**.

10. On the **Domain Name and DNS Servers** screen, enter **Parent domain**, **Server name**, and **IP address** values, and click **Next**.

11. On the **WINS Servers** screen, click **Next**.

12. On the **Activate Scope** screen, leave the default setting of **Yes, I want to activate this scope now**, then click **Next**.

13. Click **Finish** to complete the **New Scope Wizard**.

14. In the DHCP console, the scope will be seen as follows:

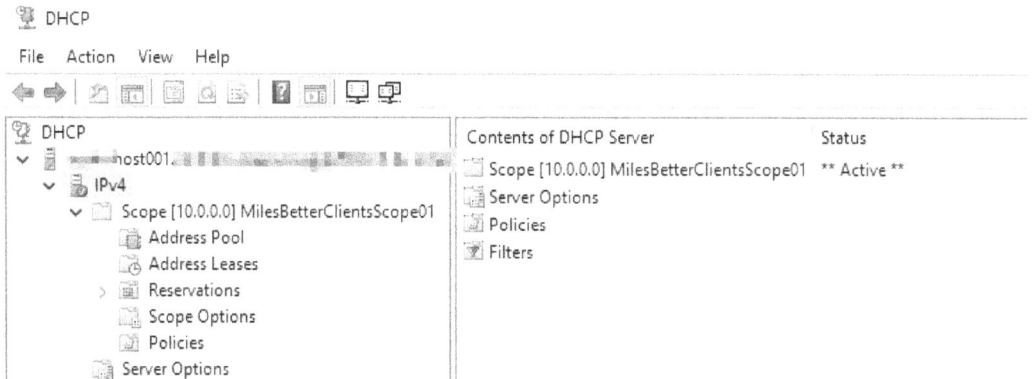

Figure 5.37 – Created scope

In this task, you created a scope.

Task – using commands to manage DHCP

This task will cover running commands to manage DHCP and retrieve information. Proceed as follows:

1. Run the following command to release an IP address from the DHCP server: `ipconfig / release`.

2. Run the following command to renew an IP address from the DHCP server: `ipconfig / renew`.

3. We can use the `netsh` command-line environment for DHCP server info. From **Command Prompt**, enter the following commands:

    *   `netsh dhcp show server`

    *   `netsh dhcp server show all`

    *   `netsh dhcp show version`

    *   `netsh dhcp server show all`

    *   `netsh dhcp show scope`

    *   `netsh dhcp show dnsconfig`

    *   `netsh dhcp show optionvalue`

With this, we have completed this exercise. This exercise taught us the skills needed to implement the DNS service in Windows Server and create a scope. Now, let us summarize this chapter.

Summary

This chapter provided coverage for the AZ-800 Administering Windows Server Hybrid Core Infrastructure: *Implement and manage on-premises network infrastructure* exam.

This chapter's content further developed your knowledge of and skills in on-premises network infrastructure services.

You learned about some core network infrastructure services in Windows servers, such as DHCP, DNS, IPAM, and remote access. We then finished the chapter with a hands-on exercise to develop your skills further.

You added new skills through the information provided, with the chapter's goal to take your knowledge beyond the exam objectives so that you are prepared for a real-world, day-to-day hybrid environment-focused role.

The next chapter will teach you about implementing and managing Azure network infrastructure.

Further reading

This section provides links to additional exam information and study references:

- *Microsoft Certified: Windows Server Hybrid Administrator Associate*: `https://docs.microsoft.com/en-us/learn/certifications/windows-server-hybrid-administrator/`

- *Exam AZ-800: Administering Windows Server Hybrid Core Infrastructure*: `https://docs.microsoft.com/en-us/learn/certifications/exams/az-800`

- *Exam AZ-800: skills outline*: `https://query.prod.cms.rt.microsoft.com/cms/api/am/binary/RWKI0r`

- *Microsoft Learn*: `https://docs.microsoft.com/en-us/learn/paths/implement-operate-premises-hybrid/`

Skills check

Challenge yourself with what you have learned in this chapter by trying to answer the following questions:

1. What is name resolution?
2. What is the primary method of name resolution in Windows Server?
3. What is a DNS resolver cache?
4. What is a namespace?
5. What is a zone?
6. What is the difference between a forward and reverse lookup zone?
7. What is a record?
8. List six different records and their purposes.
9. What is a forwarder?
10. What is meant by TTL?

6

Implementing and Managing Azure Network Infrastructure

This chapter covers *AZ-800 Administering Windows Server Hybrid Core Infrastructure: Implement and manage Azure network infrastructure.*

In the previous chapter, we added skills to understand the concepts of on-premises network infrastructure and services. We will start this chapter with an understanding of network services concepts in Azure. We will then look at each network service in more detail and conclude with a hands-on exercise to develop your skills further.

The following topics are included in this chapter:

- Introduction to Azure network services
- Implementing **virtual networks** (**VNets**) in Azure
- Implementing network connectivity in Azure
- Implementing network protection in Azure
- Implementing name resolution in Azure
- Hands-on exercises

In addition to the topics listed in this chapter, this chapter's goal is to take your knowledge beyond the exam objectives to prepare you for a real-world, day-to-day hybrid environment-focused role.

Introduction to Azure network services

This section introduces the functionality of the network services available in Azure, why each service is required, and how it may be used. Each network service capability will be covered in more detail in this chapter.

The following list outlines the available network services capabilities:

- **Communication between resources in Azure**: VNets provide the communication paths for network traffic between Azure resources such as **virtual machines** (**VMs**), load balancers, firewalls, and so on. Communication between VNets in the same region or another region is handled by **VNet peering** using the Microsoft backbone. **Service endpoints** and **private endpoints** are available for connecting **platform-as-a-service** (**PaaS**) resources such as database and storage services to VNets.

- **Communication with resources on-premises**: **Virtual private networks** (**VPNs**) as well as **Azure Relay**, **Azure Network Adapter**, **Azure Virtual WAN**, and **ExpressRoute** can provide hybrid connectivity scenarios, such as extending on-premises networks into Azure.

- **Communication with the internet**: VMs can communicate on the internet for both incoming and outgoing traffic and can be provided with network and application and protection services as well as **network address translation** (**NAT**) and routing.

- **Segmentation of address spaces**: VNets can be segmented with subnets like on-premises networks to optimize the address space(s) defined in the VNets. Traffic filtering and routing can also be applied at the subnet level. The address spaces will typically be *RFC 1918 private ranges*, but *public IP prefixes* are also supported. For security reasons, VM resources should have IPs assigned from the private IP address space with public IPs associated with load balancers, firewalls, and so on. In a specific scenario, the **Extended Network for Azure** provides a subnet extension, meaning an on-premises subnet can be stretched into Azure. Azure resources can also communicate with resources on the same local on-premises network.

- **Routing of network traffic**: A VNet uses default **system routes** to determine traffic paths. **User-Defined Routes** (**UDRs**) are much like static routes in on-premises networking; these allow the default system routes to be *overridden* with a more preferred and specific **next-hop** traffic path. This next hop could be a **Network Virtual Appliance** (**NVA**) such as a firewall.

- **Filtering of network traffic**: There is no traffic filtering in a VNet by default. A **network security group** (**NSG**) is a built-in Azure packet filter service for network protection. Network traffic inbound and outbound of a VNet can be controlled through a set of `allow` and `deny` rules. Filtering traffic can be applied based on factors such as source, destination, port, protocol, and so on. The **Azure Firewall** service can, in addition, provide *Layer 4* and *Layer 7* protection services.

- **Encryption of network traffic**: An **Azure VPN Gateway** can encrypt communication between Azure and remote networks such as on-premises locations.

- **Speed of network traffic**: Due to the inherent nature of the internet, if a *low-latency connection* is required between Azure and on-premises locations, then **Azure ExpressRoute** can be used. This provides a managed private network connection that bypasses the internet for traffic.

- **Name resolution**: Name resolution in Azure can be provided by a traditional **Windows Server DNS** running on a VM or provided by **Azure DNS**, which is a managed service that is fully VNet-integrated to provide both *public* and *private* zones.

In this section, we introduced some of the Azure network services. In the following sections of this chapter, we will cover each of these areas. The first topic we will look at in the next section is VNets.

Implementing VNets in Azure

This section will look at implementing and managing Azure VNets. We will cover VNet peering, VNet routing, implementing IP addressing, and the Extended Network for Azure.

The first skills area we will look at in this section is VNets in Azure and how they work.

Azure VNets

An **Azure VNet** is a software-defined representation of a **Local Area Network (LAN)** that you would have on-premises.

A VNet provides communication paths between resources through *network interfaces*. When resources in on-premises locations communicate with resources in Azure, this can be referred to as **hybrid connectivity** or **cross-premises connectivity**.

A VNet provides a private and isolated *single-tenant* network for your resources to communicate in. You can define private **IP address spaces** and segment them into subnets as you would for on-premises.

Each connected device on the VNet will be assigned a unique IP through the Azure **Dynamic Host Configuration Protocol (DHCP)** service. VMs are recommended to be assigned private IPs.

Public IPs can also be assigned to resources, although they are recommended for association with firewalls, load balancers, and so on for security reasons.

A VNet allows for network interfaces to be created that belong to a subnet as part of the segmentation of an IP address space associated with the VNet. These network interfaces are assigned an IP address in the range from that subnet by the Microsoft DHCP service; *you cannot bring your own DHCP servers*.

For a VM to communicate on the network, it must have one or more network interfaces associated with it. You would assign network interfaces according to the subnets you wish to send or receive traffic. A VM can have a network interface assigned from one or more subnets, with the maximum number of network interfaces supported by the VM type. The network interface is always allocated to the VM, even in the shutdown state. However, when the VM is deleted, the network interface remains a standalone resource but can be deleted as required.

A VNet in Azure exists in the boundary of an Azure region and is a communication boundary for network traffic within that region.

A VNet can span across a data center zone but cannot span a region. You can connect VNets in different regions and allow the traffic to span as though it were one network.

The default traffic behavior is that a VNet allows all resources within that VNet to communicate with each other but not with resources in another VNet without additional configuration steps.

A VNet can connect to on-premises locations for hybrid connectivity through a **VPN gateway**. Only *one* VPN gateway can be created for a VNet and can be created as a **VPN type** or an **ExpressRoute type**. Both types can exist on the same gateway if enough IP addressing is available to assign in address space; it is recommended to create a VPN subnet of /27 as a *minimum*.

VNet peering

VMs in one or more VNets can communicate with each other through **VNet peering**.

Unlike a VPN gateway where the communication traffic must route over the internet, VNet peering allows communication to flow as though they are the same logical network, and traffic passes over Microsoft's *backbone* and bypasses the internet. This is important in terms of *speed*, *reliability*, and *security*. VNet peering is available as **regional VNet peering**, which connects VNets in the *same region*, and **global VNet peering**, which connects VNets in *different regions*.

This is represented in the following diagram:

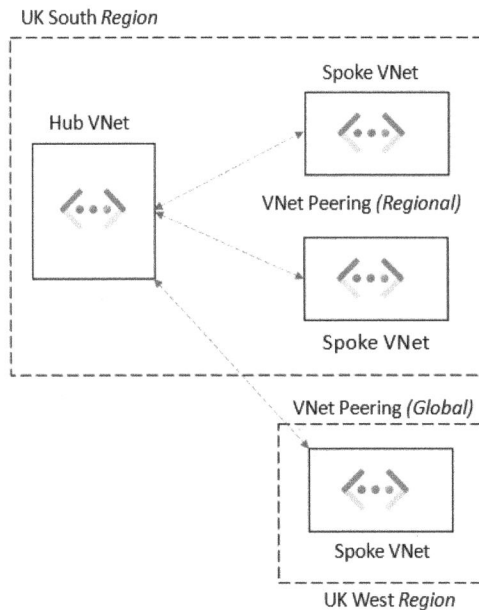

Figure 6.1 – VNet peering

In this section, we looked at what an Azure VNet was. In this next section, we will look at IP addressing for VMs.

How to implement IP addressing

When a VNet is created, you define one or more IP address spaces. This can then be further broken down into subnets to meet your requirements. Each resource can be assigned an IP address from the subnet so that other resources can communicate with it on the network.

This is represented in the following diagram:

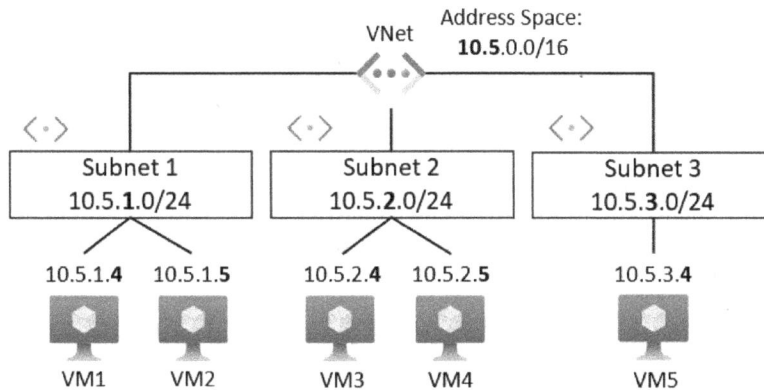

Figure 6.2 – IP address subnetting

The preceding diagram represents a VNet with an address space defined using the **classless inter-domain routing** (**CIDR**) format of 10.5.0.0/16. Subnetting has been used to further segment into smaller subnets that use a /24 notation, and each resource is allocated an IP from this range. This approach allows us to optimize the IP address allocation from the defined address /16 space. A VNet can have more than one IP address space assigned, which can be a public IP address prefix but is more commonly a private address range.

The following are the recommended private non-routable (*RFC 1918*) address ranges:

- 192.168.0.0 to 192.168.255.255 (192.168/16 prefix)

- 172.16.0.0 to 172.31.255.255 (172.16/12 prefix)

- 10.0.0 to 10.255.255.255 (10/8 prefix)

The private IP addresses are static or dynamically assigned to the network interfaces and can be changed between the two at any time. Typically, static IP addresses will be used, especially in the case of VMs that will perform roles such as domain controllers, DNS servers, or NVAs, where it is critical that the IPs remain assigned and not change. Private IPs (static and dynamic) can also be assigned to resources such as load balancers and network interfaces of storage services, for example, that use *private endpoints*.

When planning IP addressing in Azure, it is essential to note that Azure reserves *five IP addresses* for its own use in each subnet range, which cannot be allocated to any resources. These five reserved IP addresses are the *first four IP addresses* and the *last IP address* in a range, with the IP address ranges starting at 0, not 1.

The reserved five IP addresses are utilized as follows:

- x.x.x.0: This first IP address is used for the *network address*

- x.x.x.1: This second IP address is used for the *default gateway*

- x.x.x.2: This third address is used for *Azure DNS*

- x.x.x.3: This fourth IP address is used for *Azure DNS*

- x.x.x.255: This fifth IP address is used for the *broadcast address*

You will notice from the preceding list that the first IP assigned to a created VM in the subnet will be the .4 address.

Public IP addressing prefixes can be added to Azure VNets and can be assigned to resources such as network interfaces, VPN gateways, NVAs, and internet-facing load balancers. There are *basic-* and *standard*-type **stock-keeping units** (**SKUs**).

IPv6 can also be used alongside IPv4 in the same VNet and for the same resource. When a resource such as a VM has requirements for IPv6 and IPv4 communication paths, that is known as a **dual stack**.

Azure Extended Network

The **Extended Network for Azure** capability provides a subnet extension, meaning an on-premises subnet can be stretched into Azure to meet the requirements of specific scenarios, and vice versa. Azure resources can be part of your on-premises local network, akin to making the same IP broadcast domain available in separate locations. It should be noted that this is not necessarily a good practice but may be required for a specific scenario that dictates this requirement.

This is achieved by extending the on-premises Layer 2 network with a Layer 3 overlay network using a **Virtual Extensible LAN** (**VXLAN**) solution. The advantage is that when VMs move to Azure, they can retain their existing IPs. Ordinarily, this is not possible as there cannot be overlapping IP address space between the on-premises networks and Azure.

This is represented in the following diagram:

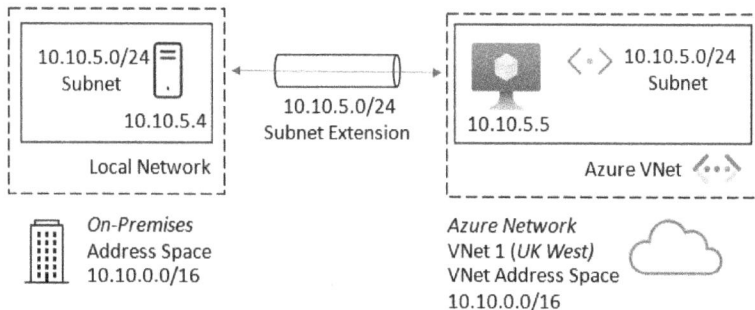

Figure 6.3 – Azure Extended Network subnet extension

This section looked at how to implement IP addressing in an Azure VNet. In the next section, we will look at implementing VNet routing.

How to implement VNet routing

As we learned earlier, an Azure VNet is a software-defined representation of on-premises networks. When a VNet is created, an *RFC 1918* private address space can be configured, and subnets created for the segmented network design are required. In an earlier example, we saw a /16 address space that contained /24 subnets for IP assignment to network interfaces that could be attached to VMs to allow them to communicate within the VNet.

Azure has a set of default system routes that automatically provide the Azure built-in routing capability; this allows VMs in the same VNet to communicate with each other with no specific network configuration required. The default gateway address is pre-defined at .1 in any subnet, and the system routes determine the next hop paths. These routes cannot be *edited* or *removed*, although you can provide custom routes that allow you to override the system routes with more specific routes that take priority over the system routes. These custom routes are referred to as UDRs. UDRs are applied through a **route table** resource. This provides specific network packet control of where the next hop should be for any packet leaving a subnet where it is not desired to use the default .1 gateway next hops.

The next-hop type can be one of the following:

- Virtual appliance
- VNet gateway
- VNet
- Internet

Implementing UDRs in subnets is particularly beneficial when we want the next hop for traffic leaving a subnet to be filtered and inspected by an NVA such as a *firewall*. This allows us to control east-west traffic and provides network protection against **lateral attacks**.

This concept is represented in the following diagram:

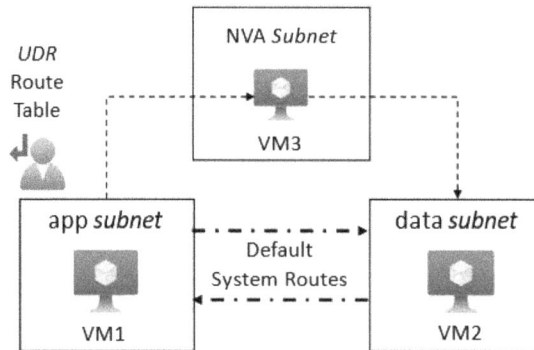

Figure 6.4 – Network routing in Azure

In the preceding diagram, we see that the default system routing shows direct communication possible between the **VM1** and **VM2** VMs. This allows **VM2** to communicate with **VM1** using the default system route.

Due to the UDR route table being associated with the app subnet, all packets leaving the app subnet will use the custom routes provided by the UDR route table.

In this case, the route table has an entry that specifies that for all traffic with a destination of the data subnet, its next hop must be **VM3**.

This means **VM1** cannot communicate with **VM2** directly, but **VM1** can only communicate with **VM2** via **VM3**, a firewall appliance VM; if **VM1** were to be compromised, this approach would protect VM2 from a lateral movement attack.

We should also discuss in this section the topic of **egress routing**. This is where we can determine the path of traffic leaving an Azure VNet, based on cost, quality, and performance metrics.

The two options available for egress routing are set out here:

- **Cold-potato routing**: With this routing method, traffic remains on the global Microsoft backbone for as long as possible. The final destination gets the traffic once the Microsoft backbone has handed off to a downstream ISP edge **Point of Presence** (**POP**) to perform this *last-mile* traffic delivery.

- **Hot-potato routing**: With this routing method, the traffic leaves the Microsoft global network at the earliest opportunity and travels to its destination via many hops across the internet, hence the term *hot potato*, as the traffic is not held on to for long before being handed off to the *next hop* or *hand-off* for the packet.

This concept is represented in the following diagram:

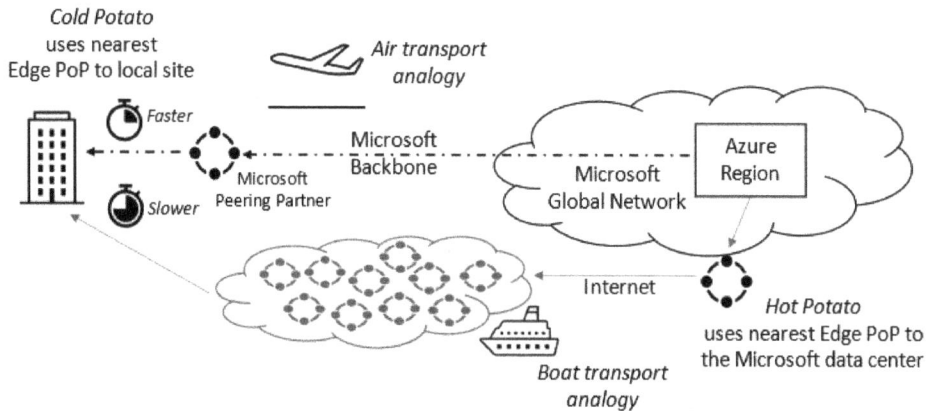

Figure 6.5 – Egress routing in Azure

The preceding diagram visualizes the cold-potato versus hot-potato egress routing methods; the benefits and disadvantages are as follows:

- Cold-potato routing is the most expensive form of egress routing but has the benefit of the lowest latency, best quality, and fastest performance. Think of this as a *courier service* versus a *regular mail* delivery service.

- Hot-potato routing is the cheapest option, but the disadvantage is that it has the slowest and least reliable delivery. Think of this as a *regular mail* delivery service versus a *courier service*.

We can also link this traffic delivery method to *air transport* versus *ground/boat shipping*.

In this section, we looked at VNet routing. The following section will look at implementing hybrid network connectivity in Azure.

Implementing network connectivity in Azure

This section will look at network connectivity in Azure between VNets as well as *cross-premises/hybrid* options available in Azure.

While there are many third-party vendor solutions, we will look to introduce only the first-party Microsoft services available in Azure and for the exam skills objectives.

VPNs are the first skills area we will look at in this section.

VPNs

The Microsoft **VPN Gateway** is a network connectivity service that provides traffic encryption through private and secure tunnels across the internet. These tunnels can connect on-premises networks to Azure VNets or connect VNets in the same region or across different regions.

When a VPN gateway is implemented, it can be created as two types:

- VPN connection type
- ExpressRoute connection

Only one VNet gateway can be created per VNet, with a **high availability** (**HA**) option; VNets *can't share* a gateway.

The gateway type must be deleted and created again with the correct type to change it.

VPN connection type gateways can be created as two VPN types:

- **Policy-based** (**static routing**) VPN gateway
- **Route-based** (**dynamic routing**) VPN gateway

The VPN gateway can create **Point-to-Site** VPNs and **Site-to-Site** VPNs.

The **Internet Protocol Security** (**IPsec**) protocol **Internet Key Exchange** (**IKE**) **versions 1** and **2** can be used; **pre-shared** keys are used for authentication.

The following connectivity types require a route-based VPN gateway:

- Point-to-Site VPN
- Inter-VNet connections
- Multi-site connections
- ExpressRoute coexistence

This section introduced the VPN Gateway service in Azure. The following sections will look at the Point-to Site and Site-to-Site VPN types.

Point-to-Site VPN

A Point-to-Site VPN type provides an encrypted connection from a single device to Azure resources by creating a secure tunnel across the internet for communication.

The protocols used for this VPN type are as follows:

- OpenVPN

- **Secure Socket Tunneling Protocol (SSTP)**

- IKEv2

This Point-to-Site VPN type is represented in the following diagram:

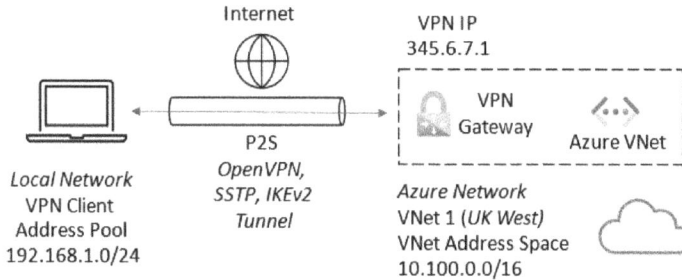

Figure 6.6 – Point-to-Site VPN

The use case for a Point-to-Site VPN is when remote workers, such as home workers or those traveling, need access to Azure resources accessible via VNets and cannot use a Site-to-Site VPN.

Site-to-Site VPN

A Site-to-Site VPN type provides an encrypted connection from *on-premises locations* to Azure resources by creating a secure tunnel across the internet for communication to support hybrid connectivity/cross-premises scenarios.

The Azure VPN gateway supports policy-based and route-based VPNs and the IPsec v1/v2 protocols; pre-shared keys are used for authentication.

This Site-to-Site VPN type is represented in the following diagram:

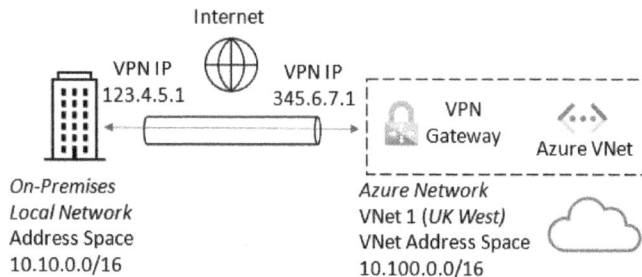

Figure 6.7 – Site-to-Site VPN

In the preceding diagram, the on-premises location must have a VPN device with a public IP address. Note that without the deployment of **Azure Extended Network**, the on-premises and Azure VNet address spaces *cannot overlap*; Layer 3 routing over the public internet is required.

A VPN gateway can also establish VPN tunnels to multiple on-premises locations using a single VNet gateway.

This multi-site VPN connection type is represented in the following diagram:

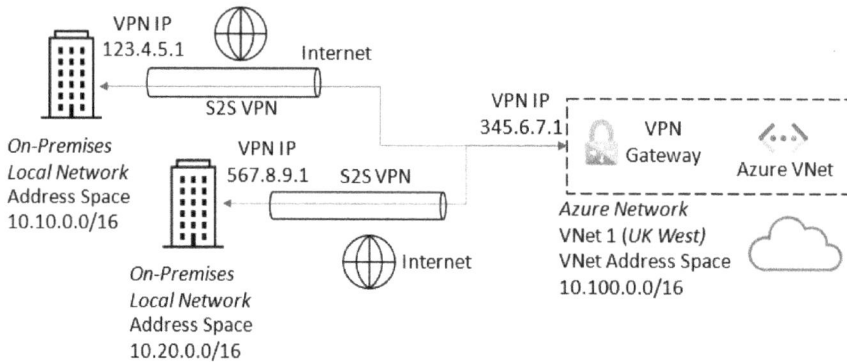

Figure 6.8 – Multi-site VPN connection

For multiple VPN connection tunnels to co-exist on a single VNET gateway, you must use a **route-based VPN** to support this.

VNet-to-VNet VPN

VPNs can also connect two Azure VNets to allow encrypted communications.

This VNet-to-VNet VPN type is represented in the following diagram:

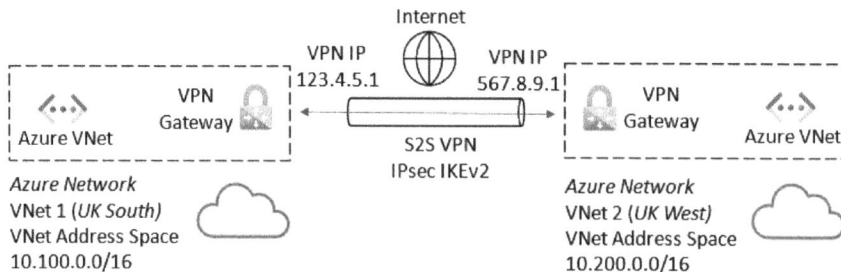

Figure 6.9 – VNet-to-VNet VPN connection

To interconnect Azure VNets, we can also use the alternative method of VNet peering; however, this does not encrypt the traffic, bypasses the internet, and stays on Microsoft's backbone.

Implementing Azure VNet gateways

As we learned earlier, a VNet gateway is an Azure resource that connects on-premises networks to Azure VNets over the internet. Secure and private tunnels are created through which encrypted traffic can be sent. This allows resources in both cloud and on-premises locations to communicate as though they were on the same local network.

A VNet gateway requires the following resources to be implemented:

- **VNet**: The VNet gateway resource can only be associated with one VNet; a VNet cannot have more than one VPN gateway. There should be no overlap of IP address space ranges between the Azure VNet and the on-premises networks—that is, Azure networks should be created with unique CIDR IP address space ranges.

- **Gateway subnet**: A dedicated subnet resource is required to be created in the VNet and must be called **GatewaySubnet**. The VNet address space used must have sufficient IPs available to assign. It is recommended this should be /27, /26, or larger to ensure enough IP addresses are available, especially if you wish ExpressRoute to co-exist on the same VNet gateway.

- **Public IP address**: This resource is used by the remote VPN device to connect to the VNet gateway.

- **Local network gateway**: This resource represents the remote network location; be careful of the *local* terminology: this means the network at the other end of the tunnel from the VNet gateway. This identifies the on-premises VPN device used to connect with the Azure VPN gateway. You specify the network ranges you will allow a connection from.

- **Connection**: This resource represents a logical connection between the VPN and the local network gateway.

The preceding resources and their relationships are represented in the following diagram:

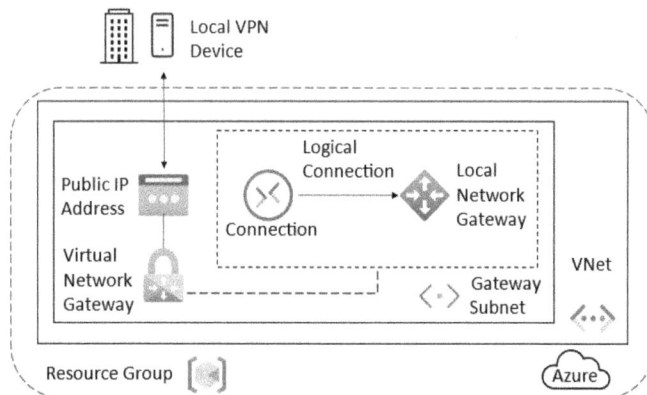

Figure 6.10 – VPN gateway resources

VNet gateway is also used when implementing ExpressRoute, which we will look at in the next section.

ExpressRoute

An **ExpressRoute circuit** is an extension of your on-premises network into Azure without the need for VPNs or traversing the internet. It provides *cross-premises/hybrid connectivity* using a reliable, **low-latency**, fast private network, with traffic routed directly to Microsoft's data centers, bypassing the internet. This private network is facilitated by a network carrier provider, which acts as the connectivity provider.

This is represented in the following diagram:

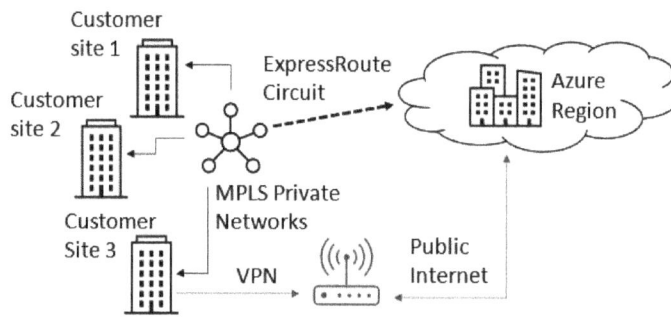

Figure 6.11 – ExpressRoute

Working with a network carrier provider, ExpressRoute can make Azure data centers appear as a site on a **Multiprotocol Label Switching** (**MPLS**) network. This provides global reach and capacity extensions of on-premises locations into Azure Data centers.

With ExpressRoute being an enterprise-grade network solution, redundancy is built in using **Border Gateway Protocol** (**BGP**) and **dynamic routing**.

It should be noted that by ExpressRoute does not *encrypt* or *filter* traffic; you should build that into a design if that is important. Typically, this will be achieved by implementing **NVAs** into the solution.

To implement an ExpressRoute circuit, the following steps outline the actions required:

1. Create an ExpressRoute gateway.
2. Create a circuit by working with a connectivity provider.
3. Create a peering.
4. Connect the ExpressRoute circuit to a VPN.

Microsoft provides two peering types, as follows:

- **Private peering**: This type provides connections to Azure VNets
- **Microsoft peering**: This type provides connections to Microsoft public services such as **Microsoft Office 365 (M365)** and **Dynamics 365 (D365)**

You may have a Site-to-Site VPN and ExpressRoute on the same VNet gateway, but you must use a **route-based VPN** to support this. It is recommended to allocate a /27 subnet to ensure adequate IPS can be assigned.

In this section, we looked at ExpressRoute. In the next section, we will look at Azure Network Adapter.

Azure Network Adapter

Windows Server 2019 introduced Azure Network Adapter as a new hybrid connectivity solution. To connect to networks and VMs running in Azure, there is the option of VPNs and ExpressRoute, which can add significant complexity to what is often a simple need to connect resources on-premises to resources in Azure.

With Azure Network Adapter and through Windows Admin Center, you can provide a one-click solution to connect your Azure VNets to on-premises networks using a Point-to-Site VPN.

Azure Relay

Azure Relay is the service that used to be called **Service Bus Relay**.

Azure Relay is a hybrid solution that provides the ability to expose internal application resources to public cloud environments. It works based on *listeners* and does this without the need for inbound firewall ports to be open and without intrusive network infrastructure changes to the on-premises environment. This is a different solution from VPNs, which work at the network integration level. This is like an application VPN where we can scope connection access to a single on-premises machine's application rather than a VPN (whose scope is very wide to a network range) and is more about connecting networks than connecting applications.

Azure Virtual WAN

Azure Virtual WAN provides a **global network transit architecture** solution.

It is a virtualized representation of a **Wide Area Network (WAN)** that would traditionally be in place on-premises, such as an MPLS solution. In the case of Virtual WAN, Microsoft now becomes your connectivity provider.

In essence, the solution is a scaled-out, any-to-any connection type VNet gateway solution. It can handle multiple connection types, such as **P2S**, **S2S**, **ExpressRoute**, and **Software-Defined WAN (SD-WAN)**.

A virtual WAN enables **endpoints** to *transit connectivity* across a **hub**. These endpoints make up **distributed spokes** that can be global.

This is represented in the following diagram:

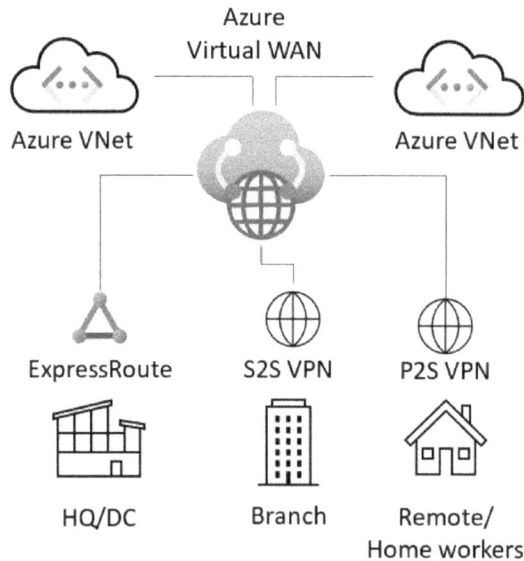

Figure 6.12 – Azure Virtual WAN

Azure Virtual WAN supports the following capabilities:

- **Point-to-Site (P2S)** VPN connectivity
- **Site-to-Site (S2S)** VPN connectivity
- ExpressRoute connectivity
- VPN and ExpressRoute interconnectivity
- SD-WAN connectivity
- Intra-cloud connectivity
- Intra-VNet connectivity
- Encryption
- Routing and traffic control
- Azure Firewall

This can be thought of as an **any-to-any**-type connectivity solution.

To implement an Azure virtual WAN, the high-level steps are as follows:

1. Create a Virtual WAN resource.
2. Create a Virtual WAN hub.
3. Create a Virtual WAN site.
4. Connect spoke endpoints to the hub.
5. Download a configuration file.
6. Validate connectivity through the hub and spoke endpoints.

In this section, we looked at Azure Virtual WAN and concluded the topic of network connectivity in Azure. The following section will look at implementing network protection in Azure.

Implementing network protection in Azure

This section will look at the network protection options available in Azure. This will cover NSGs, **application security groups** (**ASGs**), and the Azure Firewall; we will look at the use cases, how they work, and their capabilities.

This section will look at NSGs as the first skill area.

NSGs

An NSG is a *packet filter* approach controlling traffic flow into and out of resources such as VMs.

A set of **inbound** and **outbound** rules filters network traffic, denying all traffic unless *explicitly allowed*.

An NSG evaluates whether access is *allowed* or *denied* based on five data points (the *5-tuple* method); these data points are as follows:

- Source
- Source port
- Destination
- Destination port
- Protocol

NSGs are represented in the following diagram:

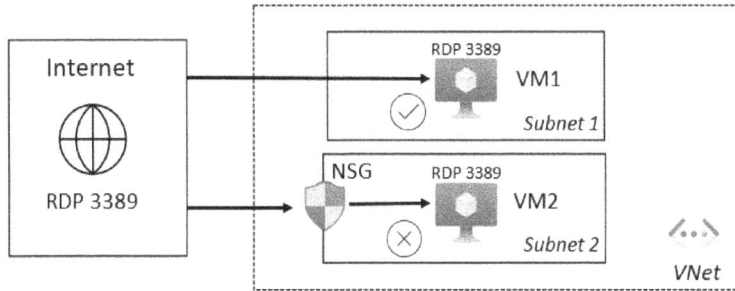

Figure 6.13 – NSGs

NSGs can be associated with a **network interface** or a **subnet**, but not a VNet. A subnet or network interface can only have one NSG assigned; an NSG can be associated with multiple subnets or network interfaces.

NSG association is represented in the following diagram:

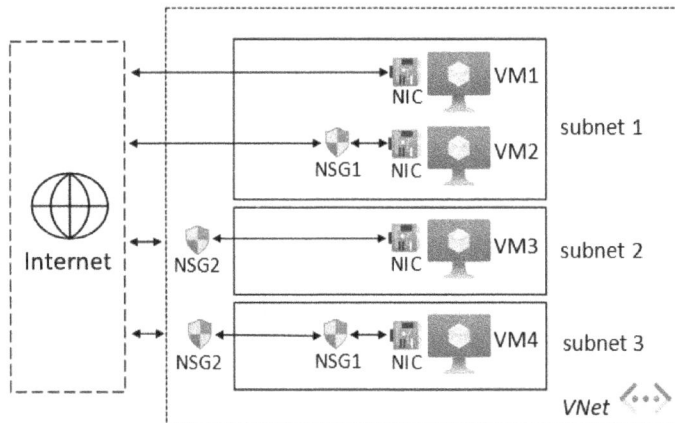

Figure 6.14 – NSG association

Each NSG rule is numbered; the lowest number will be evaluated first and will determine based on the data points if a connection to the destination is *allowed* or *denied*.

Each NSG has a set of **default rules** that cannot be modified or removed; you should add **custom rules** with a lower number to have these overrides evaluated before the default rules. The final default rule that will be enforced is a **deny-all** rule.

An NSG can only be used to protect resources in the **same subscription** and **region** as the NSG; if you need to perform centralized network protection across highly distributed VNets and across multiple regions, Azure Firewall can be used.

This section looked at NSGs. In the next section, we will look at ASGs.

ASGs

An ASG provides a set of application-/workload-specific rules that can be associated with an NSG. In a nutshell, we base ASG traffic filter rules on business logic layer names rather than IP addresses or ranges. This is intended to simplify network protection rules by determining access at the business logic layer.

An ASG provides simplified network security micro-segmentations. We can apply a single NSG to all subnets on the NSG for full traffic analytics via flow logs. We then apply ASGs, with each VM being made a member of the appropriate ASG based on its role or network segmentation; this allows granular VM network interface layer control for both east/west and north/south traffic.

This approach is achieved through traffic filter rules based on the business logic and your application topology rather than the network address space topology. It takes a declarative/intent base approach. Rather than defining specific network IPs to be included in the rule, you define the business logic name layer groups you intend to filter traffic for and let Microsoft handle the complexities of the IP rules required. This is represented in the following diagram:

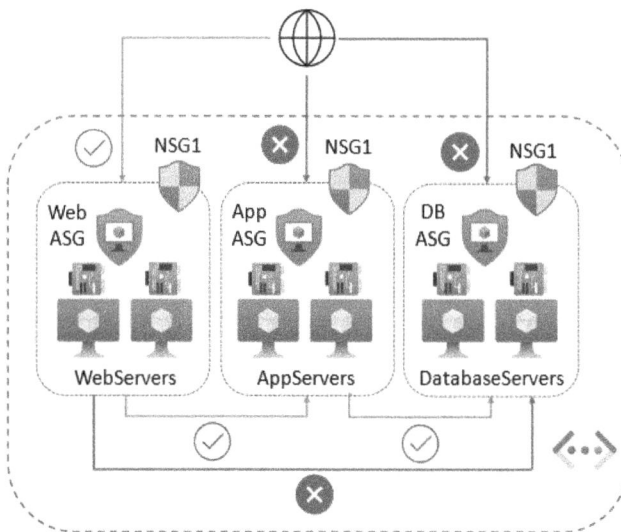

NSG 'IP' Rules Example

Action	Source	Destination
Allow	Internet	10.1.1.0/24
Allow	10.1.1.0/24	10.1.2.0/24
Allow	10.1.2.0/24	10.1.3.0/24
Deny	Any	Any

ASG 'Name' Rules Example

Action	Source	Destination
Allow	Internet	WebServers
Allow	WebServers	AppServers
Allow	AppServers	DatabaseServers
Deny	Any	Any

Figure 6.15 – ASG association

We can link this approach to the DNS concept, which uses names instead of IP addresses. In DNS, we may not know the IP address of a server, and its IP address may change, so we focus on business logic, meaning whatever the resource name is (that is constant), return the IP address (that can be variable). We can therefore use the same approach in ASGs to assign an *alias* or *nickname*, referred to as a *moniker*, to a group of VM resources and set a traffic filter rule based on the *monikers* rather than a network address.

This section looked at ASGs. In the next section, we will look at Azure Firewall.

Azure Firewall

Azure Firewall is a cloud-based Microsoft-managed **firewall-as-a-service** (**FWaaS**) solution. It provides centralized Layer 3/Layer 7 protection and control policies. Unlike an NSG, it works across subscriptions and regions.

Traffic control is through UDRs and can provide segmentation of networks for traffic control.

A typical Azure Firewall network topology is represented in the following diagram:

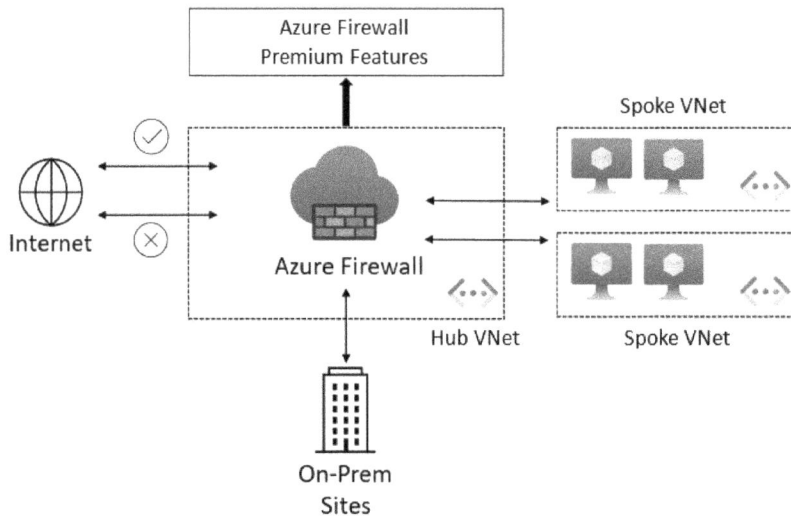

Figure 6.16 – Azure Firewall

Some of the capabilities of Azure Firewall are outlined here:

- **Destination NAT (DNAT)**
- **Source NAT (SNAT)**
- **Intrusion Prevention System (IPS)**

- **Transport layer security** (**TLS**) inspection
- **Uniform Resource Locator** (**URL**) filtering

Firewall Manager is another service that can be utilized. It provides centralized policy configuration and management for multiple deployed Azure firewalls across different Azure regions and subscriptions.

In this section, we looked at Azure Firewall and concluded the topic of network protection in Azure. The following section will look at implementing name resolution in Azure.

Implementing name resolution in Azure

This section will look at the name resolution options available in Azure; we will also cover the topic of cross-premises/hybrid name resolution.

Azure DNS is the first skill area we will look at in this section.

Azure DNS

Azure DNS is a Microsoft-managed **DNS-as-a-service** solution for name resolution of Azure resources. You do not need to provide your own DNS servers for name resolution.

The Azure DNS service provides hosting for both your **public DNS zones** and **private DNS zones**, for which Azure can be authoritative, as detailed here:

- **Public DNS**: Hosts your DNS domains and provides name resolution for DNS domains that are internet-facing
- **Private DNS**: Allows hostname resolution within a VNet and between VNets

Once your DNS zones are *migrated* to Azure DNS, Microsoft's **DNS name servers** will respond to queries for resources in these zones. **Anycast DNS** is used by the DNS service, meaning the query will be responded to by the DNS server that is closest geographically to the query.

The **Azure portal**, **Azure PowerShell**, or the **Azure CLI** can be used to create and manage Azure public and private zones.

These are the high-level steps to implement a public zone:

1. Create a public zone.
2. Add records to the zone.
3. Validate name resolution for the zone.

These are the high-level steps to implement a private zone:

1. Create a private zone.
2. Link (publish) the zone to the VNets for name resolution using the Azure DNS service.
3. Add records to the zone; if required, enable *auto-registration*.
4. Validate name resolution for the zone.

For Azure DNS private zones, the following capabilities are provided:

- Automatic registration from a private zone linked to the VNet of VMs
- DNS resolution forwarding across private zone-linked VNets
- Reverse DNS lookup within the scope of the VNet

The following are some limitations of Azure DNS:

- There is no support for **conditional forwarding**; you should use your own DNS servers in place of this if this capability is required
- There is no support for **Domain Name System Security Extensions** (**DNSSEC**); you should use your own DNS servers in place of this if this capability is required
- There is no support for **zone transfers**; you should use your own DNS servers in place of this if this capability is required
- Reverse DNS is only possible with private address spaces linked to the VNet
- You can only link one VNet to a private zone
- There is a limit to the number of zones and records per subscription

Next, we will look at Windows Server DNS on Azure VMs and integration with Azure DNS.

Windows Server DNS on Azure VMs

The DNS server role can be installed on Azure VMs running the Windows Server OS using Server Manager in the same way as for an on-premises installation.

You can specify that a VNet uses your own Windows DNS servers instead of Azure DNS; these DNS servers could be Azure VM DNS servers or on-premises DNS servers connected to the Azure VNet.

You may decide to implement Windows Server DNS on Azure VMs to replace or in addition to Azure DNS in the following scenarios:

- A need for name resolution between VNets
- A need for name resolution of Azure from on-premises

- A need for conditional forwarders

- A need for zone transfers

Your own Windows DNS servers can be set in the Azure VNet **DNS servers | Custom** setting, as shown in the following diagram:

Figure 6.17 – Custom DNS server setting

The following diagram represents hybrid name resolution using Windows Server DNS and Azure DNS:

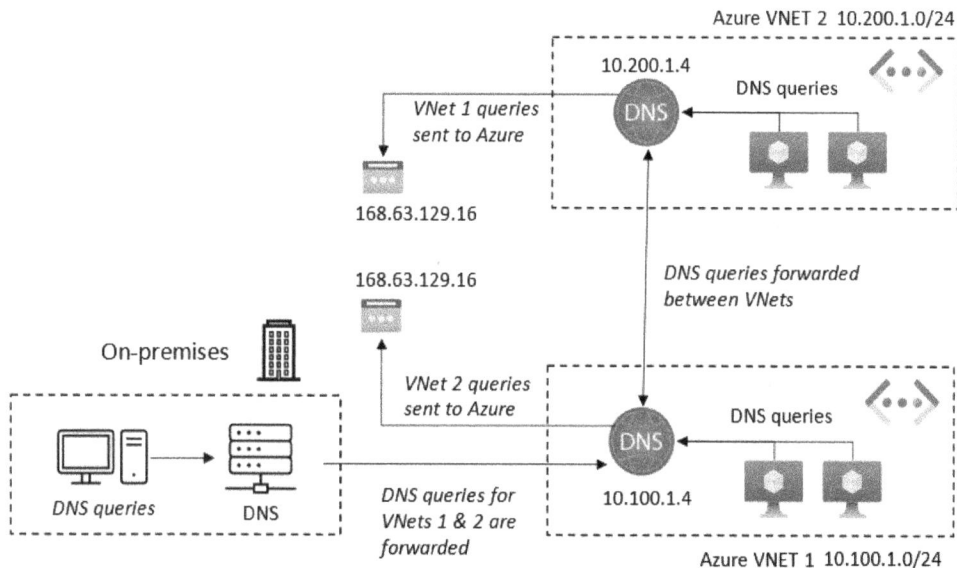

Figure 6.18 – Hybrid name resolution

In the preceding diagram, Windows Server DNS servers attached to a VNet can respond to queries for its on-premises domain, such as its own authoritative zones, and use the recursive resolvers in Azure for forwarded queries that need to resolve Azure hostnames that it will not have a record of.

The address to use for Azure DNS recursive DNS resolvers is 168.63.129.16.

Next, we will look at the split-horizon DNS in Azure.

Split-horizon DNS in Azure

Split-horizon DNS uses the same name to resolve the names of *internal* and *external* resources. It is also known as *split-brain DNS*.

This works by having two zones for the same domain; one is used for internal resource queries, and the other for external resource queries.

In Azure DNS, this is implemented by creating two zones of the same name, such as `milesbetter.solutions`.

The `milesbetter.solutions` zones should be created and configured as follows:

- **Private zone records**:

Host	Record type	IP address
`pizzapp`	CNAME	`mbs1.milesbetter. solutions`
`mbs1`	A	`10.5.1.4`

- **Public zone records**:

Host	Record type	IP address (public)
`pizzapp`	A	`123.456.789`

This configuration returns the correct lookups and information based on whether clients are on the internal or external network.

In this section, we looked at Windows Server DNS on Azure VMs and concluded the topic of implementing name resolution in Azure. In the next section, we will complete a hands-on exercises section to re-enforce some of the concepts covered in this chapter.

Hands-on exercises

To support your learning with some practical skills, we will utilize concepts and understanding gained from the earlier sections of this chapter and put them into practical application.

We will look at the following exercises:

- Exercise—Implementing an Azure VNet
- Exercise—Implementing an NSG
- Exercise—Implementing Azure DNS
- Exercise—Implementing an Azure VPN gateway
- Exercise—Implementing VNet peering

Getting started

To get started with this section, if you do not already have an Azure subscription, you can create a free Azure account at `https://azure.microsoft.com/free`. This free Azure account provides the following:

- 12 months of free services
- USD $200 credit to explore Azure for 30 days
- 25+ services that are always free

Let's move on to the exercises for this chapter.

Exercise – Implementing an Azure VNet

In this exercise, we will look to create an Azure VNet.

Follow these steps to create a VNet:

1. Log in to the Azure portal: `https://portal.azure.com`. You can alternatively use the Azure desktop app: `https://portal.azure.com/App/Download`.

2. In the search bar, type `virtual networks`; click on **Virtual networks** from the list of services shown:

Figure 6.19 – Searching for the Virtual networks resource

3. From the **Virtual networks** blade, click on the **+ Create** option from the top menu of the blade, or use the **Create virtual network** button at the bottom of the blade. Optionally, click the **Learn more** hyperlink for further information and learning:

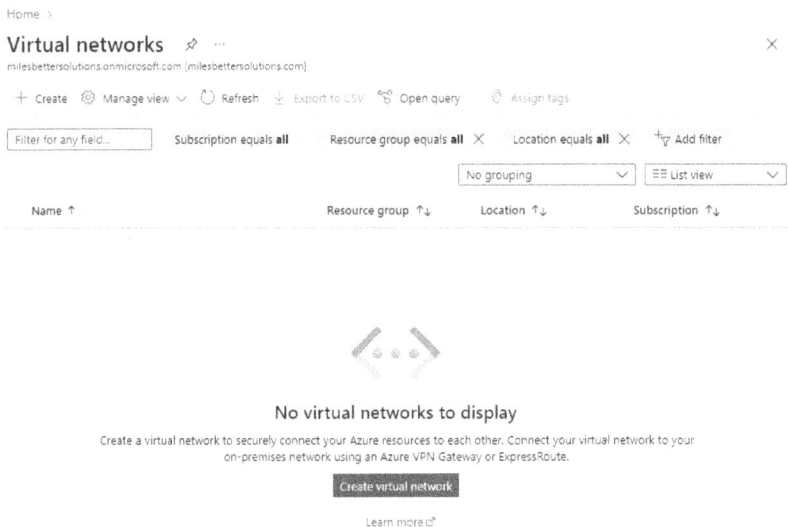

Home >

Virtual networks

milesbettersolutions.onmicrosoft.com (milesbettersolutions.com)

+ Create   Manage view   Refresh   Export to CSV   Open query   Assign tags

Filter for any field...   Subscription equals **all**   Resource group equals **all**   Location equals **all**   Add filter

No grouping   List view

Name ↑	Resource group ↑↓	Location ↑↓	Subscription ↑↓

No virtual networks to display

Create a virtual network to securely connect your Azure resources to each other. Connect your virtual network to your on-premises network using an Azure VPN Gateway or ExpressRoute.

Create virtual network

Learn more

Figure 6.20 – Creating a VNet

4. Complete the **Project details** and **Instance details** settings as required on the **Basics** tab:

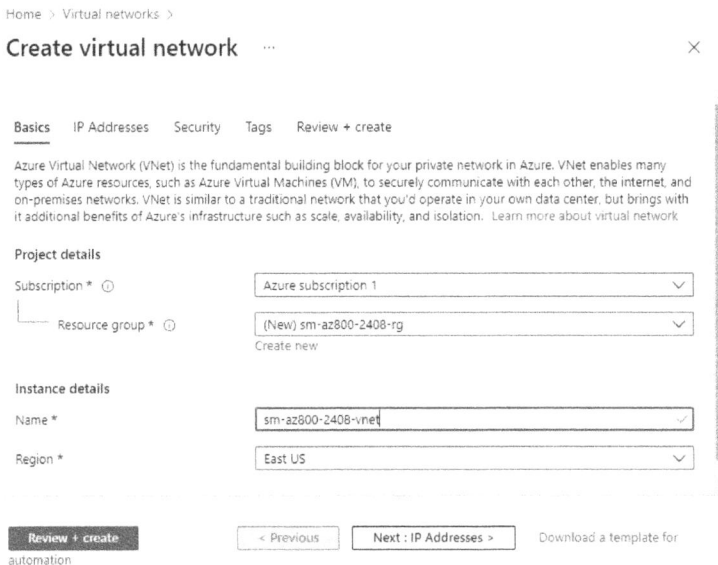

Home > Virtual networks >

Create virtual network   ···

Basics   IP Addresses   Security   Tags   Review + create

Azure Virtual Network (VNet) is the fundamental building block for your private network in Azure. VNet enables many types of Azure resources, such as Azure Virtual Machines (VM), to securely communicate with each other, the internet, and on-premises networks. VNet is similar to a traditional network that you'd operate in your own data center, but brings with it additional benefits of Azure's infrastructure such as scale, availability, and isolation. Learn more about virtual network

Project details

Subscription *   Azure subscription 1

Resource group *   (New) sm-az800-2408-rg
Create new

Instance details

Name *   sm-az800-2408-vnet

Region *   East US

Review + create   < Previous   Next : IP Addresses >   Download a template for automation

Figure 6.21 – VNet Basics tab

5. Click **Next: IP Addresses**.

6. Create an IPv4 address space and subnets; use existing defaults for this exercise:

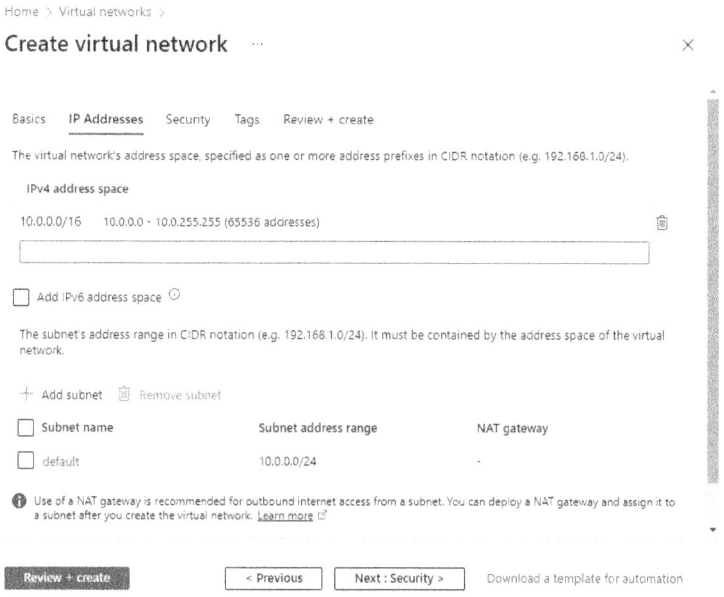

Figure 6.22 – VNet IP Addresses tab

7. Click **Next: Security**.

8. Leave the default setting of **Disable** for **BastionHost**, **DDoS Protection Standard**, and **Firewall**:

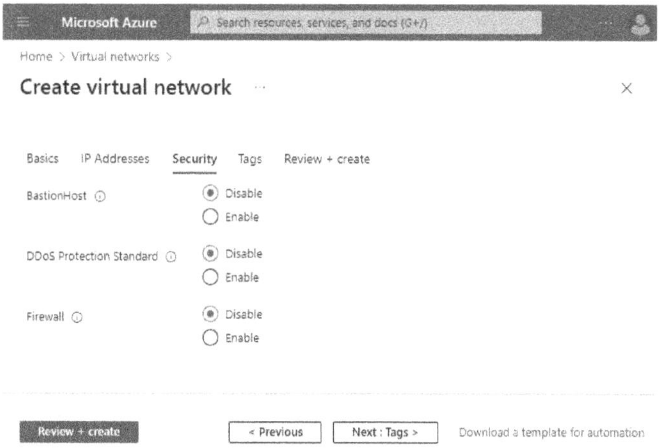

Figure 6.23 – VNet Security tab

9.   Click **Next: Tags** and skip entering any information for this exercise. Then, click **Next: Review + create**:

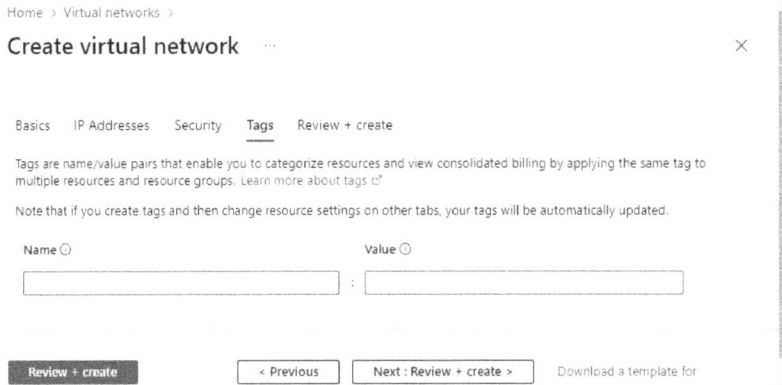

Figure 6.24 – VNet Tags tab

10.   On the **Review + create** tab, *review your settings*; you may go back to the previous tabs and make any edits if required. Once you have confirmed your settings are as needed, you can click **Create**:

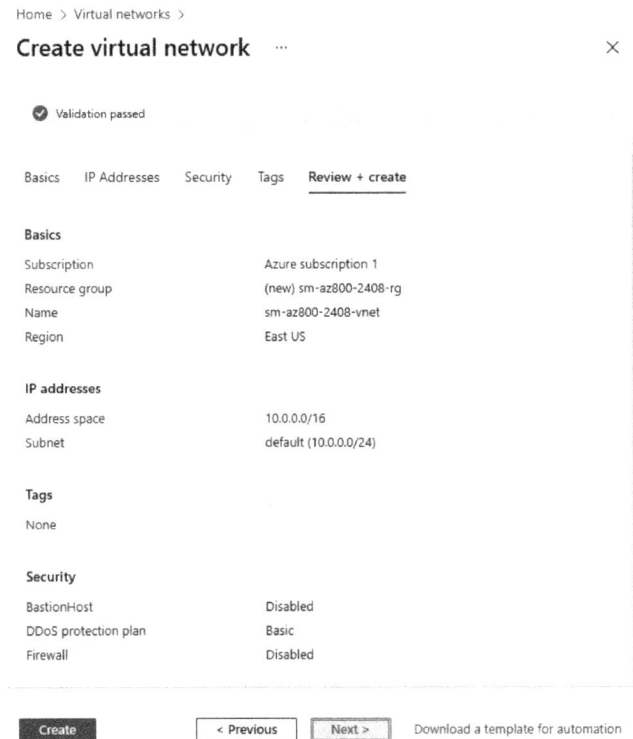

Figure 6.25 – VNet Review + create tab

11. You will receive a **Deployment succeeded** notification. Click on **Go to resource** or navigate to the **Virtual networks** blade:

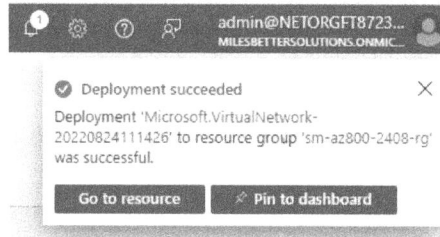

Figure 6.26 – VNet deployment succeeded

12. On the **Virtual networks** blade, the created VNet can be seen:

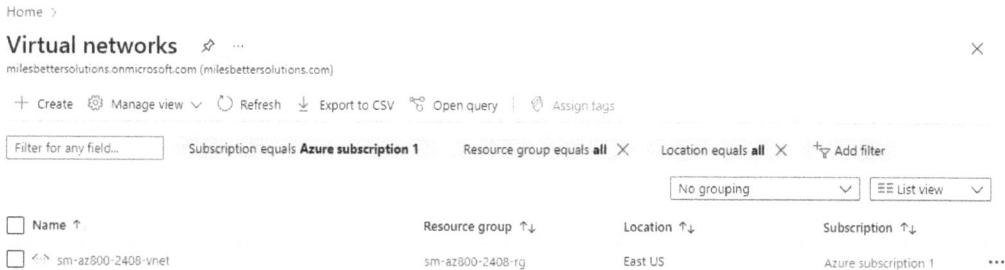

Figure 6.27 – Created VNet

With this, we have completed this exercise. It taught us the skills needed to implement an Azure VNet. The following exercise will look at implementing an NSG.

Exercise – Implementing an NSG

In this exercise, we will look to create a VM and associate an NSG at the subnet level. We will then add rules to allow **Remote Desktop Protocol** (**RDP**) access and deny outbound internet access.

In the following sub-sections, you can see the procedure to complete the exercise, segregated into tasks for a better understanding.

Task 1 – Accessing the Azure portal

1. Log in to the Azure portal: `https://portal.azure.com`. You can alternatively use the Azure desktop app: `https://portal.azure.com/App/Download`.

Task 2 – Creating a VM

1. In the search bar, type in `virtual machines`; click **Virtual machines** from the list of services:

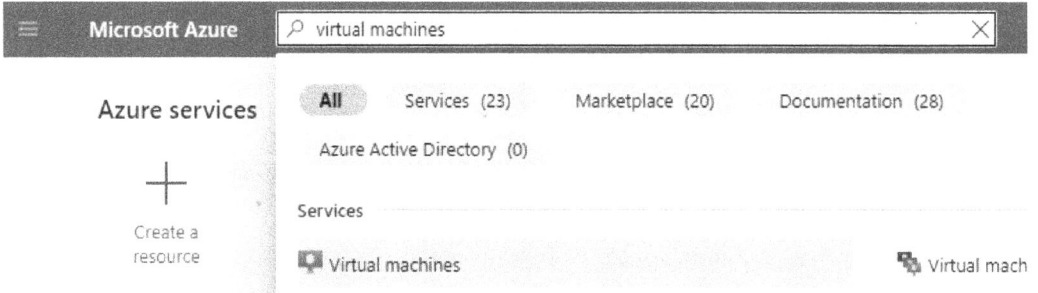

Figure 6.28 – Searching for the VM's resource

2. From the **Virtual machines** blade, click **+ Create** and then **Virtual machine** from the top menu of the blade. Optionally, click the **Learn more about** hyperlinks for further information and learning:

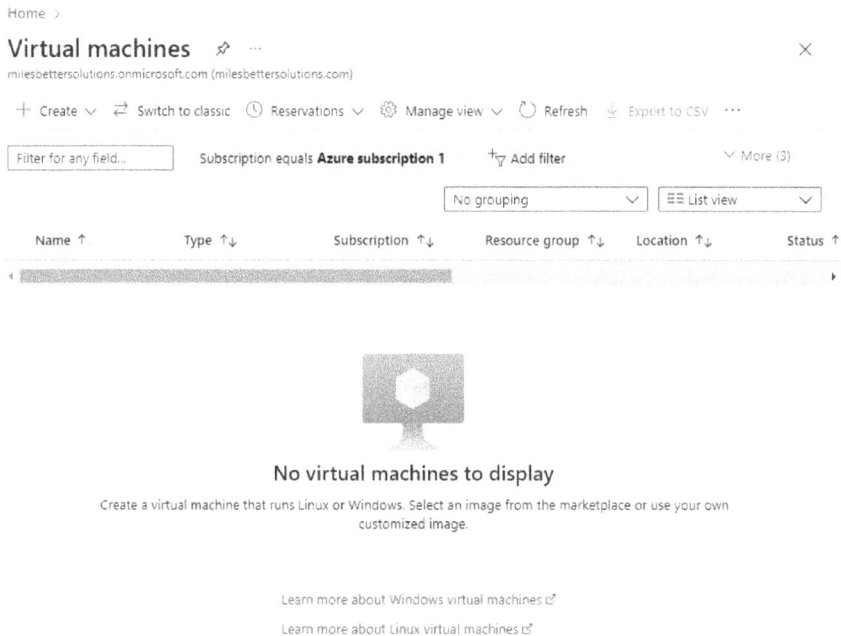

Figure 6.29 – Creating a VM

3. Set **Project details** values of **Subscription** and **Resource group** as required.

4. Set **Instance details** values as follows:

- **Virtual machine name**: *Enter a name*

- **Region**: *Select a region*

- **Availability options**: *Leave default selected*

- **Security type**: *Leave default selected*

- **Image**: *Select a Windows Server image*

- **Size**: *Leave default selected or select as required*:

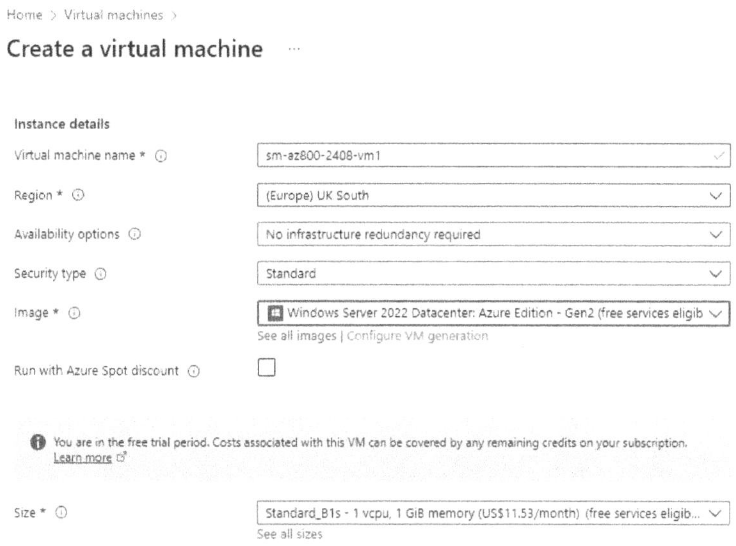

Home > Virtual machines >

Create a virtual machine ···

Instance details

Virtual machine name * ⓘ	sm-az800-2408-vm1
Region * ⓘ	(Europe) UK South
Availability options ⓘ	No infrastructure redundancy required
Security type ⓘ	Standard
Image * ⓘ	Windows Server 2022 Datacenter: Azure Edition - Gen2 (free services eligib

See all images | Configure VM generation

Run with Azure Spot discount ⓘ ☐

ⓘ You are in the free trial period. Costs associated with this VM can be covered by any remaining credits on your subscription. Learn more

Size * ⓘ Standard_B1s - 1 vcpu, 1 GiB memory (US$11.53/month) (free services eligib...

See all sizes

Figure 6.30 – Creating a VM: settings

5. Set the **Administrator account** *username* and *password* as required:

Administrator account

Username * ⓘ	sm8002408
Password * ⓘ	••••••••••••
Confirm password * ⓘ	••••••••••••

Figure 6.31 – Creating a VM: settings (continued)

6.   Set **Public inbound ports** rules to **None**:

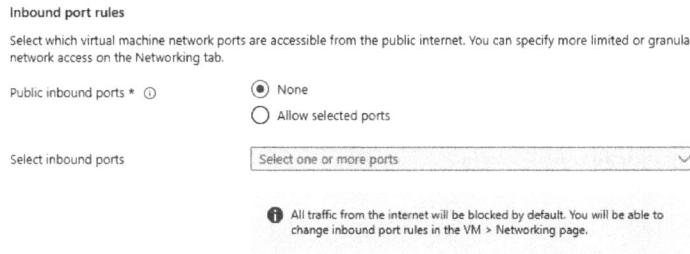

Figure 6.32 – Creating a VM: settings (continued)

7.   For this exercise, set **Licensing** at the unchecked default; that is, **No use of an existing Windows Server license**.

8.   Click **Next: Disks**.

9.   For this exercise, leave the **Disks** tab at the *default settings*:

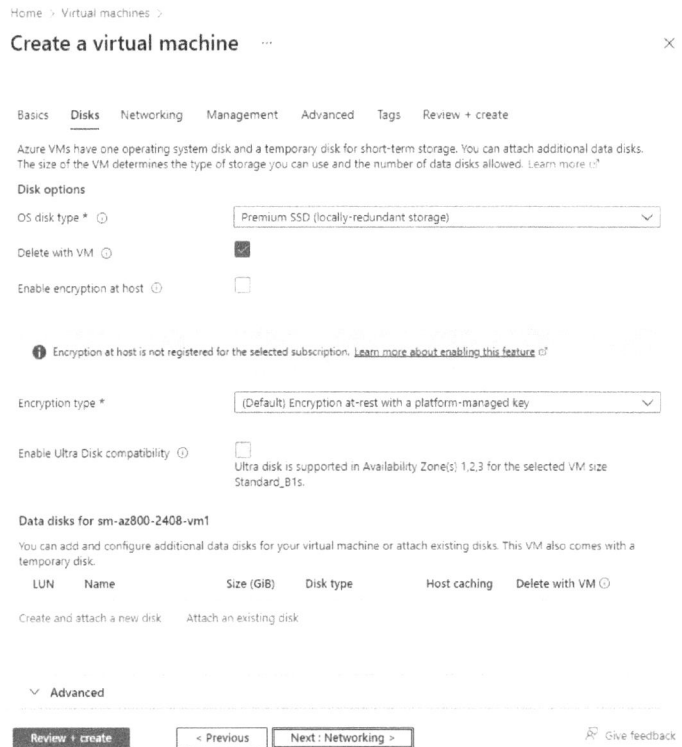

Figure 6.33 – Creating a VM: settings (continued)

10. Click **Next: Networking**.

11. From **Network interface**, set the following:

    - **Virtual network**: *Use the default provided*

    - **Subnet**: *Use the default provided*

    - **Public IP**: Click **Create new**, *enter a name*, and click **OK**

    - **NIC network security group**: Select **None**

    - **Delete public IP and NIC when VM is deleted**: *Check to enable/allow*

    - Leave all other settings at their defaults:

| Basics | Disks | **Networking** | Management | Advanced | Tags | Review + create |

Define network connectivity for your virtual machine by configuring network interface card (NIC) settings. You can control ports, inbound and outbound connectivity with security group rules, or place behind an existing load balancing solution. Learn more ☐

Network interface

When creating a virtual machine, a network interface will be created for you.

Virtual network * ⓘ   (new) sm-az800-2408-rg-vnet   ⌄
    Create new

Subnet * ⓘ   (new) default (10.1.0.0/24)   ⌄

Public IP ⓘ   (new) sm-az800-2408-vm1-ip   ⌄
    Create new

NIC network security group ⓘ   ◉ None
    ○ Basic
    ○ Advanced

    ⚠ All ports on this virtual machine may be exposed to the public internet. This is a security risk. Use a network security group to limit public access to specific ports. You can also select a subnet that already has network security groups defined or remove the public IP address.

Delete public IP and NIC when VM is deleted ⓘ   ☑

Enable accelerated networking ⓘ   ☐   The selected VM size does not support accelerated networking.

[Review + create]   [< Previous] [Next : Management >]   ⅋ Give

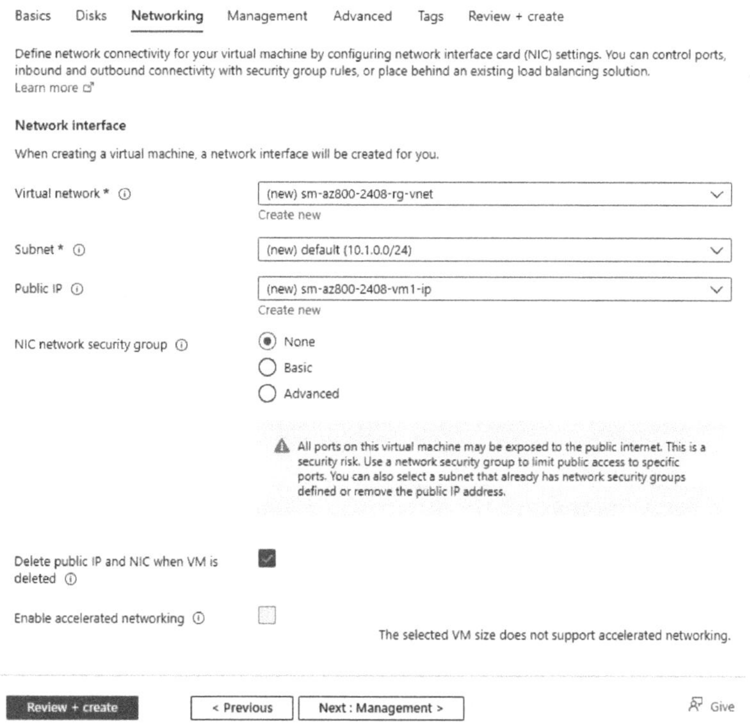

Figure 6.34 – Creating a VM: settings (continued)

12. Click **Next: Management**.

13. Leave the settings at their *defaults*.

14. Click **Next: Advanced**.

15. Leave the settings at their *defaults*.

16. Click **Next: Tags** and skip entering any information for this exercise. Then, click **Next: Review + create**.

17. On the **Review + create** tab, *review your settings*; you may go back to the previous tabs and make any edits if required. Once you have confirmed your settings are as needed, you can click **Create**:

Figure 6.35 – Creating a VM: Review + create tab

18. You will receive a notification that the resource deployment succeeded:

Figure 6.36 – VM deployment succeeded

Task 3 – Creating an NSG

1. In the search bar, type in `network security groups`; click **Network security groups** (*not* **Network security groups (classic)**) from the list of services:

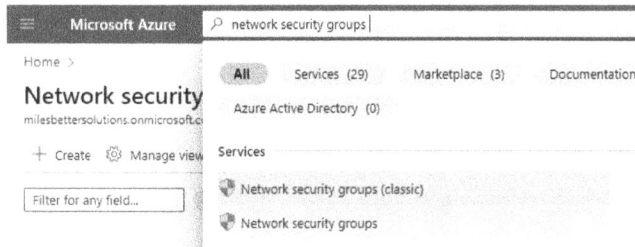

Figure 6.37 – Searching for the NSG resource

2. From the **Virtual machines** blade, click + **Create** from the top menu of the blade. Optionally, click the **Learn more** hyperlink for further information and learning:

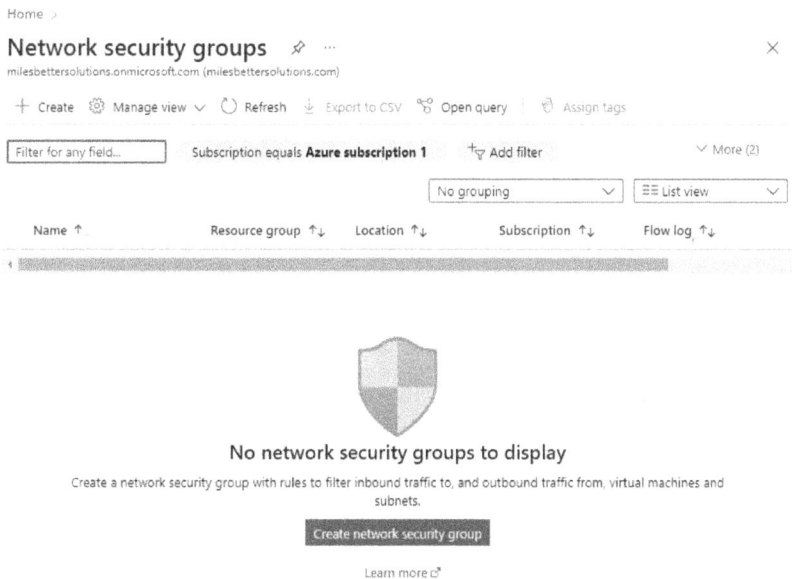

Figure 6.38 – Creating an NSG

3. Complete the **Project details** and **Instance details** settings as required. Set **Region** the same as where the VM was created:

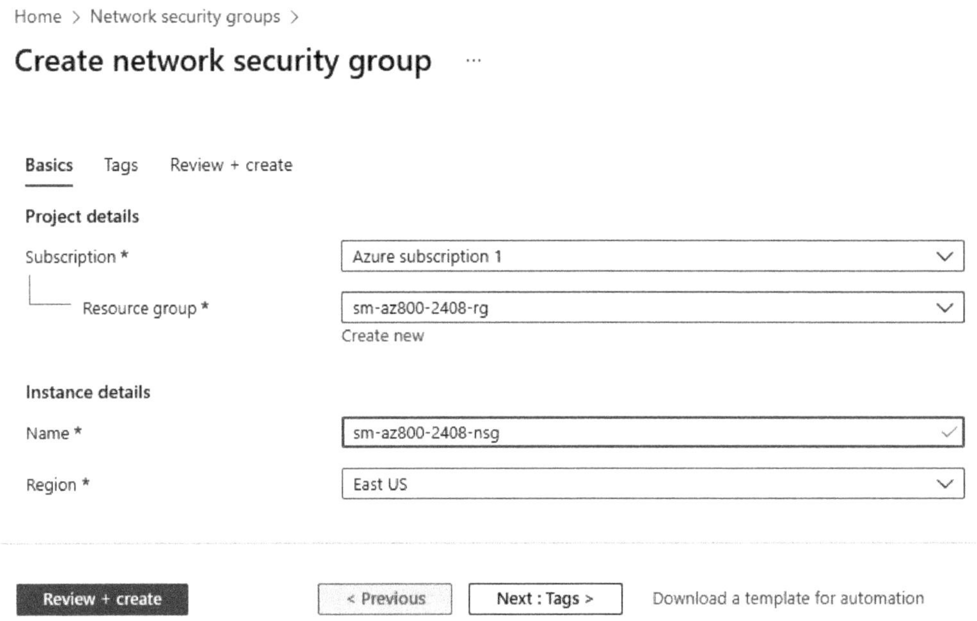

Home > Network security groups >

Create network security group   ...

Basics   Tags   Review + create

Project details

Subscription * | Azure subscription 1 |

            Resource group * | sm-az800-2408-rg |
                         Create new

Instance details

Name * | sm-az800-2408-nsg |

Region * | East US |

[Review + create]   [< Previous] [Next : Tags >]   Download a template for automation

Figure 6.39 – Creating an NSG: settings

4. Click **Next: Tags** and add any tags as required, then click **Next: Review + create**.

5. On the **Review + create** tab, *review your settings*; you may go back to the previous tabs and make any edits if required. Once you have confirmed your settings are as needed, you can click **Create**.

6. You will receive a notification that the resource was *created successfully*.

7. Click **Go to resource** from the **Deployment** blade; alternatively, navigate to the **Azure NSG** instance.

Task 4 – Associating an NSG with a subnet

1. From the created **Network security group** blade of the instance created, click **Subnets** under **Settings**:

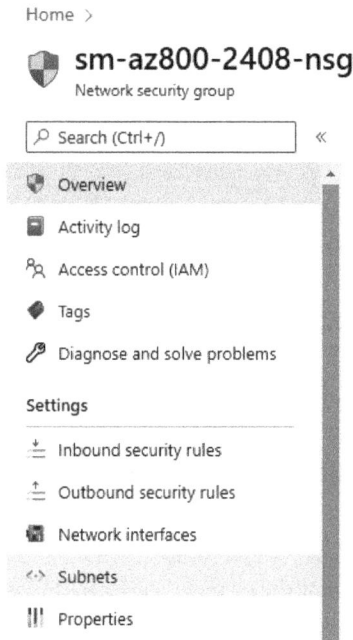

Figure 6.40 – Created NSG

2. Click + **Associate** from the top toolbar:

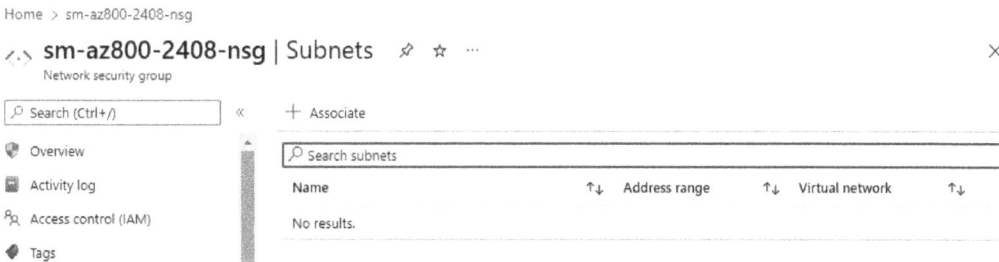

Figure 6.41 – Associating an NSG

3. Select the **Virtual network** and **Subnet** configurations of the VM you created in the previous exercise from the **Associate subnet** blade:

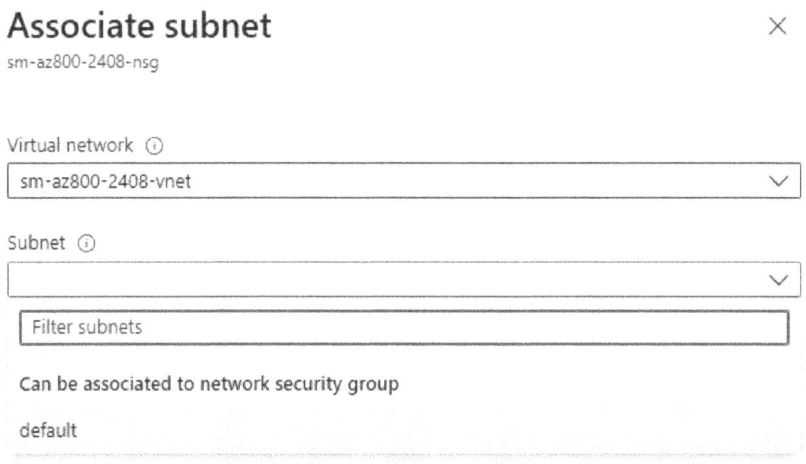

Figure 6.42 – Associating an NSG (continued)

4. Click **OK**.

5. You will receive a notification that the changes were *saved successfully*.

Task 5 – Adding an inbound rule to allow RDP access

1. Navigate to the NSG created in this exercise. Notice that all *inbound connections* are **denied** unless their source is **VirtualNetwork** or **Azure LoadBalancer**. We will address this in the next task.

2. From the **Network security groups** blade, click **Inbound security rules** under **Settings**:

Home > Network security groups >

Network security g...   «

milesbettersolutions.onmicrosoft.com (milesbetter...

+ Create   ⚙ Manage view ∨   ···

Filter for any field...

Name ↑

🛡 sm-az800-2408-nsg   ···

🛡 **sm-az800-2408-ns**

Network security group

🔍 Search (Ctrl+/)   «

🌐 Overview

📋 Activity log

👥 Access control (IAM)

🏷 Tags

🩺 Diagnose and solve problems

Settings

📥 Inbound security rules

📤 Outbound security rules

🖥 Network interfaces

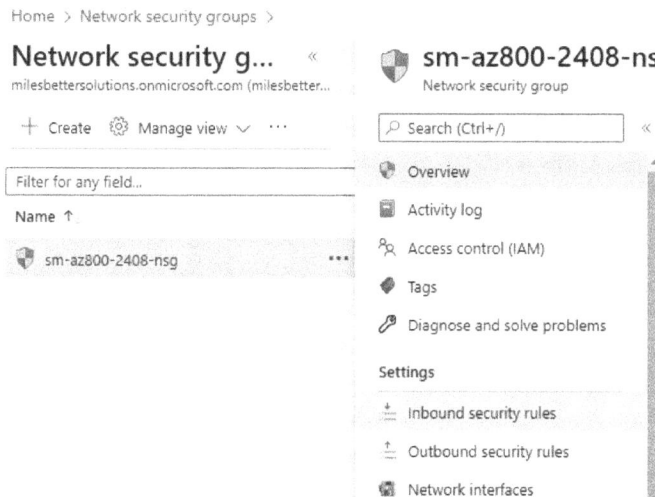

Figure 6.43 – Created NSG

3.   From the **Inbound security rules** blade, click **+ Add**:

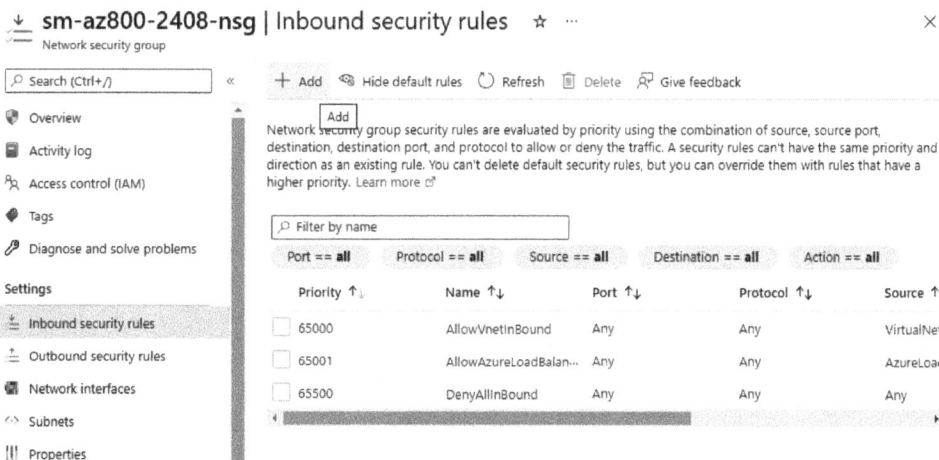

📥 **sm-az800-2408-nsg** | Inbound security rules   ☆   ···   ✕

Network security group

🔍 Search (Ctrl+/)   «   + Add   🔲 Hide default rules   ⟳ Refresh   🗑 Delete   📣 Give feedback

🌐 Overview   Add

📋 Activity log   Network security group security rules are evaluated by priority using the combination of source, source port,
                              destination, destination port, and protocol to allow or deny the traffic. A security rules can't have the same priority and
👥 Access control (IAM)   direction as an existing rule. You can't delete default security rules, but you can override them with rules that have a
                              higher priority. Learn more ⬀
🏷 Tags

🩺 Diagnose and solve problems   🔍 Filter by name

Settings   Port == **all**   Protocol == **all**   Source == **all**   Destination == **all**   Action == **all**

📥 Inbound security rules   Priority ↑↓   Name ↑↓   Port ↑↓   Protocol ↑↓   Source ↑↓

📤 Outbound security rules   ☐ 65000   AllowVnetInBound   Any   Any   VirtualNetw

🖥 Network interfaces   ☐ 65001   AllowAzureLoadBalan... Any   Any   AzureLoad

⟨⟩ Subnets   ☐ 65500   DenyAllInBound   Any   Any   Any

⫴⫴ Properties   ◄   ►

Figure 6.44 – Adding an NSG

4.   Open a browser and in Google, type what's my IP and note your IP for the next step.

5.   From the **Add inbound security rule** blade, set the following:

   • **Source**: Select **IP Addresses**

   • **Source IP addresses/CIDR ranges**: *Set this to your IP from the Google result for your IP in step 4 of this task*

- **Source port ranges**: Leave at default of * (*asterisk symbol*)

- **Destination**: Leave at default of **Any**

- **Service**: Select **RDP**

- **Action**: Ensure **Allow** is set

- **Priority**: Leave at default of **100**

- **Name**: Provide a name, such as `AllowInbound_RDP_KnownIP`

- **Description**: Enter one as required

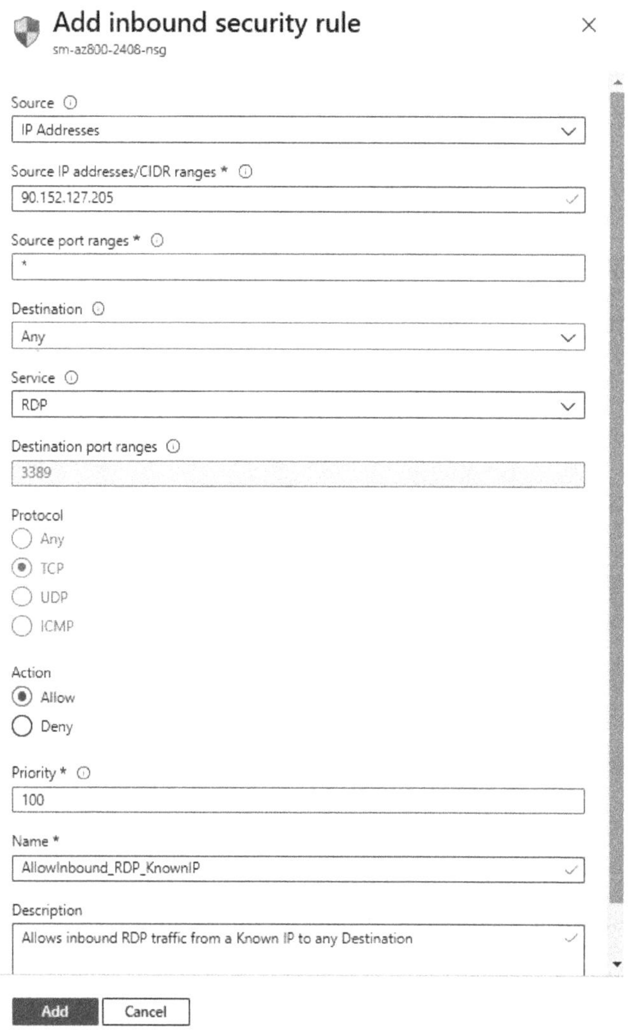

Figure 6.45 – Adding an NSG (continued)

6. Click **Add**.

7. You will receive a notification that the rule was *created successfully*:

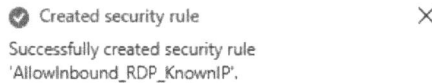

> ✓ Created security rule   ✕
>
> Successfully created security rule
> 'AllowInbound_RDP_KnownIP'.

Figure 6.46 – NSG created

Task 6 – Accessing the VM via an RDP connection

1. Navigate to the **Virtual machines** blade, and on the **Overview** blade, click on **Connect** and then **RDP**:

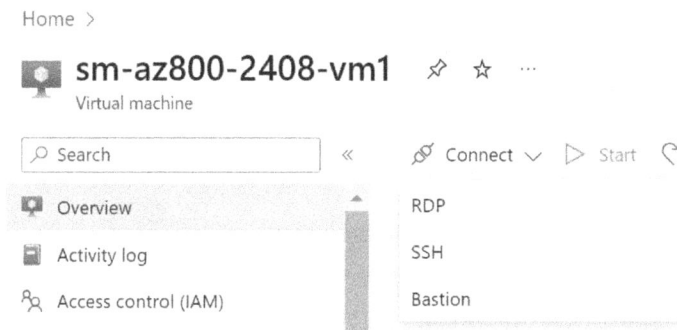

Home >

sm-az800-2408-vm1 📌 ☆ ⋯
Virtual machine

🔍 Search   «   🔗 Connect ∨   ▷ Start   ↻

💻 Overview   RDP

🗐 Activity log   SSH

👥 Access control (IAM)   Bastion

Figure 6.47 – VM connection

2. On the **Connect** blade, click the **Download RDP File** button:

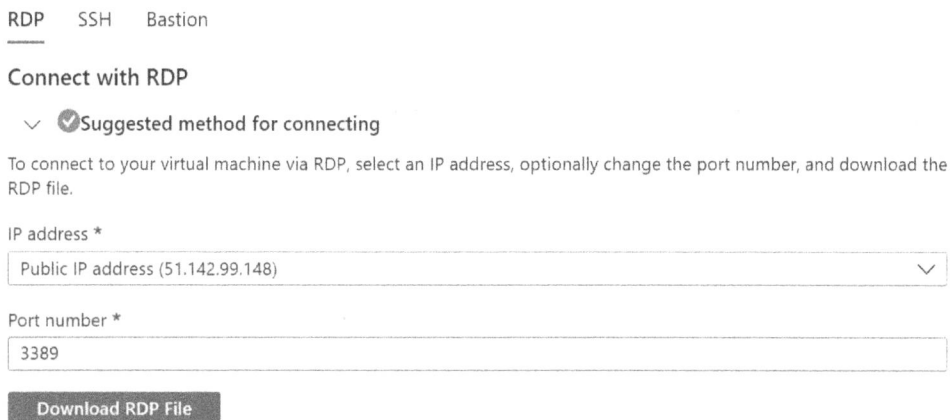

RDP   SSH   Bastion

Connect with RDP

∨ ✓ Suggested method for connecting

To connect to your virtual machine via RDP, select an IP address, optionally change the port number, and download the RDP file.

IP address *

Public IP address (51.142.99.148)   ∨

Port number *

3389

Download RDP File

Figure 6.48 – VM connection (continued)

3. Open the RDP file from where it is saved.

4. Click **Connect** to start an RDP session that is allowed by the ruleset in this task:

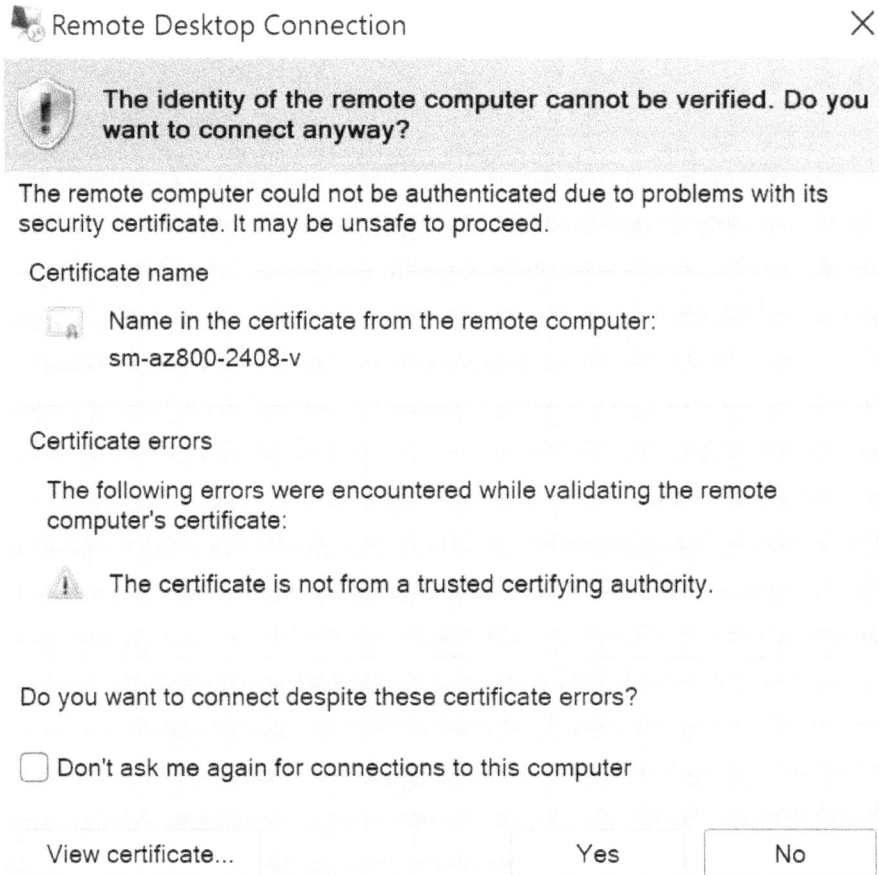

Remote Desktop Connection   ✕

> ! **The identity of the remote computer cannot be verified. Do you want to connect anyway?**

The remote computer could not be authenticated due to problems with its security certificate. It may be unsafe to proceed.

Certificate name

Name in the certificate from the remote computer:
sm-az800-2408-v

Certificate errors

The following errors were encountered while validating the remote computer's certificate:

⚠ The certificate is not from a trusted certifying authority.

Do you want to connect despite these certificate errors?

☐ Don't ask me again for connections to this computer

View certificate...   Yes   No

Figure 6.49 – Remote desktop connection

Task 7 – Accessing the internet from the VM

1. While logged in to the VM, open a browser, and confirm you reached the internet by visiting a site such as `https://docs.microsoft.com`. *You may need to adjust the browser security settings.* We will restrict this access to the internet in the next exercise:

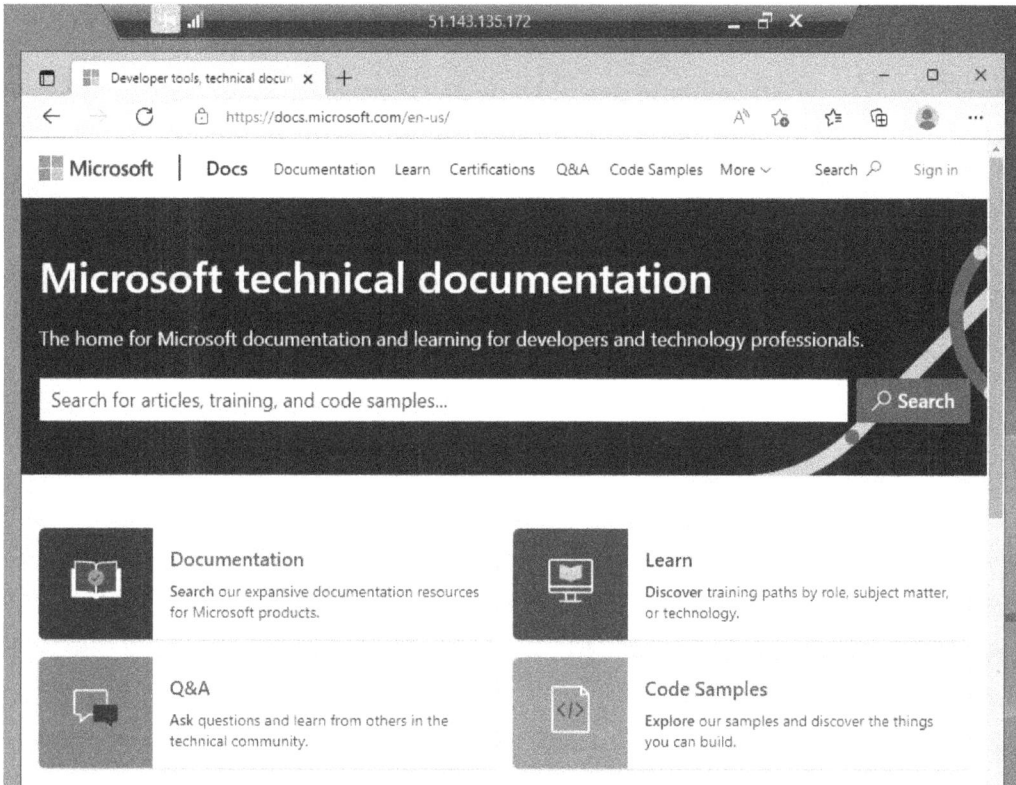

Figure 6.50 – Accessing the internet

Task 8 – Adding an outbound rule to deny internet access

1. On the **Virtual machine** blade, click **Networking** under **Settings**; you will see from the **Outbound port rules** tab that all outbound connections are allowed to the internet.

2. On the **Outbound port rules** tab, click **Add outbound port rule**.

3. On the **Add outbound security rule** blade, leave all other options at their defaults apart from the following:

    -   **Destination**: Select **Service Tag**

    -   **Destination service tag**: Select **Internet**

- **Destination port ranges**: Enter the * symbol (*asterisk symbol*)

- **Action**: Ensure **Deny** is set

- **Priority**: Leave at the default value of **100**

- **Name**: Provide a name, such as DenyOutbound_Internet

- **Description**: Enter as required:

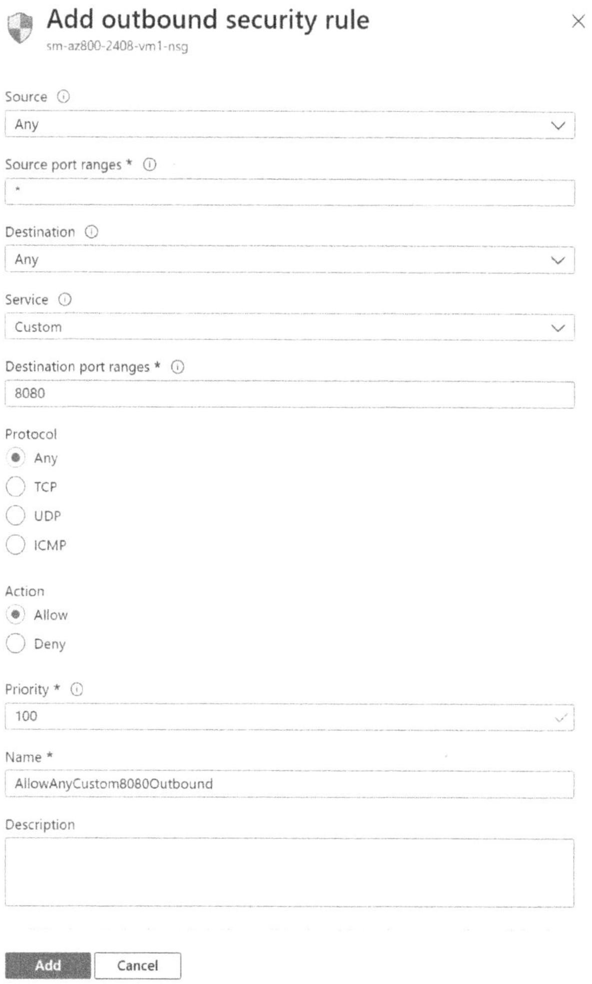

Figure 6.51 – Adding an NSG rule

4. Click **Add**.

5. You will receive a notification that the rule was *created successfully*.

6. The new rule will be shown listed:

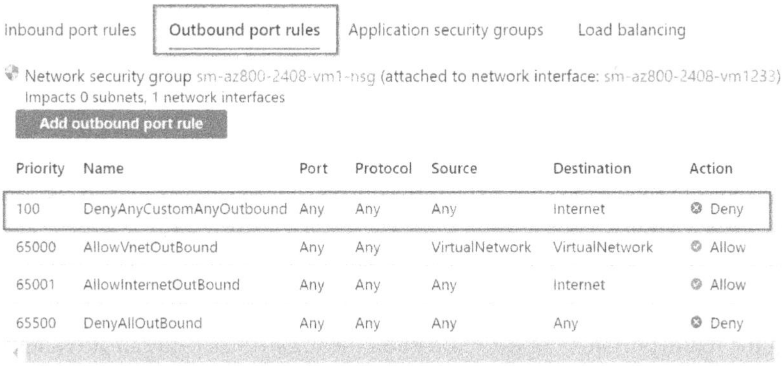
Inbound port rules
Outbound port rules
Application security groups
Load balancing
Network security group sm-az800-2408-vm1-nsg (attached to network interface: sm-az800-2408-vm1233)
Impacts 0 subnets, 1 network interfaces
Add outbound port rule
Priority Name Port Protocol Source Destination Action
100 DenyAnyCustomAnyOutbound Any Any Any Internet Deny
65000 AllowVnetOutBound Any Any VirtualNetwork VirtualNetwork Allow
65001 AllowInternetOutBound Any Any Any Internet Allow
65500 DenyAllOutBound Any Any Any Any Deny

Figure 6.52 – New NSG rule added

7. In the VM, open a browser again and confirm you can no longer reach the internet by visiting a site such as `https://docs.microsoft.com`.
8. This time, you will see a message from the browser such as **Hmmm… can't reach this page**:

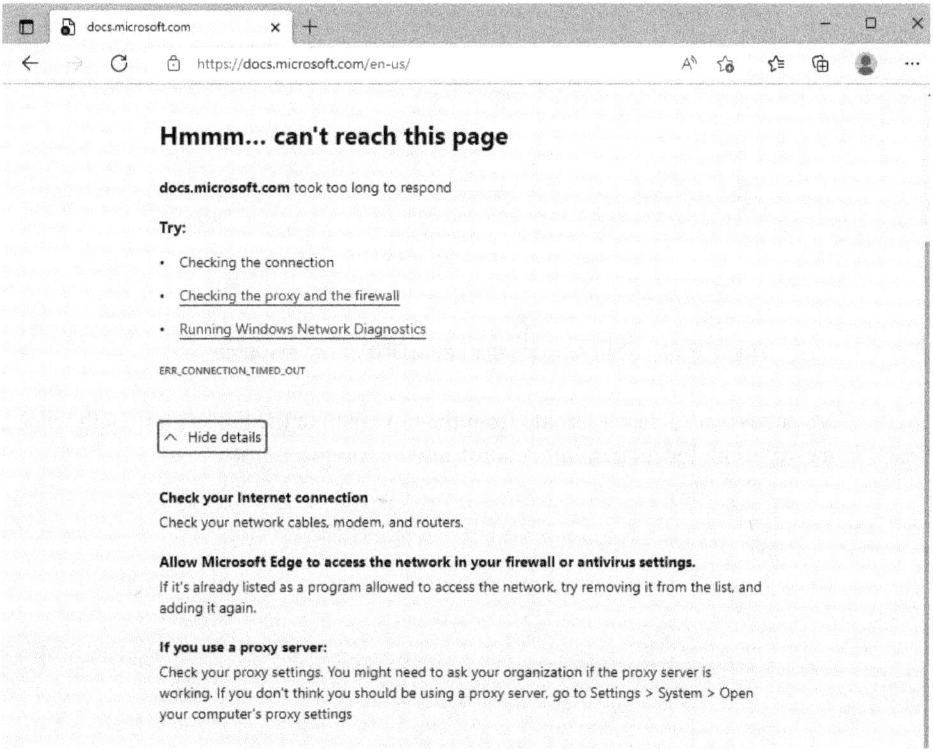
docs.microsoft.com
https://docs.microsoft.com/en-us/
Hmmm... can't reach this page
docs.microsoft.com took too long to respond
Try:
Checking the connection
Checking the proxy and the firewall
Running Windows Network Diagnostics
ERR_CONNECTION_TIMED_OUT
Hide details
Check your Internet connection
Check your network cables, modem, and routers.
Allow Microsoft Edge to access the network in your firewall or antivirus settings.
If it's already listed as a program allowed to access the network, try removing it from the list, and adding it again.
If you use a proxy server:
Check your proxy settings. You might need to ask your organization if the proxy server is working. If you don't think you should be using a proxy server, go to Settings > System > Open your computer's proxy settings

Figure 6.53 – Internet access denied

With this, we have completed this exercise. This exercise taught us the skills needed to implement and configure an NSG. The following exercise will look at implementing Azure DNS.

Exercise – Implementing Azure DNS

This exercise will look at creating and configuring public and private zones in Azure DNS.

In the following sub-sections, you can see the procedure to complete the exercise, segregated into tasks for a better understanding.

Task 1 – Accessing the Azure portal

1. Log in to the Azure portal: `https://portal.azure.com`. You can alternatively use the Azure desktop app: `https://portal.azure.com/App/Download`.

Task 2 – Creating a public zone

1. In the search bar, type in `dns zone`; click **DNS zones** from the list of services:

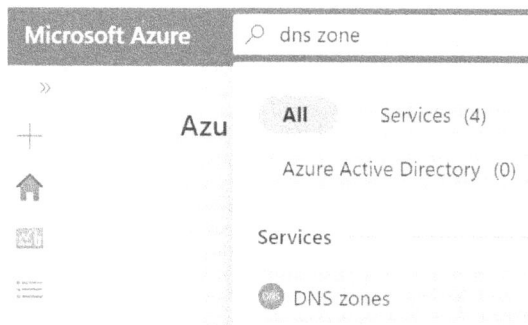

Figure 6.54 – Searching for the Azure DNS zones resource

2. On the **DNS zones** blade, click + **Create** from the top menu of the blade. Optionally, click the **Learn more** hyperlink for further information and learning:

Home >

DNS zones

milesbettersolutions.onmicrosoft.com

+ Create   Manage view ⌄   ⟳ Refresh   ↓ Export to CSV   Open query   ...

Filter for any field...	Add filter	More (3)

	No grouping ⌄	List view ⌄

Name ↑	Numb... ↑↓	Resource group ↑↓	Location ↑↓

No dns zones to display

Azure DNS is a hosting service for DNS domains that provides name resolution by using Microsoft Azure infrastructure. By hosting your domains in Azure, you can manage your DNS records by using the same credentials, APIs, tools, and billing as your other Azure services.

Create dns zone

Learn more ☐   ♀ Give feedback

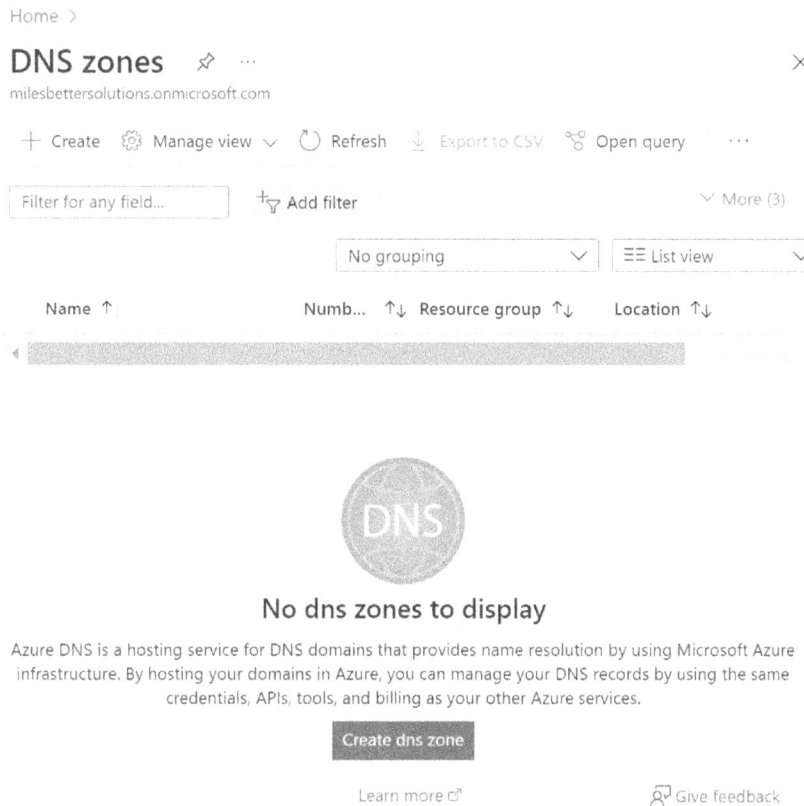

Figure 6.55 – Creating a DNS zone

3. Complete the **Project details** and **Instance details** settings as required.

4. Click **Next: Review + create**.

5. On the **Review + create** tab, *review your settings*; you may go back to the previous tabs and make any edits if required. Once you have confirmed your settings are as needed, you can click **Create**.

6. You will receive a notification that the resource was *created successfully*.

7. Click **Go to resource** from the **Deployment** blade; alternatively, navigate to the **DNS zones** blade.

Task 3 – Adding records

1. Search **All resources** for the zone created or navigate to the **DNS zones** blade and open your created zone.

2. From the **DNS zones** blade, click **+ Record set** from the top menu of the blade.

3. From the **Add record set** blade, enter the required information, and click **OK**.

4. From the **DNS zones** blade, review your record added:

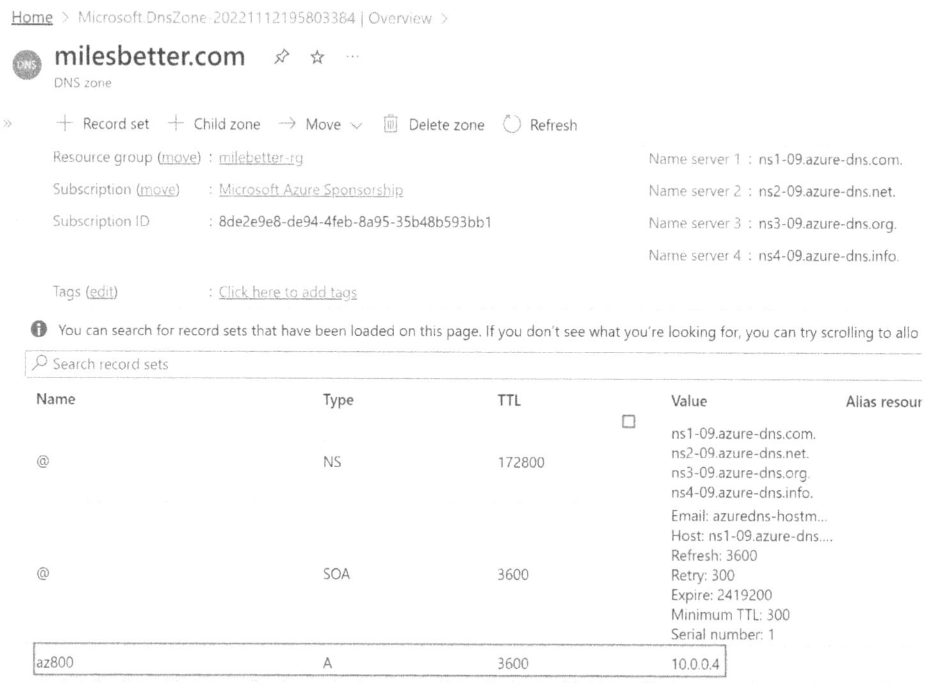

Figure 6.56 – Adding DNS records

For this exercise, we have used an example name and record; *you should use an example that meets your requirements.*

5. To test the name resolution, open Command Prompt and run the following command:

    ```
    nslookup az800.milesbetter.com ns1-01.azure-dns.com
    ```

    Replace with the information for the fully qualified domain name (FQDN) and name server that is relevant to your environment.

6. You should see a response as follows:

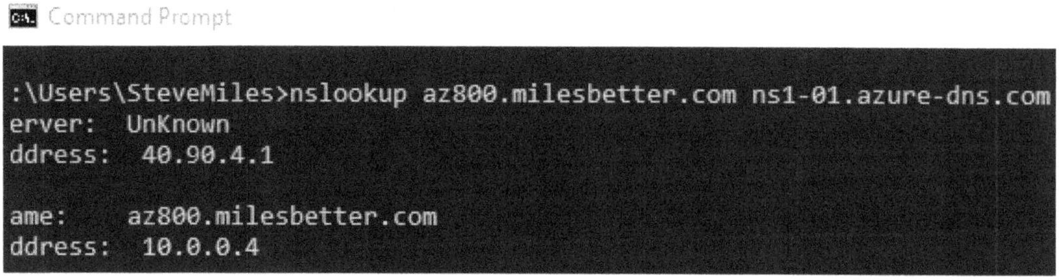

Figure 6.57 – nslookup command output

Task 4 – Creating a private zone

1. In the search bar, type in `private dns zone`; click **Private DNS zones** from the list of services.

2. On the **Private DNS zones** blade, click **+ Create** from the top menu of the blade. Optionally, click the **Learn more** hyperlink for further information and learning.

3. Complete the **Project details** and **Instance details** settings as required.

4. Click **Next: Review + create**.

5. On the **Review + create** tab, *review your settings*; you may go back to the previous tabs and make any edits if required. Once you have confirmed your settings are as needed, you can click **Create**.

6. You will receive a notification that the resource was *created successfully*.

7. Click **Go to resource** from the **Deployment** blade; alternatively, navigate to the **Private DNS zones** blade.

Task 5 – Linking VNets

1. Search **All resources** for the zone created or navigate to the **Private DNS zones** blade and open your created zone.

2. From the **Private DNS zones** blade, under **Settings**, select **Virtual network links**.

3. From the **Virtual network links** blade, click **+ Add** from the top menu of the blade.

4. Enter a **Link name** value.

5. Select a **Subscription** type.

6. Select the **Virtual network** configuration you will link the zone to.

7. Optionally, select whether to enable auto registration.

8. Select **OK**.

9. You will receive a notification that the resource was *created successfully*.

10. Click **Go to resource** from the **Deployment** blade; alternatively, navigate to the **Private DNS zones** blade.

11. Adding and testing records is the same process as the steps we completed in the *Creating a public zone* task.

With this, we have completed this exercise. This exercise taught us the skills needed to implement Azure DNS. This next exercise will look at creating an Azure VPN gateway.

Exercise – Implementing an Azure VPN gateway

In this exercise, we will look to create an Azure VPN gateway for hybrid connectivity to on-premises resources.

For this exercise, you will require the following in place:

- An Azure VNet; you may use an existing VNet or the VNet from the previous exercise in this chapter on VNet creation

In the following sub-sections, you can see the procedure to complete the exercise, segregated into tasks for a better understanding.

Follow the steps in the following tasks to create a VPN gateway (VNet gateway).

Task 1 – Accessing the Azure portal

1. Log in to the Azure Portal: `https://portal.azure.com`. You can alternatively use the Azure Desktop App: `https://portal.azure.com/App/Download`.

Task 2 – Creating a VNet gateway

1. In the search bar, type `virtual network gateways`; click on **Virtual network gateways** from the list of services shown:

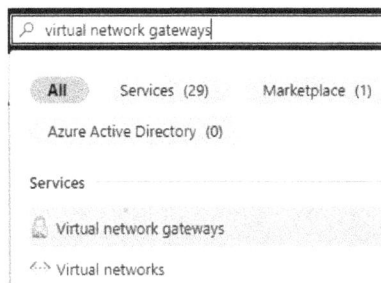

Figure 6.58 – Searching for the VNet gateway resource

2. From the **Virtual network gateways** blade, click on the + **Create** option from the top menu of the blade, or use the **Create virtual network gateway** button at the bottom of the blade. Optionally, click the **Learn more about Virtual network gateway** hyperlink for further information and learning.

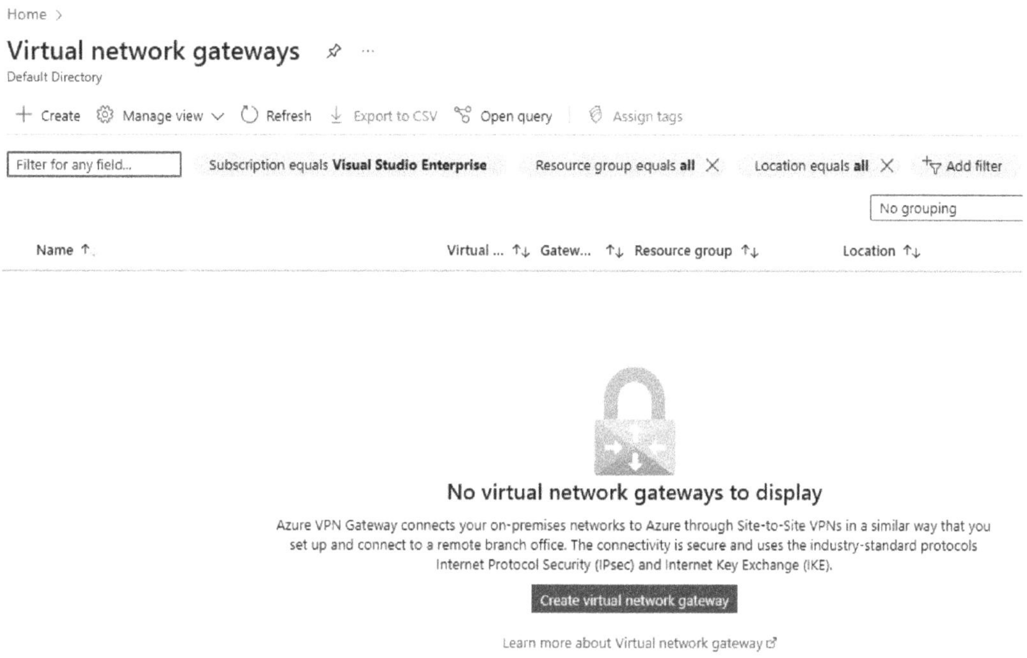

Home >

Virtual network gateways

Default Directory

+ Create   ⊙ Manage view ∨   ↻ Refresh   ↓ Export to CSV   ⁰ᣖ Open query   ⊘ Assign tags

| Filter for any field... | Subscription equals **Visual Studio Enterprise** | Resource group equals **all** ✕ | Location equals **all** ✕ | ⁺⁊ Add filter |

No grouping

| Name ↑. | Virtual ... ↑↓ | Gatew... ↑↓ | Resource group ↑↓ | Location ↑↓ |

No virtual network gateways to display

Azure VPN Gateway connects your on-premises networks to Azure through Site-to-Site VPNs in a similar way that you set up and connect to a remote branch office. The connectivity is secure and uses the industry-standard protocols Internet Protocol Security (IPsec) and Internet Key Exchange (IKE).

Create virtual network gateway

Learn more about Virtual network gateway ⧉

Figure 6.59 – Creating a VNet

3. On the **Create virtual network gateway** blade, select your subscription to use for the resources for this exercise, then select your VNet; this will auto-populate the **Resource group** field:

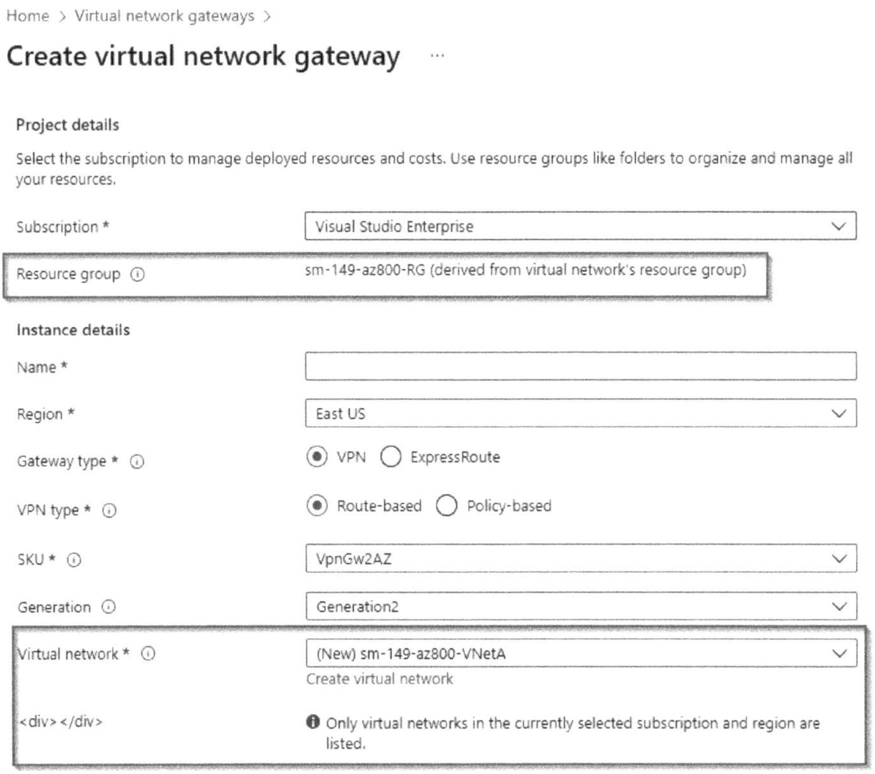

Figure 6.60 – Creating a VNet (continued)

4.   Complete the following settings, and leave all other entries at their defaults:

- **Name**: *Enter your chosen name*

- **Region**: *This must be left at the region for the VNet being used*

- **SKU**: VpnGw1

- **Public IP address name**: *Enter your chosen name*:

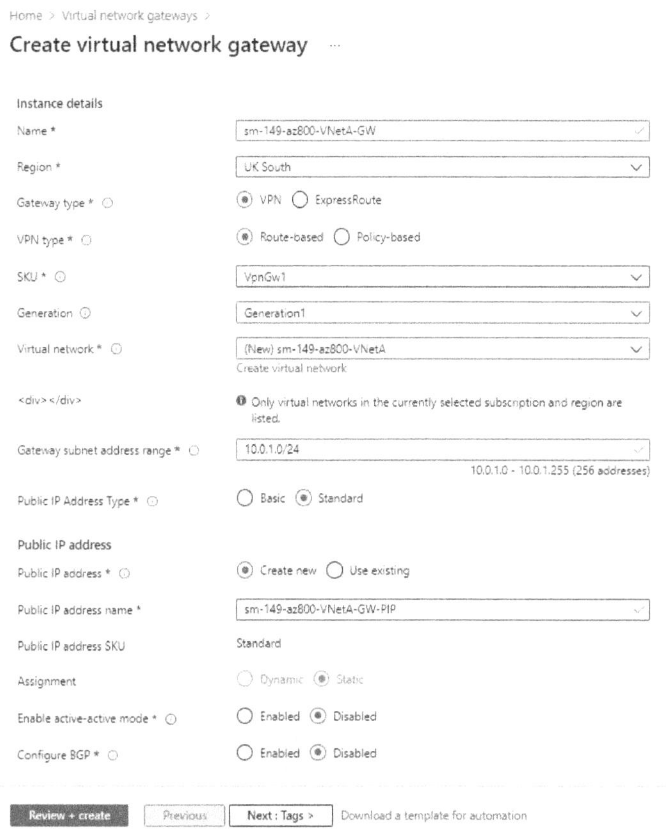

Figure 6.61 – Creating a VNet (continued)

5. Click **Next: Review + create**.

6. On the **Review + create** tab, review your settings; you may go back to the previous tabs and make any edits if required. Once you have confirmed your settings are as needed, you can click **Create**:

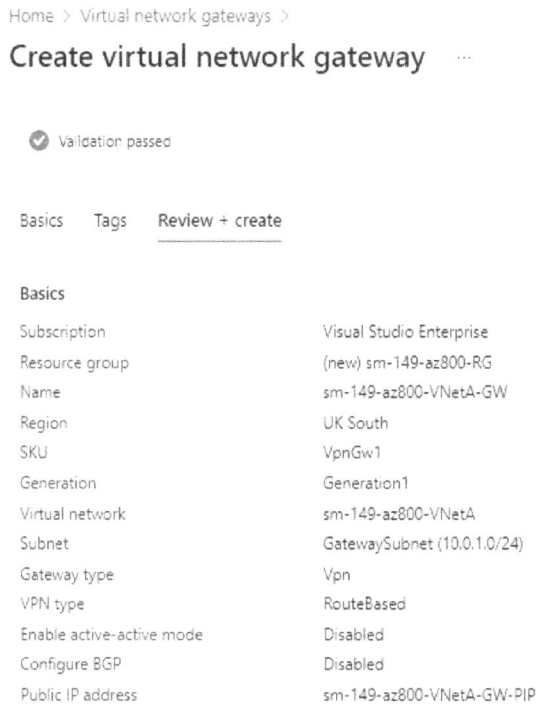

Figure 6.62 – Reviewing settings

7.   The creation process may take up to *1 hour*.

8.   You will receive a notification that the resource was *created successfully*.

9.   Click **Go to resource** from the **Deployment** blade; alternatively, navigate to the **Virtual network gateways** blade to review the resource created.

With this, we have completed this exercise. This exercise taught us the skills needed to implement an Azure VPN gateway. In the next exercise, we look at implementing VNet peering.

Exercise – Implementing VNet peering

In this exercise, we will look to add a peering relationship between two VNets so that resources can communicate across the Microsoft backbone without traversing the public internet.

For this exercise, you will require the following in place:

- Two Azure VNets; you may use existing VNets or the VNet from the previous exercise in this chapter on VNet creation and create another to peer with

Follow these steps to add VNet peering.

Task – Add VNet peering

1. Log in to the Azure portal: `https://portal.azure.com`. You can alternatively use the Azure desktop app: `https://portal.azure.com/App/Download`.

2. Navigate to the **Virtual networks** blade and click your first VNet to set up, peering from the list of VNets. Then, click **Peerings** under the **Settings** section:

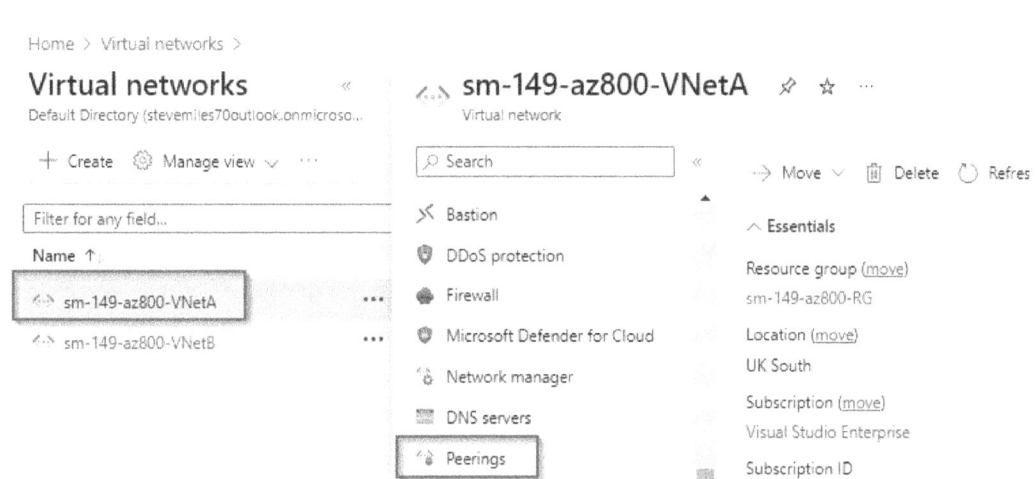

Figure 6.63 – Selecting the first VNet to peer

3. From the **Virtual networks peerings** blade, click on the + **Add** option.

4. From the **Add peering** blade, set the following; leave all others at defaults:

   - **This virtual network - Peering link name**: *Enter your chosen name*

   - **Remote virtual network – Peering link name**: *Enter your chosen name*

   - **Virtual Network**: *Select the VNet you wish to establish a peer with; that is, the remote VNet*

5. You will receive a notification that peerings were *added successfully*:

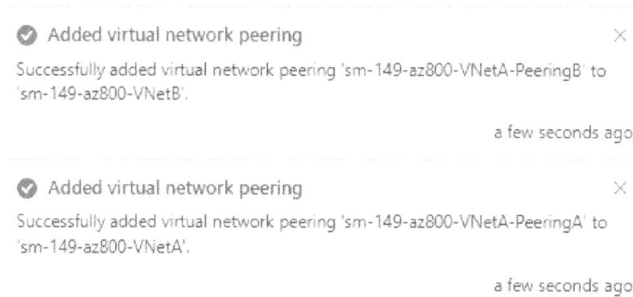

Figure 6.64 – Peering added successfully

6. From the **Peerings** blade, you will see that the peering has been added, the peering status is **Connected**, and the remote VNet peer is shown:

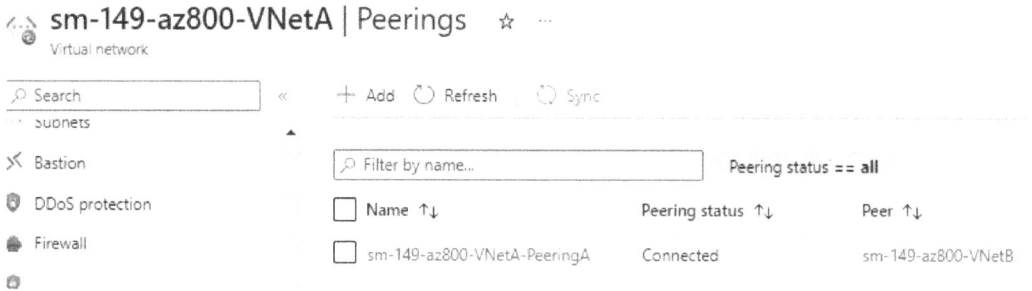

Figure 6.65 – VNet peerings

With this, we have completed this exercise. This exercise taught us the skills needed to implement Azure VNet peering. Now, let's summarize this chapter.

Summary

This chapter aimed to take your knowledge beyond the exam objectives; we added new skills and learning with the content provided. This further develops your knowledge and skills for on-premises network infrastructure services and enables you to be prepared for a real-world, day-to-day hybrid environment-focused role.

The content in this chapter covered information for the *Implement and manage Azure network infrastructure* skills outline section for the *AZ-800 Administering Windows Server Hybrid Core Infrastructure* exam.

We learned about some of the core network services components in Azure, such as implementing VNets and DNS, as well as looking at connectivity options and network protection aspects. A hands-on exercise then finished the chapter to develop your skills further.

The next chapter teaches about implementing and managing Azure network infrastructure.

Further reading

This section provides links to additional study references and additional exam information:

- *Microsoft Certified: Windows Server Hybrid Administrator Associate*: `https://docs. microsoft.com/en-us/learn/certifications/windows-server-hybrid- administrator/`

- *Exam AZ-800: Administering Windows Server Hybrid Core Infrastructure*: `https://docs. microsoft.com/en-us/learn/certifications/exams/az-800`

- *Exam AZ-800*: Skills outline: `https://query.prod.cms.rt.microsoft.com/cms/ api/am/binary/RWKI0r`

- *Microsoft Learn*: `https://docs.microsoft.com/en-us/learn/paths/implement- operate-premises-hybrid/`

Skills check

Challenge yourself with what you have learned in this chapter:

1. What is an Azure VNet?

2. Which IP types are supported to be created?

3. How many IP addresses in a VNet subnet are reserved, and what are their purposes?

4. How does a VM communicate with other resources in Azure?

5. What is VNet peering?

6. Which VPN types are supported to connect to Azure?

7. What is a VPN gateway, and which types can be created?

8. What is ExpressRoute, and what are its benefits?

9. What is Azure Virtual WAN?

10. What is Azure Extended Network, and what is a use case?

11. Explain VNet routing in Azure and the purpose of NVAs.

12. What are NSGs and ASGs?

13. What is Azure Firewall?

14. What is Azure DNS?

15. What are the limitations of Azure DNS, and when might Windows Server DNS still be required?

Part 3:
Hybrid Storage

This part will provide complete coverage of the knowledge and skills required for the skills measured in the *Manage storage and file services* section of the exam.

This part of the book comprises the following chapters:

- *Chapter 7, Implementing Windows Server Storage Services*
- *Chapter 8, Implementing a Hybrid File Server Infrastructure*

7
Implementing Windows Server Storage Services

In the previous chapter, we added skills to understand the concepts of Azure network infrastructure and services.

This chapter covers content from *AZ-800 Administering Windows Server Hybrid Core Infrastructure: Manage storage and file services*.

We will start this chapter by understanding the basics of the Windows Server filesystem and protocols; then, we will learn how to implement and manage Windows Server-based storage services and solutions. We will conclude with a hands-on exercise to develop your skills further.

The following topics are included in this chapter:

- Managing Windows Server file servers
- Implementing Windows Server iSCSI
- Implementing Storage Spaces and Storage Spaces Direct
- Implementing Windows Server Data Deduplication
- Implementing Windows Server Storage Replica
- Implementing a Distributed File System
- Hands-on exercise

In addition to the topics listed in this chapter, this chapter's goal is to take your knowledge beyond the exam objectives to prepare you for a real-world, day-to-day hybrid environment-focused role.

Managing Windows Server file servers

This section will look at the file-based storage solutions in Windows Server. We will start this chapter by looking at definitions and concepts to help set a baseline and foundation of knowledge that you can build from.

What is a Windows Server filesystem?

When a disk is presented to a computer, for it to be used for stored data, volumes must be created for it, which you then format with a filesystem. In this section, we will examine the question of what is a filesystem.

A **filesystem** is a mechanism or structure that allows you to hierarchically organize files and their access. It contains directories and folders created on a volume. It allows you to control a naming convention and format that can be stored. It also provides security and protection through access control.

The following filesystem types can be used with Windows Server storage systems:

- **File Allocation Table (FAT)**, FAT32, **Extended File Allocation Table (exFAT)**
- **New Technology File System (NTFS)**
- **Resilient File System (ReFS)**

In the following sections, we will look at these filesystems and the other filesystem technologies available on Windows Server.

What are the FAT, FAT32, and exFAT filesystems?

FAT is the simplest but least secure filesystem used by Windows. It uses a *volume-level table* for filesystem objects; the root directory and tables are fixed on a formatted disk. For resiliency, two copies of a table are kept.

The maximum volume size for FAT is 4 GB; FAT32 and exFAT support a maximum of 64 GB for partitions.

There is no security mechanism with FAT or FAT 32, so you should not use these volume types on any Windows Server OS computers. External media such as USB flash drives are the intended use case for the FAT, FAT32, and exFAT filesystems.

What is the NTFS filesystem?

NTFS is the standard Windows Server OS filesystem.

Unlike FAT, FAT32, and exFAT, NTFS security is built-in and provides file and folder **access control lists (ACLs)** to provide granular access to and control of the files and folder structure. NTFS also provides *encryption* and *compression*; a file or folder can only be encrypted or compressed, but not both.

The intended use case for NTFS is for the Windows Server OS **boot** and **data volumes**.

Windows Server roles such as **Active Directory Domain Services** (**AD DS**), **Volume Shadow Service** (**VSS**), and **Distributed File System** (**DFS**) will require NTFS as the filesystem.

What is ReFS?

Resilient File System (**ReFS**) was introduced with the Windows Server 2012 OS. As its name suggests, the primary enhancement over NTFS is its data corruption enhancements. ReFS supports a maximum of 35 **Petabytes** (**PB**).

Like NTFS, security is built-in and provides file and folder ACLs to provide granular access to and control of the files and folder structure. But unlike NTFS, it provides no encryption or compression.

The intended use case for ReFS is **data volumes**, not **boot volumes** or **removable media**.

What is FSRM?

File Server Resource Manager (**FSRM**) is a feature that's included in the Window Server File and Storage Services server role.

It is used to manage and classify data stored on file servers. Quotas can be set on folders, reports can be created for storage usage monitoring, and files can automatically be classified based on those performing tasks. The FSRM management console can be seen in the following screenshot:

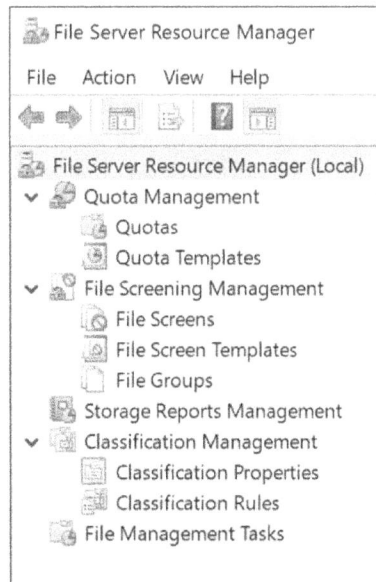

Figure 7.1 – FSRM console

The features of FSRM are as follows:

- **Quota Management**
- **File Screening Management**
- **Storage Reports Management**
- File classification infrastructure
- Access-denied assistance

FSRM is configured and managed using the **File Server Resource Manager** app or **PowerShell**. FSRM only supports NTFS-formatted filesystems and does not support ReFS.

Quotas can be created to limit volume or folder size; auto-generated quotas can also be applied to existing volumes or folders' subfolders and any that are created in the future. Quota templates can also be created for volumes and folders to enforce company-wide governance. Email notifications can be sent for quota limits that are met or exceeded.

The **File Screening Management** capability allows the file types a user can save to be controlled, such as no music files or executables. New volumes and files can have file screen templates applied so that company-wide governance can be enforced; file screen exceptions can be created. Notifications can be generated when users attempt to save unauthorized files.

What is SMB?

Server Message Block (**SMB**) is a file-sharing protocol used on a TCP/IP-based network. The protocol is used by an application to access files on a remote server.

SMB 3.0 was introduced in Windows Server 2012 and included the following features:

- SMB Multichannel
- SMB Direct
- SMB Transparent Failover
- SMB Scale-Out
- SMB Encryption
- PowerShell management commands

SMB 3.1.1 was introduced in Windows Server 2016 and provided improved encryption, preauthentication integrity to provide **man-in-the-middle attack** protection, and removal of the `RequireSecureNegotiate` setting with the ability to disable **Secure Negotiate** where required.

What is VSS?

Windows Server uses VSS to perform backups. For the data to be backed up, VSS coordinates activities to ensure consistent copies of data. These copies of data are called **shadow copies** and are also known as **snapshot** or **point-in-time** copies. A VSS solution has the following components:

- **VSS service**: This is the OS component that coordinates and orchestrates the operations.

- **VSS requester**: A requester is typically a backup software such as Windows Server Backup that requests the VSS service. It requests shadow copy operations such as *create*, *import*, and *delete*.

- **VSS writer**: This component provides a guaranteed consistent dataset for the backup operation. The writer is typically a part of the workload application that is backed up, such as Microsoft SQL Server. The Windows OS has VSS writers for various components.

- **VSS provider**: This component creates the shadow copies and maintains them. They act as proxies for the VSS service.

These VSS components can be seen in the following diagram:

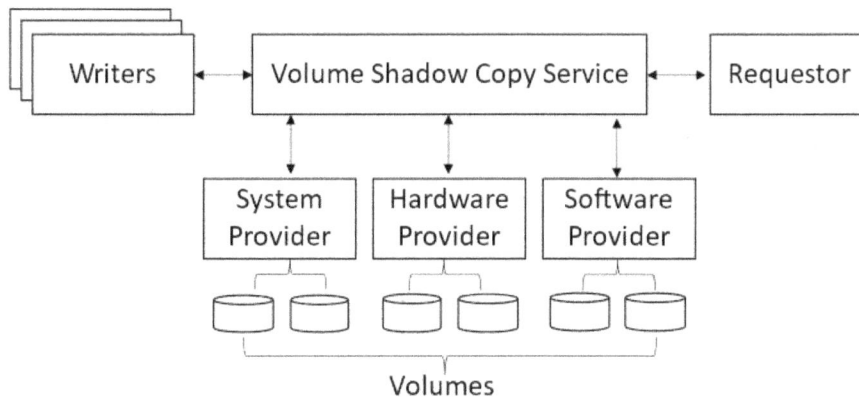

Figure 7.2 – VSS components

VSS supports volumes with a maximum size of 64 TB. SMB 3.0 must be used if you need SMB file shares to work in conjunction with VSS.

How does VSS work?

The **VSS requestor** (*your backup software*) will query the workload for a backup (snapshot copy) to determine the installed **VSS writers**.

The **VSS writer** is instructed to perform a **data quiesce** task. The snapshot copy is then requested to be created by the **VSS provider**.

When the backup has been completed, the VSS writer performs any other post-backup tasks required and then returns the workload to normal operations.

In this section, we introduced the topic of Windows Server filesystems and their related technologies. In the next section, we will look at implementing Windows Server iSCSI.

Implementing Windows Server iSCSI

In this section, we will look at **Internet Small Computer System Interface** (**iSCSI**), which is pronounced as *scuzzy* (SKUZ-ee), like *fuzzy*.

What is iSCSI?

To set the scene, **Small Computer System Interface** (**SCSI**) is defined as a set of standards that connects and transfers data between computers and peripheral connected devices, such as disks.

iSCSI is a **storage area network** (**SAN**) technology protocol, and its communication with storage systems is done over Ethernet networks using TCP/IP. It emulates the local bus SCSI disk drive interconnect, making remote storage systems appear to the storage clients as local bus storage systems.

iSCSI is an alternative to the **Fibre Channel** (**FC**) and **Fibre Channel over Ethernet** (**FCoE**) protocol storage systems. iSCSI is cheaper and less complex to implement but has lower reliability and performance as the communications are sent using TCP/IP.

Now, let's look at its components.

What are the iSCSI components?

iSCSI uses the following physical and logical components:

- Physical components:

    - **Ethernet network**: This is the physical transport medium that connects iSCSI clients (computers) to iSCSI storage systems. Standard network switches and RJ45 copper Ethernet cables are used; no specialist FC SAN switches or cabling is required.

    - **iSCSI initiator**: This is the client (computer) interface component. It is responsible for packaging the SCSI commands to be sent to the storage system and sending them over the Ethernet network as TCP/IP packets. This is included in the Windows Server OS. To implement it, the service needs to be started and configured.

- **iSCSI target**: This is the storage system interface component. The iSCSI target represents the storage system where SCSI commands will be processed; the target will appear as local storage to the initiator side server. The initiator scales to 256 iSCSI targets, with 512 virtual disks per server, provides VHD or VHDX files for virtual disks, and provides authentication. In the Windows Server OS, it is implemented as a server role under **File and Storage Services**. This role is installed and managed using Server Manager, Windows Admin Center, or PowerShell.

- Logical components:

  - **iSCSI Qualified Name (IQN)**: This is a logical name component. It provides a unique worldwide identifier for identifying an iSCSI node. It uses the IQN and does not use an IP address for identification and communication on the network. The node can be either an initiator or a target. The iSCSI initiators must be configured with an IQN and then use the IQN to contact and communicate with the iSCSI targets.

  - **Internet Storage Name Service (iSNS)**: This is a logical name component. It is a protocol used by iSCSI initiators to locate storage resources through discovery (such as DNS for iSCSI) and allows iSCSI devices to be configured and managed.

These components are shown in the following diagram:

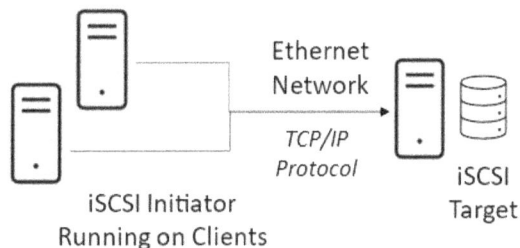

Figure 7.3 – iSCSI components

Both the *initiator* and *target* components are included in the Windows Server OS. The iSCSI *initiator* is included in all supported Windows OS versions.

In this section, we looked at the components of an iSCSI solution. In the next section, we will look at how it works.

How does iSCSI work?

As a storage protocol, iSCSI operates within the **Open Systems Interconnection (OSI)** network stack model.

iSCSI commands sent between a server and a storage system are encapsulated in TCP/IP packets and sent over an Ethernet network as any regular network packet; the payload can be a *SCSI command*. The packets are then passed back up through the network stack, and the SCSI commands are stripped out and interpreted as local device SCSI commands by the storage system. Data transmitted between the storage system and client computers follow the same TCP/IP encapsulation.

This concept is shown in the following diagram:

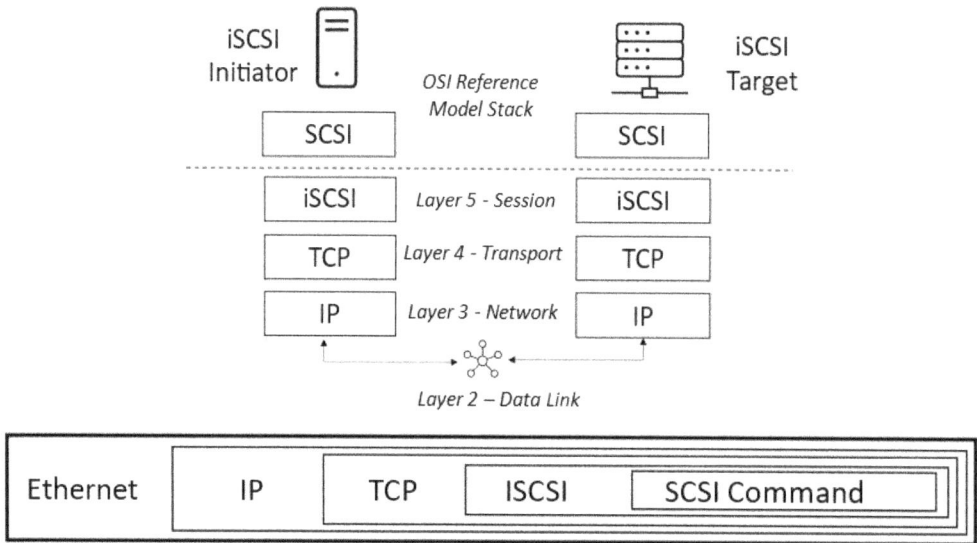

Figure 7.4 – iSCSI encapsulation

This contrasts with the *FC* and *FCoE SAN* protocols, which are alternative methods of transporting the *block* data.

To identify initiators and targets, FC uses **World Wide Names**. For iSCSI addressing, **IQN iSCSI Qualified Names**, as defined in **RFC-3271**, are used.

The IQN has a limit of 255 characters, and the format is as follows:

- `iqn.yyyy-mm.naming-authority:unique name`
- `iqn.2004-07.com.example:MBS_Target_1_SN_WC786556`

Next, let's look at how it is implemented.

How is iSCSI implemented?

iSCSI transmits information in IP packets over an Ethernet network infrastructure. This can be dedicated to storage traffic or sharing the data network but segregated by using **virtual local area networks (VLANs)** and **Quality of Service (QoS)** for performance.

To increase performance, an iSCSI **host bus adapter (HBA)** or **TCP Offload Engine (TOE)** card can be used to offload the IP packet processing from the server's processors.

The following should be considered when implementing iSCSI:

- Workloads
- Network topology
- Performance
- Security
- High availability
- The skill sets of staff and cross-function team collaboration

We will look at some of these considerations in the following sections.

Workloads

You should consider the use cases and types of workloads you need a storage system for in your environments, such as *block*, *file*, and *object-based*.

iSCSI target servers can provide access to **block-based** storage systems that are required by applications such as SQL Server, Exchange, and others.

The data is accessed at the **block level**, which refers to the data blocks and the SCSI commands; these commands and data are encapsulated in packets and sent over an Ethernet network to iSCSI initiators for client access.

iSCSI-connected storage systems can be heterogenous; the iSCSI initiator is not dependent on the Windows OS, meaning support for multi-platform storage clients.

Diskless and *network boot* scenarios are also supported using differencing disks to save storage space.

Network topology

You should consider whether a **shared** storage and data network topology or a network dedicated to storage will be implemented. This is important to evaluate in the planning stage as this can impact network performance and has security considerations. For shared data and storage networks, *QoS* should be configured for the network.

The *minimum* network speed should be 1 Gbps; a 10 Gbps, 40 Gbps, or 100 Gbps infrastructure can be used.

The following is a reference architecture for a shared storage and data network topology:

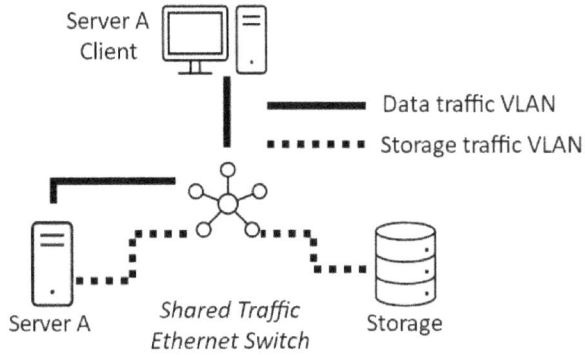

Figure 7.5 – iSCSI Ethernet network topology

In this preceding topology, an Ethernet network infrastructure is *shared* between storage and data traffic but *segregated logically* with separate VLANS for data traffic and storage traffic.

The following is a reference architecture for a dedicated storage network topology:

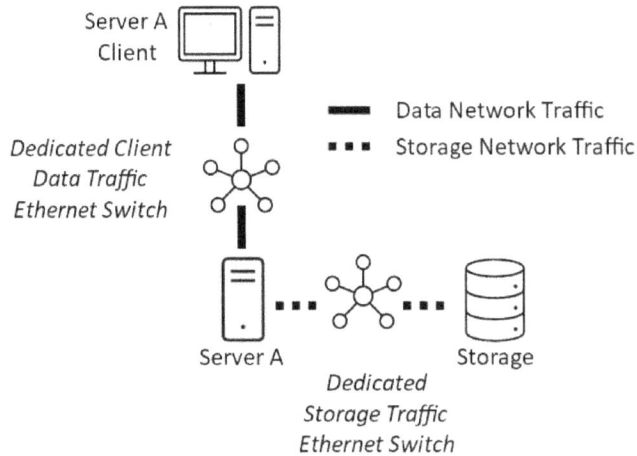

Figure 7.6 – iSCSI Ethernet network topology

In this preceding topology, an Ethernet network infrastructure is dedicated to storage traffic. The storage traffic is isolated physically by the dedicated storage switches.

Security

You should consider using a dedicated network that is isolated. iSCSI does not support **zoning**, so the initiator and target are configured with **password-based authentication** for spoofing attack protection. To authenticate initiator connections, **Challenge-Handshake Authentication Protocol (CHAP)** is used; reverse CHAP is used to authenticate the iSCSI target.

Logical unit number (LUN) masking is the same as FC, but the **IQN** is used for client identification instead of a **WWPN**. To enhance security, **IPSec** can provide end-to-end encryption.

High availability

You should consider how redundancy will be provided. You can use **Microsoft Multipath Input/ Output (MPIO)** or **Multiple Connections per Session (MCS)** as follows:

- **MPIO**: This multipathing software can provide redundant network paths to a target for multiple sessions to the same target. The multipathing software is responsible for choosing the paths to take.

- **MCS**: This provides the target with multiple TCP/IP connections from the initiators for the same iSCSI session.

In this section, we looked at how iSCSI is implemented and concluded the topic of implementing Windows Server iSCSI. The following section will look at implementing Storage Spaces and Storage Spaces Direct.

Implementing Storage Spaces and Storage Spaces Direct

This section will introduce storage virtualization using Microsoft Storage Spaces and Storage Spaces Direct. The first technology we will introduce in this section will be **Storage Spaces**; we will compare it with **Storage Spaces Direct** in the next section and understand the differences.

What is Storage Spaces?

Storage Spaces is a Microsoft storage virtualization technology that allows the local disks of a computer to be aggregated into an array of disks that can be used to create virtual disks. It is also a way to provide storage system redundancy through protection against drive failures. This can be used as an alternative to the traditional disk failure protection method of a **Redundant Array of Independent Disks (RAID)** solution.

When selecting a storage technology for a given scenario, Storage Spaces provides an inexpensive storage solution that is simple to implement and manage while being scalable and reliable.

How does Storage Spaces work?

Storage Spaces is similar in concept to the **traditional SAN** storage model. In this model, physical disks from a storage enclosure are grouped/aggregated to provide a RAID pool. From this, we create LUNs, which are presented to a server as one or more volumes by initializing them, assigning a drive letter, and then formatting them with a filesystem such as NTFS. This can be seen in the following diagram:

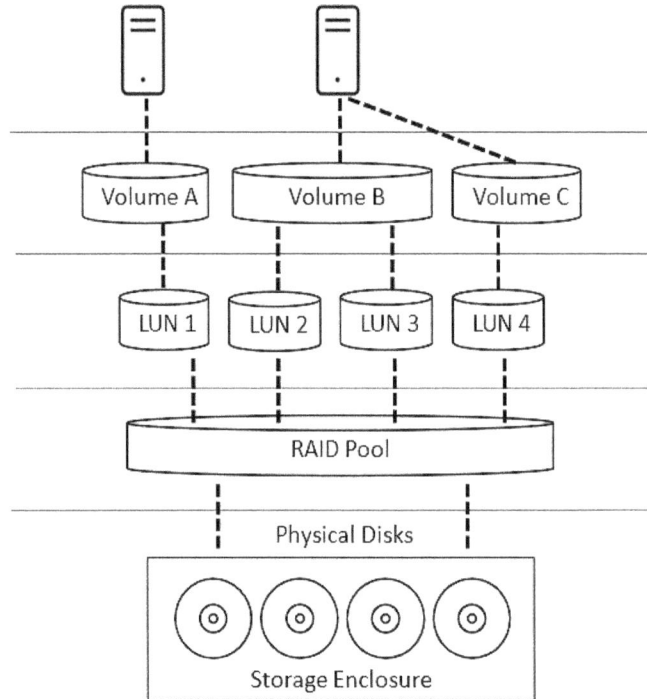

Figure 7.7 – Traditional SAN concept

In the **Storage Spaces** model, physical disks are grouped from a pool of disk resources, and in this storage pool of physical disks, virtual disks are created; a virtual disk is equivalent to a LUN. These virtual disks can be presented to a computer, then initialized and brought online so that they can be formatted and have a drive letter assigned. A Storage Spaces virtual disk can be formatted with FAT32, NTFS, and ReFS.

Storage Spaces has the following components:

- **Storage pools**: These are physical disks that are combined to create a pool of disk resources from which virtual disks can be created. The disks in the pool can be of different types and sizes; the pools abstract the underlying physical disks, and much like in compute virtualization, the hardware layer is abstracted. The physical disk can only be a part of one pool.

- **Storage spaces**: These are equivalent to RAID sets (pools) and are virtual disks in this terminology. They are created on the physical disks from the available space in the storage pools. A volume is a virtual disk partition and can have a drive letter assigned; volumes are the equivalent of LUNs on a SAN.

The relationship between these components can be seen in the following diagram:

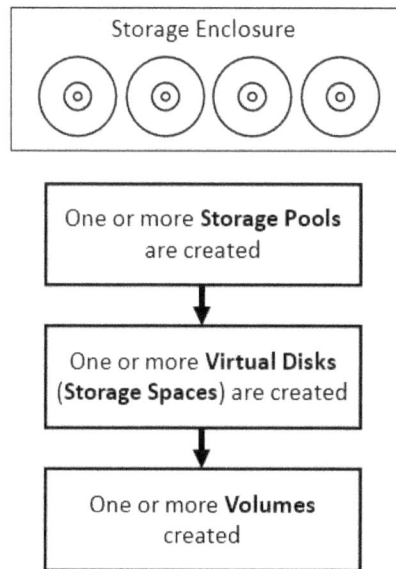

Figure 7.8 – Storage Spaces components

Storage Spaces can provide the following functionality:

- **Storage tiers**: Different disk types and speeds can be used in a pool, such as **hard disk drives (HDD)** and **solid-state drives (SSD)**; USB drives can even be used to create a storage pool. Storage Spaces uses tiering to achieve capacity and performance balance. It automatically moves data that is more frequently accessed to faster disks and to slower disks for less frequently accessed data. Data is moved once a day at 01:00 A.M. You can configure which storage is used by files and pins as required.

- **Provisioning models**: Storage Spaces provides two provisioning models for virtual disks, as follows:

  - **Thick (fixed) provisioning**: This creates a fixed size at the point of creation.

  - **Thin provisioning**: This does not pre-allocate a size at the point of creation and utilizes just-enough and just-in-time creation. A trim process can reclaim unused allocated space.

  A storage pool can have both *thin* and *fixed* provisioned virtual disks.

- **Resiliency levels**: Storage Spaces provides three resiliency models, as follows:

  - **Simple**: Maintains one copy of the data. This level is suitable for workloads that have no requirement for resiliency through storage spaces.

  - **Parity**: Maintains two copies of the data with a single parity and three copies with dual parity. This level is suitable for sequential large read/write operations in pattern workloads.

  - **Mirror**: Maintains two copies of the data in a two-way mirror and three copies in a three-way mirror. This level is suitable for all workloads.

- **Write-back caching**: This allows you to write data to disks so that it can be optimized and is enabled by default. This works with tiering to write to faster disks automatically when a peak in disk writing is detected.

- **Management**: Storage Spaces can be managed in Server Manager using the File and Storage Services role, Windows Admin Center, PowerShell, and APIs.

Storage Spaces also has some limitations, as follows:

- No support for boot or system volumes

- No support where the physical disks are abstracted from the storage layer, such as VM pass-through disks, and where a separate RAID layer is introduced

- No support for FC, FCoE, or iSCSI

- Formatted or partitioned disks cannot be added to a storage pool

- The same sector size must be used for all disks in a storage pool

- Only **Serial Attached SCSI (SAS)** storage can be used with failover clusters

In this section, we looked at how Storage Spaces works. In the next section, we will introduce Storage Spaces Direct, which has a similar name, but you should be aware that it is a different solution. We will look at how it compares to Storage Spaces.

What is Storage Spaces Direct?

Storage Spaces Direct is a software-defined storage solution; it enables multiple compute cluster nodes in a converged or hyper-converged technology stack to have their local disks presented and aggregated into a single scalable and resilient shared storage system. It can be likened to a virtual SAN. It was first introduced as a Windows Server storage technology with Windows Server 2012.

Storage Spaces Direct can be found as a core technology of Azure Stack HCI, an Azure service that is Microsoft's data center modernization and private cloud OS platform. Storage Spaces Direct is also supported as an **Infrastructure-as-a-Service (IaaS)** storage solution to be implemented in the Azure public cloud.

To recap on the previous sections, Storage Spaces, in contrast, was a much simpler solution for aggregating local disks on a single computer and using these local resources to create virtual disks on which you could create volumes. Storage Spaces is not redundant and is still a direct component of Storage Spaces Direct, which we will look at in the next section.

Storage Spaces Direct provides several capabilities and benefits, some of which are as follows:

- Scales to 4 PB per cluster.

- USB flash drive witness.

- ReFS volumes compression and deduplication. The first 4 TB of a file is deduplicated, with 64 TB supported per volume.

- Mirror accelerated parity.

In this section, we introduced Storage Spaces Direct. In the next section, we will look at how it works.

How does Storage Spaces Direct work?

The Storage Spaces Direct software-defined solution works by creating a cluster by connecting several servers (nodes) with local storage drives over Ethernet; no storage fabric, special cables, or adapter is required.

Storage Spaces Direct uses the following components:

- Failover clustering role

- **Cluster Shared Volume** (**CSV**) filesystem

- Storage spaces

- Software Storage Bus

- SMB 3

A core component of Storage Spaces is CSVs, which allow all nodes' NTFS and ReFS volumes to have reads and writes simultaneously. There is no locking, masking, change of drive ownership, or dismounting and mounting volumes; this allows for faster failovers and provides load balanced distribution of requests to the storage from the compute.

To implement Storage Spaces Direct, you will need to enable Storage Spaces Direct on a failover cluster that you have created as a prerequisite step. The next stage is to aggregate the storage drives from each server in the cluster and create a software-defined pool of virtual shared storage. Volumes will be presented as drives to the VMs where you store data, such as **virtual hard disk** (**VHD**) files, which will then be created in the storage pool. The volumes use the CSV filesystem so that each server will see the volumes as locally mounted and can be accessed and perform operations simultaneously.

In this section, we looked at how Storage Spaces Direct works. In the next section, we will look at its deployment options.

What are the Storage Spaces Direct deployment options?

Storage Spaces Direct provides deployment options that support both converged and hyper-converged infrastructure systems. The hyper-converged deployment is the only one supported by Azure Stack HCI.

Converged deployment

The converged deployment uses separate clusters for storage and compute, also called disaggregated deployment. Compute and storage clusters can be scaled independently to meet your needs. This deployment model is shown in the following over-simplified diagram:

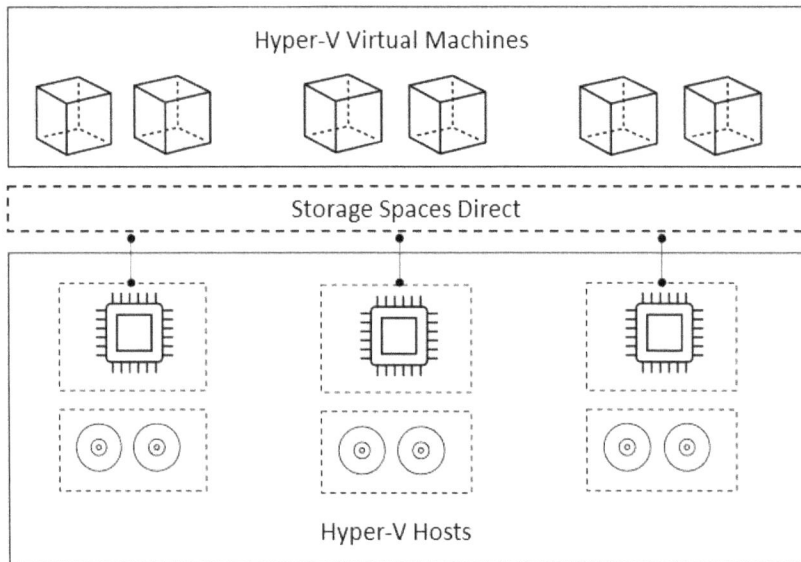

Figure 7.9 – Converged deployment model

This model provides a **Scale-out File Server** (**SoFS**) layer above Storage Spaces Direct to provide SMB 3 network shares. This may be a use case for a **Virtual Desktop Infrastructure** (**VDI**) scenario.

Hyper-converged deployment

The hyper-converged deployment uses a single cluster for both compute and storage. This deployment model is shown in the following over-simplified diagram:

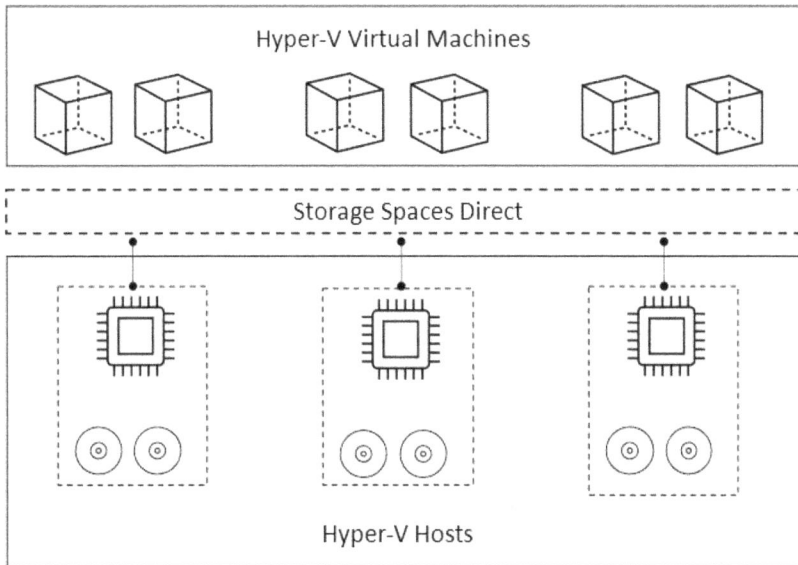

Figure 7.10 – Hyper-converged deployment model

Storage Spaces Direct can be deployed as part of a Windows Server cluster or an Azure Stack HCI cluster.

Storage Spaces Direct is not a complex solution to implement, as would be the case with a traditional SAN, due to the software-defined nature of the solution. Server Manager, Window Admin Center, and PowerShell can be used.

In this section, we looked at the deployment options for Storage Spaces Direct, which has concluded the topic of implementing Storage Spaces and Storage Spaces Direct. In the next section, we will look at Windows Server Data Deduplication.

Implementing Windows Server Data Deduplication

This section introduces **Data Deduplication** in the context of the technology component included in Windows Server.

What is Windows Server Data Deduplication?

Windows Server includes **Data Deduplication** as an installable and configurable server role. The premise of Data Deduplication as a technology is to use less storage space by removing duplicated data while the integrity of the system is still maintained.

The components of Windows Server Data Deduplication are as follows:

- Deduplication service:

  - Deduplication and compression

  - Garbage collection

  - Scrubbing

- Filter driver

- Unoptimization

The performance impact of deduplication is minimized using multi-threaded post-processing and utilizes a variable-size chunk store that can maximize space by using optional compression.

How does Windows Server Data Deduplication work?

The Windows Server implementation of Data Deduplication works based on two principles:

- Optimization should not affect *disk writes*. A post-processing model is used; all data is *first* written to disk and *then* optimized.

- Optimization should not affect *user access*. It must be transparent to the user that they are accessing deduplicated files on an optimized volume.

The deduplication process consists of the following five steps:

1. The filesystem is scanned for files that match an optimization policy

2. Files are broken into variable-size chunks

3. Unique chunks are identified

4. Chunks are moved to the chunk store and, optionally, compressed

5. The original file data is replaced with a chunk store reparse point

Before implementing deduplication, you should assess your environment to determine whether there is enough duplicate information to make the task worthwhile. The following are workloads that are recommended as good candidates for deduplication due to the duplicate data that's often stored:

- General-purpose file servers; shares such as user home folders, work folders, and software repository shares

- VDI

- Virtualized backup applications such as **Microsoft Data Protection Manager** (**DPM**)

The following will have limited benefits from deduplication:

- General-purpose Hyper-V hosts
- SQL servers
- **Line-of-business (LOB)** servers

Data Deduplication cannot process the following data on volumes:

- System state files
- Encrypted files
- Extended attribute files
- Files that are 32 KB or smaller
- Files that have been explicitly excluded from the scope of deduplication
- Files that don't meet the configured policy for deduplication

Data deduplication is supported on NTFS. For ReFS volumes, the max volume size supported is 64 TB, and files with a maximum size of 4 TB.

What disk size savings can be expected?

The following indicates disk space savings that are typical in deduplication scenarios:

- **User documents**: 30 to 50 percent saving
- **Cab files, software binaries, images, and updates**: 70 to 80 percent saving
- **Virtual hard disk files**: 80 to 95 percent saving
- **General file share mix of preceding data types**: 50 to 60 percent saving

The disk size savings that can be achieved will depend on the data type and mix factors.

How is Windows Server Data Deduplication implemented?

Data Deduplication is implemented by installing it as a server role using Server Manager, Windows Admin Center, or PowerShell.

The following PowerShell cmdlet can be used:

```
Add-WindowsFeature -Name FS-Data-Deduplication
```

Once the server role has been added, Data Deduplication needs to be enabled; this can be done with Server Manager, Windows Admin Center, or PowerShell.

The following PowerShell cmdlet can be used:

```
Enable-DedupVolume –Volume *VolumeLetter* –UsageType
*StorageType*
```

The parameter value for -UsageType should be one of the following three types:

- **Hyper-V**: This type is used for a volume hosting Hyper-V storage
- **Backup**: This type is used for a volume hosting virtualized backup servers
- **Default**: This type is used for general-purpose volumes

The default settings, if not sufficient, can be changed to meet your scenario, as follows:

- Process the optimization of files sooner
- Changes to garbage collection and scrubbing schedules
- Volume directories or file types don't need to be deduplicated

To edit the default settings, you can use the following PowerShell cmdlet:

```
Set-DedupVolume
```

Deduplication jobs can be run on-demand or by configuring schedules; we will cover this in the following sections.

How to run jobs on demand

The following on-demand job types can be run:

- **Optimization**: `Start-DedupJob –Volume *VolumeLetter* -Type Optimization`
- **Data scrubbing**: `Start-DedupJob –Volume *VolumeLetter* -Type Scrubbing`
- **Garbage collection**: `Start-DedupJob –Volume *VolumeLetter* -Type GarbageCollection`
- **Unoptimization**: `Start-DedupJob –Volume *VolumeLetter* -Type Unoptimization`

Due to the intensive nature of processing the garbage collection job, this should only be run outside regular operating hours and only when the deletion load threshold has been reached.

How to schedule jobs

Three schedules are set by default when the server is enabled for Data Deduplication for volumes; these are as follows:

- **Optimization**: This job is scheduled to run every hour
- **Garbage collection**: This job is scheduled to run once a week
- **Scrubbing**: This job is scheduled to run once a week

All jobs run on all volumes; to have a job run on a particular volume, a new job must be created.

Job schedules can be managed with the following PowerShell cmdlets:

- `Get-DedupSchedule`
- `New-DedupSchedule`
- `Set-DedupSchedule`
- `Remove-DedupSchedule`

The `Get-DedupSchedule` cmdlet cannot be used for Windows Task Scheduler custom job schedules.

In this section, we concluded the topic of implementing Windows Server Data Deduplication. In the next section, we will look at Windows Server Storage Replica.

Implementing Windows Server Storage Replica

This section introduces Storage Replica in the context of the technology component included in Windows Server.

What is Windows Server Storage Replica?

Storage Replica is an installable Windows Server feature that's used with the file server role to provide high availability and disaster recovery of volumes for physical or virtual machines. It can be used for standalone servers or clusters and stretched failover clusters between physical sites.

Data Deduplication is supported on *NTFS* and *ReFS* volumes; *Windows Server OS volumes* cannot be replicated using Windows Server Storage Replica.

How does Windows Server Storage Replica work?

It works at the block level using SMB 3 and thin provisioning and provides unidirectional replication with automatic failover.

Kerberos AES 256 is used for authentication, and delegated administration is supported. Designated network adapters can be configured to control replication traffic.

Two NTFS or ReFS volumes are required at each source and destination. One pair is used to replicate data between the source data volume (the primary) and the target or destination (the secondary); the other pair is used to replicate logs.

The following replication methods are provided:

- **Synchronous replication**: This method is used when volumes are in close network proximity and require low latency between replication nodes. It provides crash-consistent replication of volumes, which provides zero filesystem-level data loss.

- **Asynchronous replication**: This method is used in stretch cluster scenarios, where there is a greater distance between the volumes, and the network latency will be greater than 5 **milliseconds** (**ms**). This replication method is subject to data loss, the amount being dependent on the primary and secondary replication volume's lag time.

When data loss must be avoided, then synchronous replication should be used.

This should be enabled on the source server data volumes for Data Deduplication. This will automatically replicate the deduplicated content of the volumes between the primary and secondary volumes.

How is Windows Server Storage Replica implemented?

Storage Replica can be implemented with the file server role and Storage Replica feature using Server Manager, Windows Admin Center, or PowerShell.

Before Storage Replica can be implemented using one of the previous methods, there are some prerequisites; these are as follows:

- The volumes to be replicated must be on servers that are members of an AD DS forest; workgroup servers are not permitted.

- Each server requires at least 2 CPU cores and 2 GB of RAM.

- The OS must be one of the following:

  - *Windows Server 2016 Datacenter edition*

  - *Windows Server 2019 Datacenter edition*

  - *Windows Server 2022 Datacenter edition*

  - *Windows Server 2019 Standard edition*; however, this only supports a single volume of up to 2 TB for replication

- **Remote Direct Memory Access** (**RDMA**) is preferable for synchronous replication for each server; however, at a minimum, a 1-gigabit Ethernet adapter is required.

- Each source and destination server must have a volume for data and a volume for logs; the specific criteria are as follows:

  - **GUID Partition Table (GPT)** - initialized disks; not **master boot record (MBR)**

  - ReFS- or NTFS- formatted volumes

  - Matching volume sizes and sector sizes for data and log volumes

  - More performant storage should be used for the log volumes than the data volumes

  - No other workloads should access the log volumes

- Servers hosting the replicated volume must have the following bi-directional communications allowed:

  - ICMP

  - SMB (port 445) and SMB Direct (port 4445)

  - WS-MAN (port 5985)

- Less than 5 ms latency is required on the network between replication nodes for synchronous replication

The `Test-SRTopology` PowerShell cmdlet can validate whether the source and target volumes meet the preceding prerequisites; you can access the report that's been generated with `TestSrTopologyReport.html`.

In this section, we looked at the prerequisites for implementing Storage Replication, thereby concluding the topic of implementing Windows Server Storage Replication. In the next section, we will look at implementing the DFS role in Windows Server.

Implementing the DFS

The **Distributed File System (DFS)** in Windows Server comprises two server roles that can be installed as follows:

- **DFS Namespaces (DFSN or DFS-N)**: This DFS role allows you to group shared folders on a collection of file servers into logical namespaces. This allows users to navigate shared folders that exist across multiple file servers and maybe across different sites. This simplifies the user journey of finding documents and files by providing a centralized folder namespace, similar to a global UNC path. It provides this through the file server abstraction layer without knowing which server holds which share folder, file, and so on.

- **DFS Replication (DFSR or DFS-R)**: This DFS role allows you to efficiently replicate folders between multiple servers and on multiple sites. It only replicates changed blocks, not entire files; it uses the **remote differential compression (RDC)** algorithm to do so.

DFS can be implemented using Server Manager (via the *Add Roles and Features* and *File and Storage Services* roles) or PowerShell.

The following PowerShell cmdlet will install *DFSN, DFSR,* and the *DFS* management tools:

```
Install-WindowsFeature FS-DFS-Namespace, FS-DFS-Replication,
RSAT-DFS-Mgmt-Con
```

Interoperability with Azure Virtual Machines is supported. The following URL covers the requirements and limitations: `https://learn.microsoft.com/en-us/previous-versions/windows/it-pro/windows-server-2012-R2-and-2012/jj127250(v=ws.11)#interoperability-with-azure-virtual-machines.`

In this section, we looked at implementing the DFS role in Windows Server. In the next section, we will complete a hands-on exercise to reinforce some of the concepts covered in this chapter.

Hands-on exercise

To support your learning with some practical skills, we will utilize the concepts and understanding we gained from this chapter and put them to practical use.

We will look at an exercise that covers implementing Data Deduplication.

Getting started

To get started with this section, you will need access to a Windows Server; this can be physical, virtual, or an Azure IaaS VM. This exercise applies to the following Windows server OS types: Windows Server 2016, Windows Server 2019, Windows Server 2022, and Azure Stack HCI, versions 20H2 and 21H2.

It is recommended that you ensure hotfix KB4025334 is installed using Windows Server 2016.

Let's move on to the exercise for this chapter.

Exercise – Implementing Data Deduplication

In this exercise, we will learn how to enable Data Deduplication in Windows Server.

We have broken down this exercise into individual tasks for ease of understanding.

Task 1 – installing the deduplication role

In this task, we install the deduplication role in Windows Server:

1. Log into your server with an admin account, click **Add Roles and Features** from the **Manage** area of **Server Manager**, and click **Next** on the **Before You Begin** screen.
2. Select **Role-based or feature-based installation** from the **Installation Type** screen and click **Next**.

3. From the **Server Selection** screen, accept the default selection of the current server and click **Next**.

4. Select **Data Deduplication** under **File and Storage Services** from the **Roles** list of the **Server Roles** screen.

5. Click **Add Features** on the pop-up screen. Then, click **Next**:

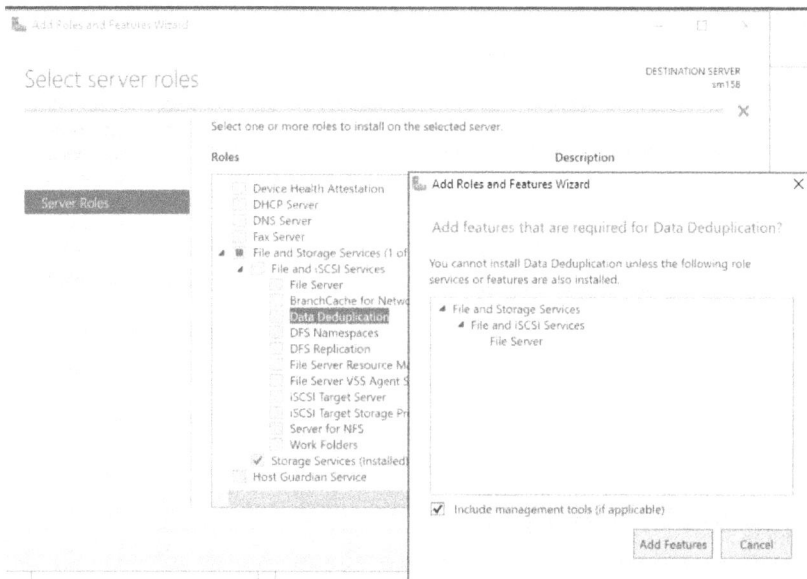

Figure 7.11 – Installing the Data Deduplication server role

6. On the **Features** screen, accept the default selections and click **Next**.

7. On the **Confirmation** screen, review the information, click **Install**, view installation signs of progress until you receive a message saying that the installation succeeded, then click **Close** on the **Add Roles and Features Wizard** popup.

8. With that, you have installed the *Data Deduplication server role*.

In the next task, we will configure deduplication.

Task 2 – configuring deduplication

In this task, we will enable and configure deduplication:

1. Log into your server with an admin account and open **Server Manager**.

2. From **Server Manager**, click **File and Storage Services**.

3. Click **Volumes** under **File and Storage Services**.

4. Select a **Volume**, then right-click and select **Configure Data Deduplication…**:

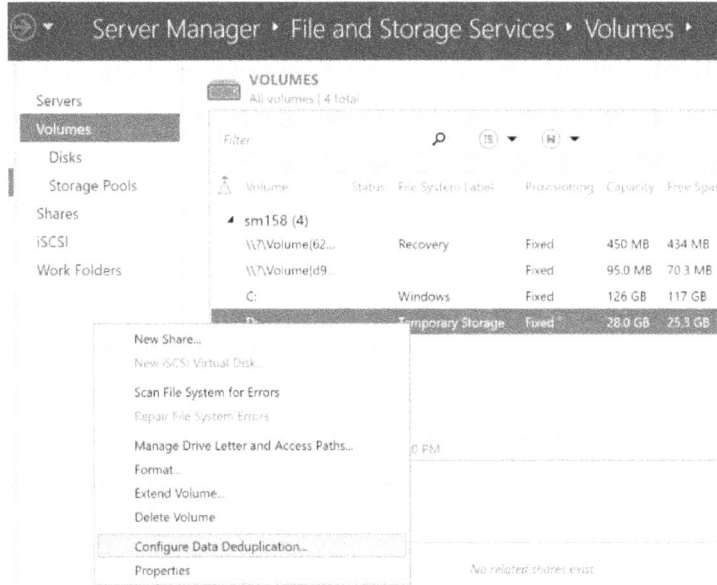

Figure 7.12 – Configure Data Deduplication…

5. Select the required Data Deduplication type from the **Deduplication Settings** screen, and then click **OK**:

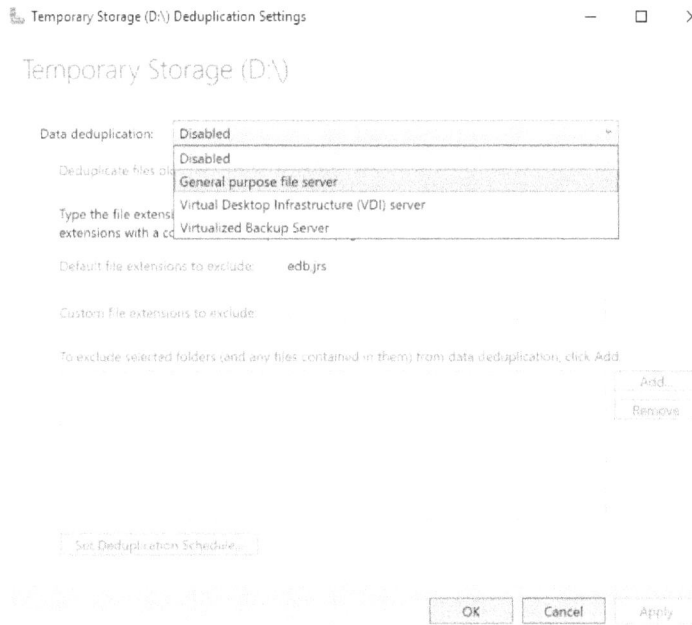

Figure 7.13 – Selecting a deduplication type

6. Configure additional settings as required, then click **OK** to close the **Deduplication Settings** screen.

With that, we have completed this exercise. This exercise taught us the skills needed to implement Data Deduplication. Now, let's summarize this chapter.

Summary

This chapter provided coverage for *AZ-800 Administering Windows Server Hybrid Core Infrastructure: Manage storage and file services*.

First, we learned about the basics of the Windows Server filesystems and protocols; then, we looked at storage solutions such as Storage Spaces and Storage Spaces Direct and understood how they differ. After that, we looked at storage optimization through Data Deduplication within Windows Server and Storage Replica, which is the high availability and disaster recovery solution for Windows Server.

We completed this chapter with a hands-on exercise to provide you with additional practical skills.

This chapter aimed to take your knowledge beyond the exam objectives; we added new skills and learning with the content provided. This further developed your knowledge and skills for on-premises network infrastructure services and enabled you to be prepared for a real-world, day-to-day hybrid environment-focused role.

In the next chapter, you will learn how to implement Azure Storage services.

Further reading

This section provides links to additional study references and additional exam information:

*   *Microsoft Certified: Windows Server Hybrid Administrator Associate*: `https://docs.microsoft.com/en-us/learn/certifications/windows-server-hybrid-administrator/`

*   *Exam AZ-800: Administering Windows Server Hybrid Core Infrastructure*: `https://docs.microsoft.com/en-us/learn/certifications/exams/az-800`

*   *Exam AZ-800: Study Guide*: `https://query.prod.cms.rt.microsoft.com/cms/api/am/binary/RWKI0r`

*   *Microsoft Learn*: `https://docs.microsoft.com/en-us/learn/paths/configure-storage-file-services/`

Skills check

Check what you have learned in this chapter by answering the following questions:

1. What filesystem types can be used with Windows Server?

2. Explain the similarities and differences between NTFS and ReFS.

3. What is FSRM?

4. Explain some of the features of the SMB protocol.

5. What is VSS?

6. What is iSCSI, and how does it differ from SCSI, FC, and FCoE?

7. What are the different iSCSI components? Describe their functionality.

8. Explain how iSCSI works.

9. What network topologies are supported for iSCSI?

10. How is iSCSI secured?

11. What is Storage Spaces?

12. Explain how Storage Spaces Direct differs from Storage Spaces.

13. What are the components of Storage Spaces Direct?

14. Explain the two Storage Spaces Direct deployment models.

15. What is Windows Server Data Deduplication?

16. What file types can't be processed for Data Deduplication?

17. What are the best use case scenarios for Data Deduplication?

18. What is Windows Server Storage Replica?

19. What two replication methods are supported?

20. List the prerequisites for implementing Windows Server Storage Replica.

8

Implementing a Hybrid File Server Infrastructure

In the previous chapter, we learned about the concepts of Windows Server storage and filesystems.

This chapter covers content from *AZ-800 Administering Windows Server Hybrid Core Infrastructure: Implement a hybrid file server infrastructure*.

We will learn about the storage services that are available in Azure and look at how they can be implemented to meet use case requirements when deploying workloads into Azure. We will conclude with some hands-on exercises to help you develop your skills even further.

The following topics are included in this chapter:

- Introduction to Azure Storage services
- Implementing Azure Files
- Introduction to Azure File Sync
- Migrating a DFS-R deployment to Azure File Sync
- Hands-on exercises

In addition to the topics listed in this chapter, this chapter's goal is to take your knowledge beyond the exam objectives to prepare you for a real-world, day-to-day hybrid environment-focused role.

Introduction to Azure Storage services

This section looks at the available Storage services in Azure. We will start this chapter by looking at various definitions and concepts to help set a baseline and foundation of knowledge that you can build from.

Storage types

The following storage types are available in Azure:

- **Files**: This provides the **Server Message Block** (**SMB**) protocol, which provides fully managed, highly available serverless network file shares. These traditional mapped drives can be communicated using port 445 and the *SMB 3.x* protocol.

- **Binary large object** (**Blob**): This provides cost-effective massive scale-out unstructured data storage. The following three blob types are available:

  - **Page Blob**: Used to store random access data objects, such as VM disks

  - **Block Blob**: Used to store ordered data objects, such as backups

  - **Append Blob**: Used to store consecutively added data objects, such as log files

- **Table**: This provides a data store for *non-relational* (**NoSQL**) *structured* and *semi-structured* data.

- **Queue**: This provides a data store for large amounts of asynchronous messages.

Selecting the right data storage that matches your workload type is a key design decision to provide an optimal cost, performance, and operational solution. You may need to implement different storage types in a single solution since no Azure Storage service will meet all your needs.

Storage accounts

These Azure resources are repositories and boundaries where you can store all your data objects. In addition to working as data planes, they can also be considered management plane consoles. Storage accounts can define the settings and configurations that can be applied to the data held in the storage account, such as tiering, redundancy, security, access control, protection, monitoring, and so on.

Storage accounts are automatically encrypted for each storage service. They can be monitored using Azure Storage Analytics and Azure Monitor.

You can create your first storage account using the Azure portal, as shown in the following screenshot:

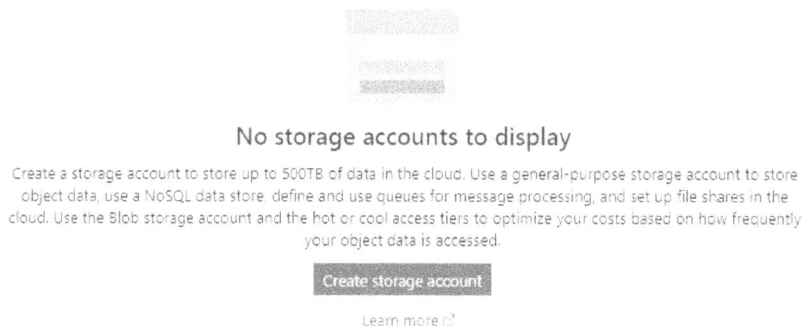

No storage accounts to display

Create a storage account to store up to 500TB of data in the cloud. Use a general-purpose storage account to store object data, use a NoSQL data store, define and use queues for message processing, and set up file shares in the cloud. Use the Blob storage account and the hot or cool access tiers to optimize your costs based on how frequently your object data is accessed.

Create storage account

Learn more

Figure 8.1 – Creating a storage account in Azure

You can also use **PowerShell**, the **Azure CLI**, or an **Infrastructure as Code (IaC)** deployment language such as **Terraform**, **Bicep**, or **Azure Resource Manager (ARM)**.

Once created, the storage account can be accessed via HTTP/HTTPS through a globally unique namespace. These are known as endpoints and can be public or private. Information about these, along with *Quickstart templates*, can be found at https://docs.microsoft.com/en-us/azure/templates/microsoft.storage/storageaccounts?pivots=deployment-language-bicep.

We will look at creating a storage account via the Azure portal method in the *Hands-on exercises* section of this chapter.

The following types of storage accounts can be created:

- **Standard: General-purpose v2**
- **Premium block blobs**
- **Premium page blobs**
- **Premium File shares**

These storage account types are represented in the following screenshot:

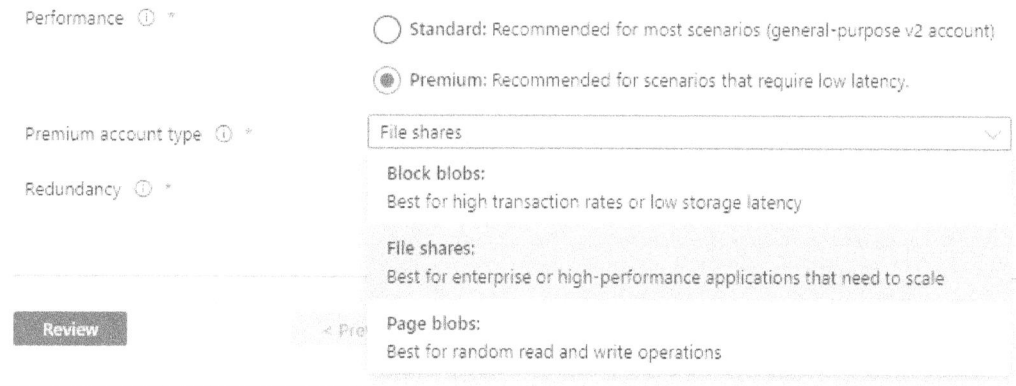

Figure 8.2 – Storage account types

You cannot change the performance type setting after creating a storage account. This is shown in the following screenshot:

Standard storage accounts are backed by magnetic drives and provide the lowest cost per GB. They're best for applications that require bulk storage or where data is accessed infrequently. Premium storage accounts are backed by solid state drives and offer consistent, low-latency performance. They can only be used with Azure virtual machine disks, and are best for I/O-intensive applications, like databases.
Learn more about storage account performance.

solve problems   Performance ⓘ
(IAM)   ○ Standard ◉ Premium

                                  ⓘ This setting cannot be changed after the storage account is created.

Figure 8.3 – Storage account type setting

You must create a new storage account of the performance type you need and then move the data to the created storage account.

This section looked at storage types, accounts, redundancy, and access. In the next section, we will look at implementing Azure Files.

Implementing Azure Files

The most common use case for Azure Files is to provide a replacement for on-premises file servers in a lift-and-shift manner to an Azure scenario. This will require you to provide access to network file shares via **Server Message Block** (**SMB**), **Network File System** (**NFS**), and **HTTP(S)** protocols. Client connections can be made by *Windows*, *Linux*, and *macOS* devices when the appropriate connectivity and access permissions are implemented.

Azure Files can also be used for backup and disaster recovery scenarios for the on-premise locations of file servers . Azure File Sync can be used to replicate file shares from on-premise file servers, providing a cloud tiering capability.

Azure Files is based on a serverless deployment, meaning there are no file servers to manage; this layer has been abstracted so that you can focus on creating shares and managing their access.

Azure Files uses the standard SMB 3.x protocol and can provide granular file permissions with support for NTFS when domain services are used to provide access. Data redundancy, performance tiering, encryption, versioning, backup, malware protection through Microsoft Defender for Storage, public and private endpoint access, and monitoring can all be configured as part of the service.

Azure Files supports a maximum file size of 4 TB and a maximum file share size of 100 TB (at the time of writing).

Access is available from anywhere, with the caveat that you must consider the following:

- **Ports and protocol**: Although the SMB 3.x protocol is considered safe and secure, many providers and organizations will block port 445, meaning your file shares will not be accessible. Port 445 should be permitted, or connectivity solutions such as ExpressRoute or a VPN will need to be implemented to work around this. You can explore this further here: `https://docs.microsoft.com/en-gb/azure/storage/files/storage-files-networking-overview#azure-networking`.

- **Permissions**: Access is another potential hurdle for file shares. Anonymous access isn't supported. The authentication methods that are supported are **access key**, **shared access signature (SAS)**, and **identity-based**. The identity-based method uses **Kerberos** and will need to be used in conjunction with hybrid identities to provide **NTFS** permissions.

- **Performance**: Once we have removed the hurdles surrounding ports, protocols, and permissions, we may find that the throughput and latency are not at the level needed to provide a workable or optimal solution. The optimal performance solution will be where Azure resources access the shared drives within the same region as the Azure Files storage account.

The storage account key will be required to gain full admin control and ownership of a file share using Azure Files; this process and its actions should be tightly controlled and governed.

Storage tiers

Azure Files provides the following storage tiers so that you get the optimal price and performance to match your needs:

- **Premium**: These file shares use **solid-state drives** (**SSD**), which provide low latency and fast and consistent performance.

- **Transaction optimized**: These file shares use **hard disk drives** (**HDD**), where low latency is not required. As the name implies, this tier is intended for workloads that perform transaction-heavy operations. This tier was first introduced as the *Standard* tier before being renamed *transaction optimized*.

- **Hot**: These file shares use HDD and are suitable for general-purpose workloads.

- **Cool**: These file shares use HDD, are the most cost-effective, and are optimized for online archives.

The chosen tier needs to match how frequently data is accessed and the rate at which data needs to be moved between tiers; the access frequency and retrieval (rehydration) process can negatively impact costs if the incorrect tier is chosen.

Storage redundancy

Several replication options are provided by Microsoft for your data. The following ones can be used with Azure Files:

- **Locally redundant storage (LRS)**: Three copies of your files are stored in a storage cluster in a single physical location in a region. This has the lowest level of availability and durability.

- **Zone-redundant storage (ZRS)**: Three copies of your files are stored per LRS and synchronously replicated across three storage clusters in different availability zones. Each zone is a physical location in the same region with one or more data centers that provide independent services such as network, cooling, and power. Data must be written to all three clusters before the writes can be accepted. This is the highest availability and durability option within a region and has the highest cost.

- **Geo-redundant storage (GRS)**: Three copies of your files are stored per LRS and then asynchronously replicated to a storage cluster in a single location in a different region. These are replicated synchronously as LRS across the cluster in the secondary region.

- **Geo-zone-redundant storage (GZRS)**: Copies are replicated as per ZRS and then asynchronously replicated to a storage cluster in a single location in a different region. These are replicated synchronously as LRS across the cluster in the secondary region as per GRS.

These options are shown in the following screenshot:

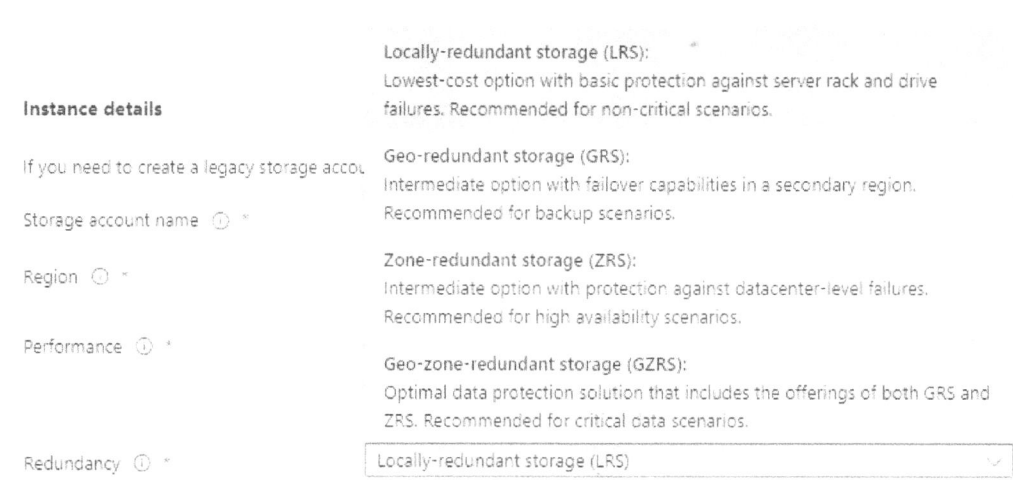

Figure 8.4 – Storage redundancy options

Note that premium file share storage accounts only support the LRS and ZRS redundancy options. There is no geo-redundancy option for a secondary region.

In addition, **Read Access GRS (RA-GRS)** and **Read Access GZRS (RA-GZRS)** are two other storage types that are available that support read access to the data in the secondary region; this can support a use case where you want to serve a copy of the data in a region closer to users, while the primary region is up. Note that not all storage types are available in all regions.

Storage encryption

Encrypting data at rest in a storage account is provided by all Azure Storage services and is enabled by default. As the data is written, it is automatically encrypted and transparently decrypted when accessed. By default, **Microsoft-managed keys (MMK)** are used. This can be seen in the following screenshot:

Figure 8.5 – Storage encryption type

As an alternative, **customer-managed keys (CMK)** can be used to encrypt the data; they can be stored in an **Azure Key Vault** as the *key store*. All storage accounts have encryption-in-transit enabled by default, as shown here:

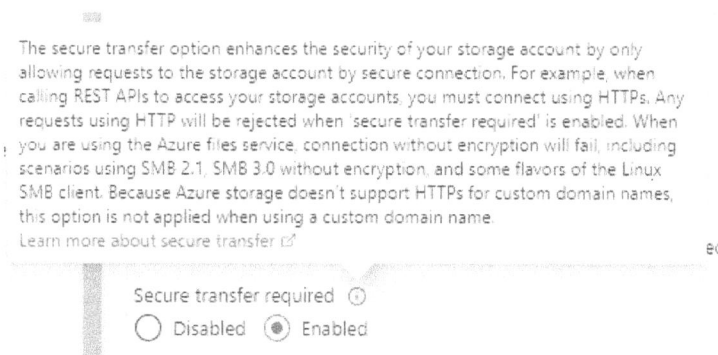

Figure 8.6 – Encryption-in-transit setting

In the default configuration of **Secure transfer required** being **Enabled**, **HTTP** connections are not allowed and will be rejected. Connections will fail if encryption is disabled, even in the case of **SMB 3.0**.

Storage protection

The following settings are enabled by default when creating a storage account:

- **Soft delete for blobs**: Retained for 7 days
- **Soft delete for containers**: Retained for 7 days
- **Soft delete for file shares**: Retained for 7 days

These options can be seen in the following screenshot:

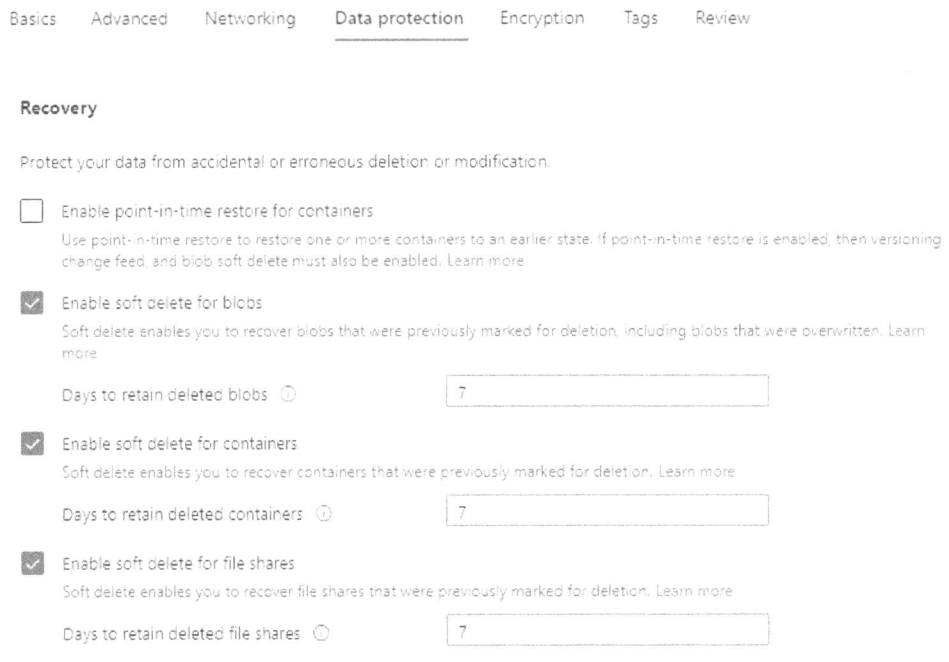

Figure 8.7 – Storage protection

In addition, snapshots can be taken of file shares for general backup operations. Snapshots are incremental, and a share can contain up to 200.

Storage network access

When a storage account is created, the default networking configuration allows public access from all networks. You can change this to one of the following:

- **Enable public access from selected virtual networks and IP addresses**
- **Disable public access and use private access**

These options are shown in the following screenshot:

Network connectivity

You can connect to your storage account either publicly, via public IP addresses or service endpoints, or privately, using a private endpoint.

Network access *

(●) Enable public access from all networks

( ) Enable public access from selected virtual networks and IP addresses

( ) Disable public access and use private access

❶ Enabling public access from all networks might make this resource available publicly. Unless public access is required, we recommend using a more restricted access type. Learn more

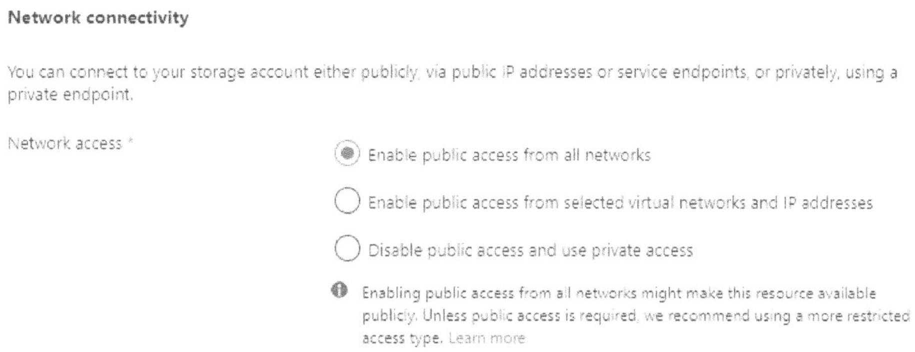

Figure 8.8 – Storage network access options

When you select the **Disable public access and use private access** option, you will be presented with the option to add a private endpoint. A private endpoint assigns a private IP to the storage account from the VNet. This can be seen in the following screenshot:

(●) Disable public access and use private access

Private endpoint

Create a private endpoint to allow a private connection to this resource. Additional private endpoint connections can be created within the storage account or private link center.

＋ Add private endpoint

Name	Subscription	Resource g...	Region	Target sub-...	Subnet	Private DN...	Private

Click on add to create a private endpoint

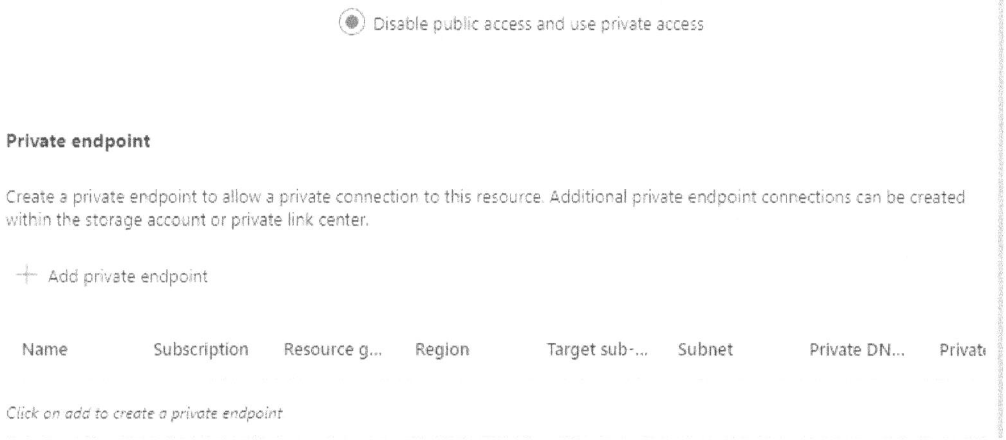

Figure 8.9 – Private endpoint

Once a private endpoint has been added, the endpoint for the storage account is only accessible from the VNet.

Storage authentication

Access to Azure Files must be authenticated; there is no support for anonymous access. The supported authentication methods for Azure Files are **identity-based**, **access key**, and **shared access signature (SAS)**.

Identity-based

This access method provides **Kerberos authentication** using identities from **Active Directory**, **Azure AD Domain Services**, and **Azure AD Kerberos**. These options can be seen in the following screenshot:

Choose the Active Directory source that contains the user accounts that will access a share in this storage account. You can set up identity-based access control for user accounts located in either one of these three domain services.

• Active Directory domain controller you host on a Windows Server (generally referred to as "on-premises AD" even though you might host these servers in Azure)
• Azure Active Directory Domain Services (Azure AD DS), a platform as a service, hosted directory service and domain controller in Azure
• Azure AD Kerberos allows using Kerberos authentication from Azure AD-joined clients. In order to use Azure AD Kerberos, user accounts must be hybrid identities.

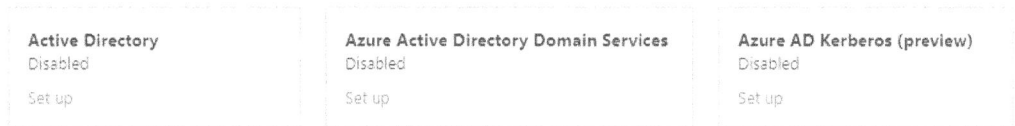

Active Directory	Azure Active Directory Domain Services	Azure AD Kerberos (preview)
Disabled	Disabled	Disabled
Set up	Set up	Set up

Figure 8.10 – Identity-based access

At the time of writing, cloud-only identities – that is, those that solely exist in Azure AD – are not supported. User accounts must be hybrid identities. Azure AD is a directory service only; it provides no direct domain controller functionality.

Access key

This static authentication method uses a key, not a password. It provides the owner full access to the file shares and bypasses all access control. This can be seen in the following screenshot:

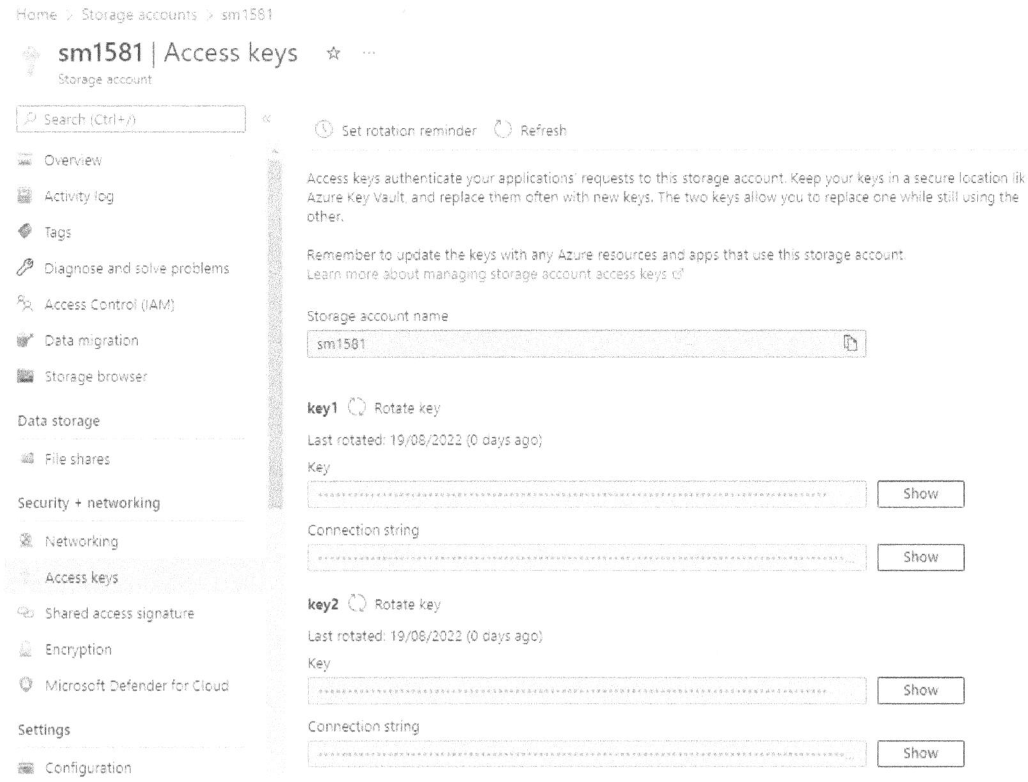

Home > Storage accounts > sm1581

sm1581 | Access keys ☆ ⋯
Storage account

🔍 Search (Ctrl+/)   «

▦ Overview
▤ Activity log
🏷 Tags
🔧 Diagnose and solve problems
🔑 Access Control (IAM)
📁 Data migration
🗂 Storage browser

Data storage

📁 File shares

Security + networking

🛡 Networking
🔑 Access keys
🔑 Shared access signature
🔒 Encryption
🛡 Microsoft Defender for Cloud

Settings

🔧 Configuration

⏱ Set rotation reminder   🔄 Refresh

Access keys authenticate your applications' requests to this storage account. Keep your keys in a secure location lik Azure Key Vault, and replace them often with new keys. The two keys allow you to replace one while still using the other.

Remember to update the keys with any Azure resources and apps that use this storage account.
Learn more about managing storage account access keys ↗

Storage account name

sm1581   📋

key1 🔄 Rotate key

Last rotated: 19/08/2022 (0 days ago)

Key

••   Show

Connection string

••   Show

key2 🔄 Rotate key

Last rotated: 19/08/2022 (0 days ago)

Key

••   Show

Connection string

••   Show

Figure 8.11 – Access keys

This access should be controlled and governed. It shouldn't be used by users, only by service owners and admins.

Shared access signature (SAS)

SAS provides a **Uniform Resource Identifier** (**URI**) that is generated dynamically. It provides access based on the storage key and can restrict access based on IP addresses, allowed protocols, permissions, and a set expiry time. These options can be seen in the following screenshot:

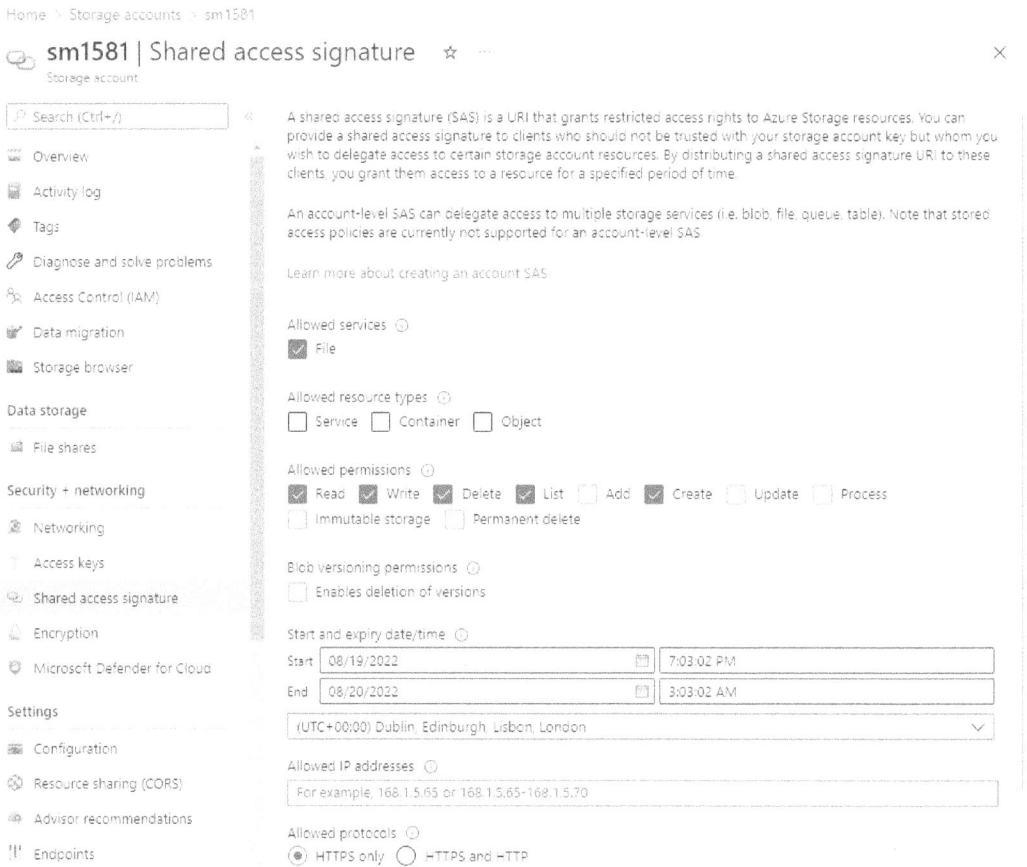

Figure 8.12 – SAS

The SAS method use case is for API access.

File share permissions

Role-based access control (**RBAC**) can manage file share permissions when configuring identity-based authentication. The following are the Azure File Share-specific built-in roles that can be assigned to users:

- **Storage File Data SMB Share Contributor**: This role allows users to read, write, and delete access to file shares

- **Storage File Data SMB Share Elevated Contributor**: This role allows users to read, write, delete, and modify NTFS permissions access for file shares

- **Storage File Data SMB Share Reader**: This role only gives users read access to file shares

These roles can be seen in the following screenshot:

Showing 3 of 349 roles

Name ↑↓	Description ↑↓
Storage File Data SMB Share Contributor	Allows for read, write, and delete access in Azure Storage file shares over SMB
Storage File Data SMB Share Elevated Cont...	Allows for read, write, delete and modify NTFS permission access in Azure Storage file shares over SMB
Storage File Data SMB Share Reader	Allows for read access to Azure File Share over SMB

Figure 8.13 – File share roles

Once users have access to a file share, NTFS permissions control the next level of access to folders and files, the same as any on-premise SMB file share.

In this section, we concluded the topic of implementing Azure Files. In the next section, we will look at implementing Azure File Sync.

Introduction to Azure File Sync

Azure File Sync uses Azure Files to keep file shares centralized in Azure but retains on-premise file servers to maintain existing operations, compatibility, and performance. Azure file shares can also be cached locally to on-premise file servers so that they are closest to the location where they are required. The file share data is still accessible locally via **SMB**, **NFS**, and **File Transfer Protocol** (**FTP**).

Azure File Sync is comprised of the following components:

- **Storage Sync Service**: This is the service resource that runs in Azure
- **Sync group**: A sync group contains endpoints that are to be kept in sync with each other
- **Registered server**: This is a trust relationship object between the Storage Sync Service and the server to sync to Azure
- **Azure File Sync agent**: This is a downloadable package that provides the sync process to Azure
- **Server endpoint**: This represents the folder or volume to sync on Windows Server – that is, the source of the sync
- **Cloud endpoint**: This represents the Azure file share, which is part of the sync group. It is the target for the sync

The use cases and capabilities of Azure File Sync for on-premise file servers are as follows:

- **Multi-site sync**: Multi-master replication is used to sync file shares across multiple sites.

- **Cloud tiering**: A volume-free space policy can be configured for server endpoints to define the amount of free space to preserve and store locally; this is defined as a percentage of the always available free space. You can also define a date policy, which you can configure so that you only locally cache files that have been accessed within a defined set of days. If a file is not accessed through a read or write operation in that period, it is tiered automatically to the Azure file share.

- **Cloud backup**: A daily backup can be scheduled for Azure file shares or a maximum of four on-demand daily backups. Snapshots are used to create backups of Azure file shares.

- **Disaster Recovery**: The entire namespace can be pulled to a server endpoint running the sync agent. This allows the folder structure files to be immediately available on the server.

Azure File Sync can be implemented using the Azure portal, PowerShell, or Windows Admin Center. You can follow these steps to implement Azure File Sync using the Azure portal:

1. Complete the prerequisites for Azure File Sync.

2. Prepare Windows Server so that it can use Azure File Sync.

3. Deploy the Storage Sync Service.

4. Install the Azure File Sync agent.

5. Register Windows Server with the Storage Sync Service.

6. Create a sync group and cloud endpoints.

7. Create a server endpoint.

8. Optional: Configure network settings to secure network access.

We will cover these steps in detail in the *Hands-on exercises* section.

Migrating a DFS-R deployment to Azure File Sync

DFS-N and **DFS-R** are compatible with Azure File Sync and can work together. To use them concurrently, File Sync cloud tiering should be disabled on volumes with DFS-R replicated folders, and server endpoints should not be created as DFSR read-only replication folders.

You can use Azure File Sync to migrate from a DFS-R deployment. The steps are as follows:

1. Create a sync group for the DFS-R topology that's being migrated from.

2. Install Azure File Sync on the server that contains the full DFS-R dataset.

3. Register the server and create a server endpoint on the share.

4. Allow Sync to complete a full upload to the cloud endpoint (Azure file share).

5. Install an Azure File Sync agent on each server that's part of the sync group.

6. Create new file shares on these servers.

7. Create server endpoints on the new file shares. Optionally, set a cloud tiering policy.

8. Allow the Azure File Sync agent to perform a full namespace rapid restore without the data. The local disk for the server endpoint will be filled with data based on the cloud tiering policy after the full namespaces sync has been completed.

9. Ensure the sync is completed and then test the outcome as required.

10. Users and applications can now be directed to the new share.

DFS-R servers in a sync group will require internet connectivity for Azure File Sync; in these cases, you don't want to migrate those servers' shares. Instead, you should use DFS-R and Azure File Sync together.

In this section, we introduced Azure File Sync and learned how to implement it. In the next section, we will complete some hands-on exercises to reinforce some of the concepts that were covered in this chapter.

Hands-on exercises

To support your learning with some practical skills, we will utilize the concepts and understanding we gained from this chapter and put them to practical use.

We will look at the following exercises:

- Exercise – creating a storage account

- Exercise – creating a standard performance Azure file share

- Exercise – creating a premium performance Azure file share

Getting started

To get started with this section, you will need access to the Azure portal with an account with Owner or Contributor access to an Azure subscription.

Let's move on to the exercise for this chapter.

Exercise – creating a storage account

In this exercise, we will create a storage account.

Follow these steps to get started:

1. Log into the Azure portal at `https://portal.azure.com`. Alternatively, you can use the Azure Desktop app: `https://portal.azure.com/App/Download`.

2. In the search bar, type `storage account`; click on **Storage accounts** from the list of **Services** shown:

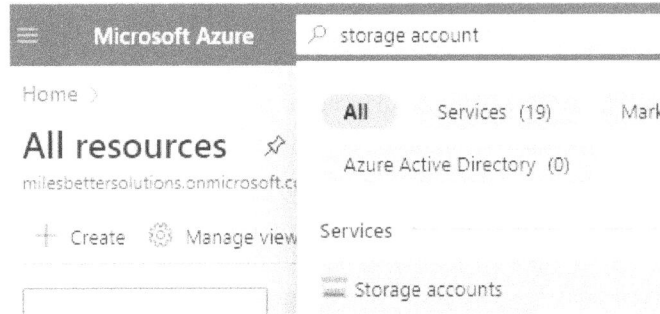

Figure 8.1 – Search storage accounts

3. Click on + **Create** or **Create storage account** from the **Storage account** blade:

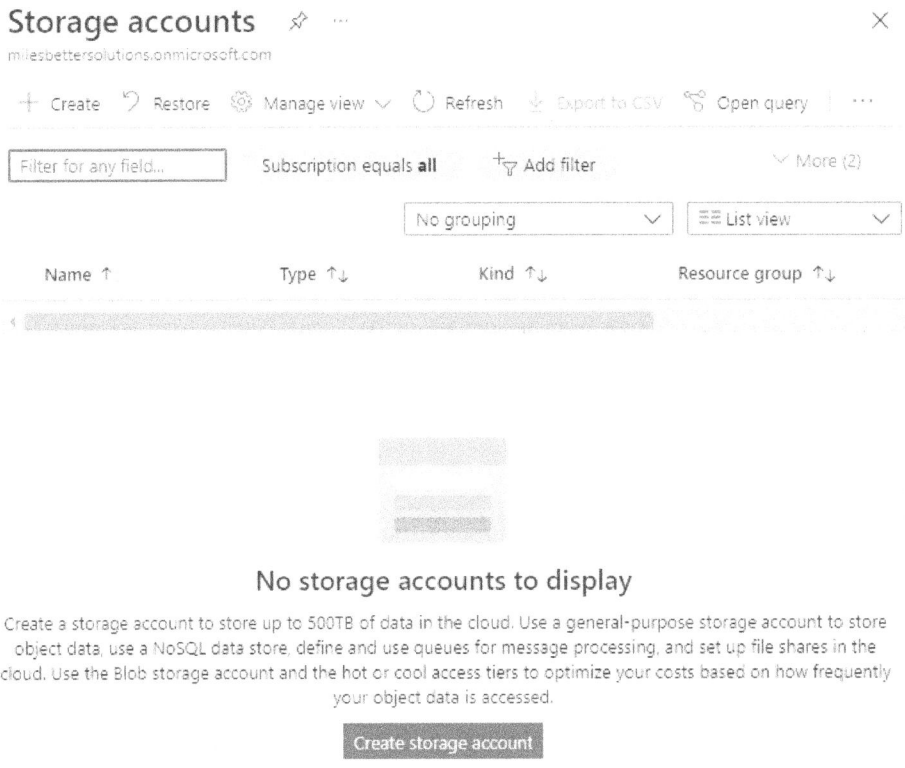

No storage accounts to display

Create a storage account to store up to 500TB of data in the cloud. Use a general-purpose storage account to store object data, use a NoSQL data store, define and use queues for message processing, and set up file shares in the cloud. Use the Blob storage account and the hot or cool access tiers to optimize your costs based on how frequently your object data is accessed.

Create storage account

Figure 8.2 – The Storage accounts screen

4. Set the **Project details** settings as required:

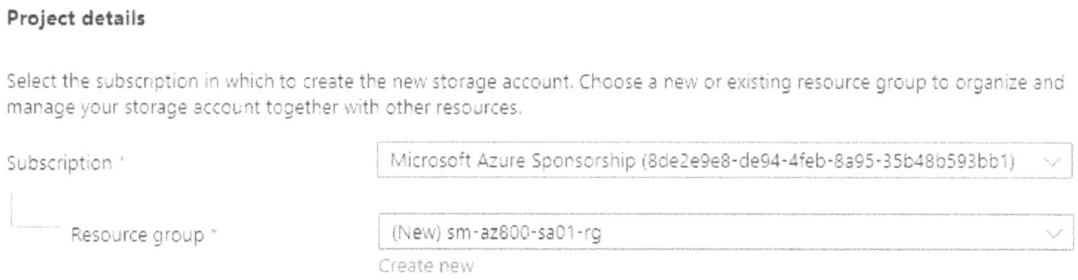

Project details

Select the subscription in which to create the new storage account. Choose a new or existing resource group to organize and manage your storage account together with other resources.

Subscription *   Microsoft Azure Sponsorship (8de2e9e8-de94-4feb-8a95-35b48b593bb1)   ∨

Resource group *   (New) sm-az800-sa01-rg   ∨
                        Create new

Figure 8.3 – Project details

5. Enter a *globally unique* **Storage account name**; *hover your mouse* over the *information* symbol for context-sensitive help choosing a name in the supported format:

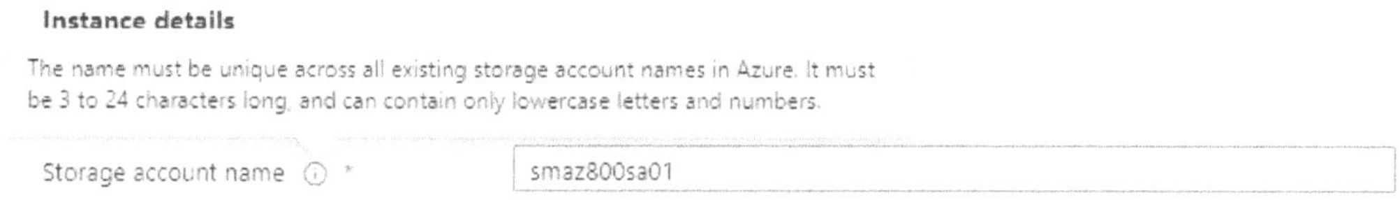

Instance details

The name must be unique across all existing storage account names in Azure. It must be 3 to 24 characters long, and can contain only lowercase letters and numbers.

Storage account name ⓘ *   smaz800sa01

Figure 8.4 – Storage accounts screen

6. Select a **Region**.

7. For **Performance**, select **Standard**:

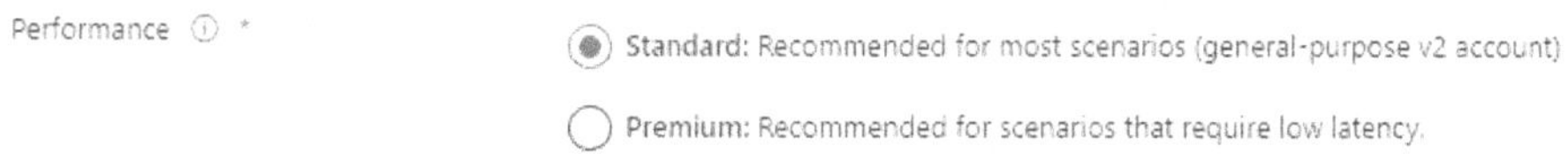

Performance ⓘ *   ⦿ Standard: Recommended for most scenarios (general-purpose v2 account)

                          ◯ Premium: Recommended for scenarios that require low latency.

Figure 8.5 – Selecting a performance type

Note

For a **Premium** storage account, you would select **Premium**, then select the required **Premium account type** for your scenario.

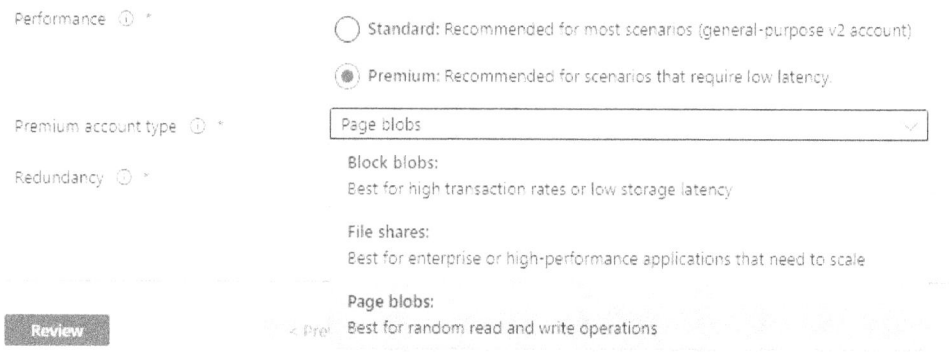

Figure 8.6 – Premium performance type

8. Review the options that are available under **Redundancy** and select **LRS**:

Figure 8.7 – Redundancy options

9. Click **Next: Advanced** > and review the options on the **Advanced** tab.

10. Click **Next: Networking** > and review the options on the **Networking** tab.

11. Click **Next: Data protection** > and review the options on the **Data protection** tab.

12. Click **Next: Encryption** > and review the options on the **Encryption** tab.

13. Click **Next: Tags** > and, optionally, add *tags* on the **Tags** tab.

14. Click **Next: Review** >. On the **Review** tab, review your settings; you may go back to the previous tabs and make any edits if required. Once you have confirmed your settings are as needed, you can click **Create**.

15. You will receive a notification that the resource deployment succeeded:

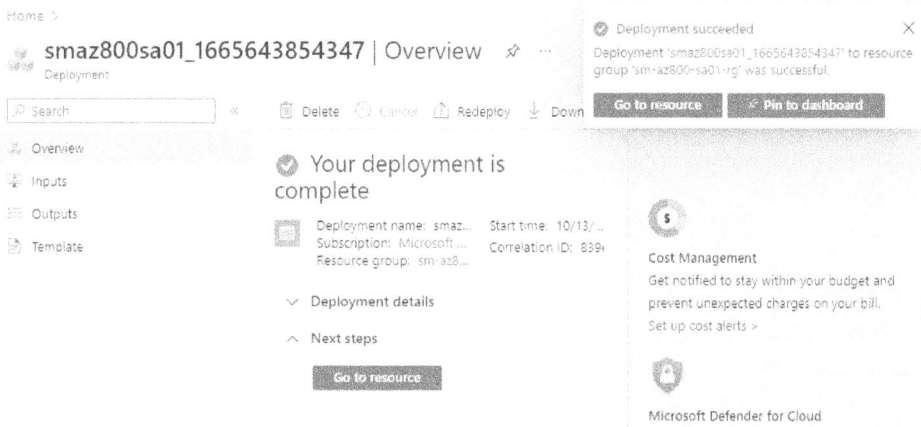

Figure 8.8 – Completed deployment

16. Click on **Go to resource** or navigate to the **Storage account** blade.

17. From the created **Storage account** blade, review the *information* from the **Overview** blade and all the available sections from the *left-hand menu*:

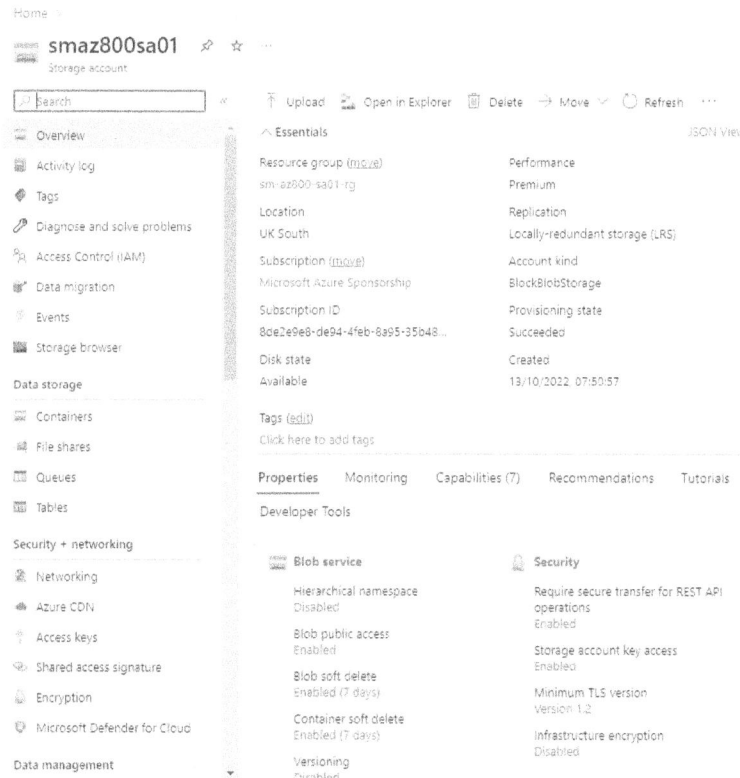

Figure 8.9 – The Storage account blade

With that, we have completed this exercise. This exercise taught us the skills needed to create a storage account; we covered the option of implementing the standard and performance tiers. In the next exercise, we will implement an Azure Files share.

Exercise – creating a standard performance Azure file share

In this exercise, we will implement an Azure file share using a standard storage account.

Follow these steps to get started:

1. Log into the Azure portal at `https://portal.azure.com`. Alternatively, you can use the Azure Desktop app: `https://portal.azure.com/App/Download`.

2. In the search bar, type `storage account`; click on **Storage accounts** from the list of **Services** shown.

3. From the **Storage account** blade, create a new storage account for this exercise to create a file share, or use the existing storage account from the previous exercise. The storage account's **Performance** type must be **Standard** for this exercise.

4. Click on **File shares** from your standard storage account in the **Data storage** section:

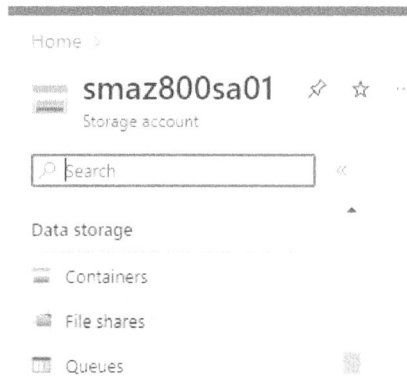

Figure 8.10 – The Storage account blade

5. Click on + **File share** via the top navigation menu:

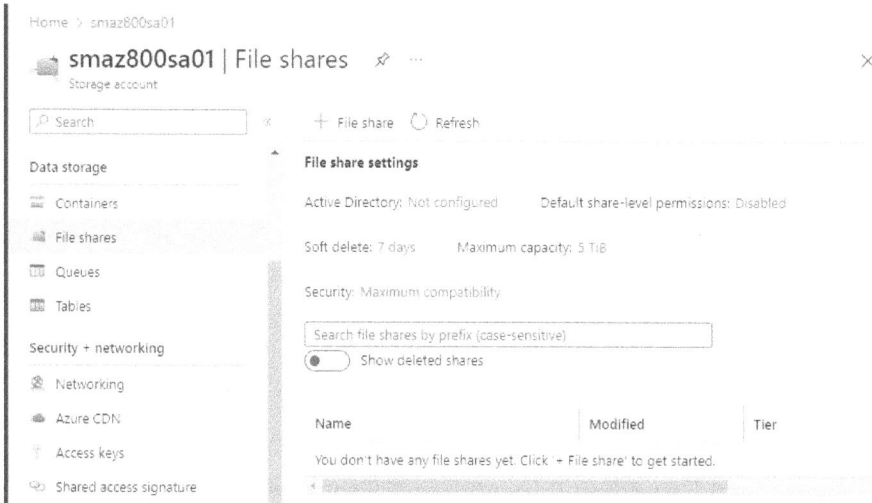

Figure 8.10 – The File shares blade

6. From the **New file share** blade, enter a **Name** for the file share.

7. For the tier, *hover your mouse* over the *information* symbol for context-sensitive help in choosing the tier. Leave **Tier** set to **Transaction optimized** and review the information presented for that tier:

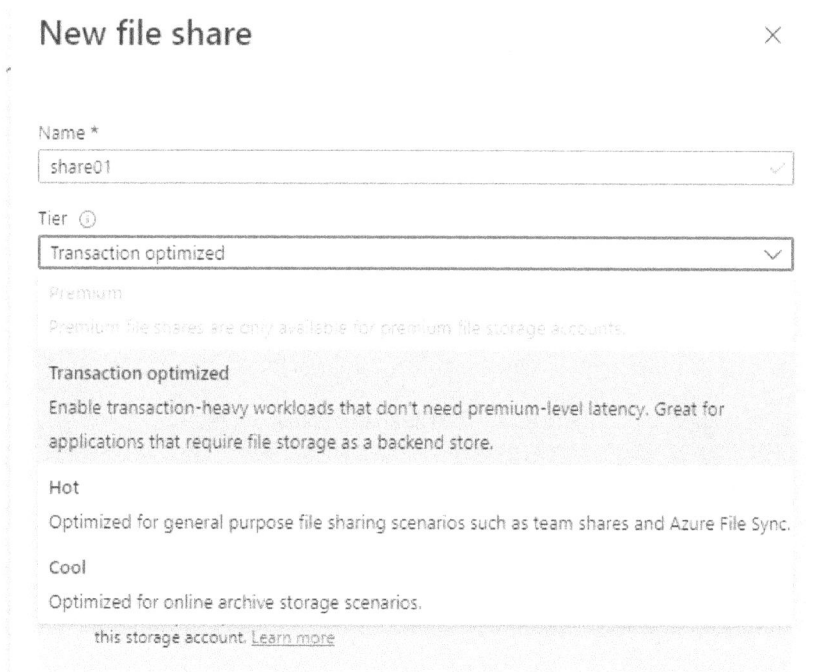

Figure 8.11 – The New file share blade

8. Click **Create**.

9. You will receive a message that the file share was successfully created.

10. The file share can now be seen in the **File shares** blade:

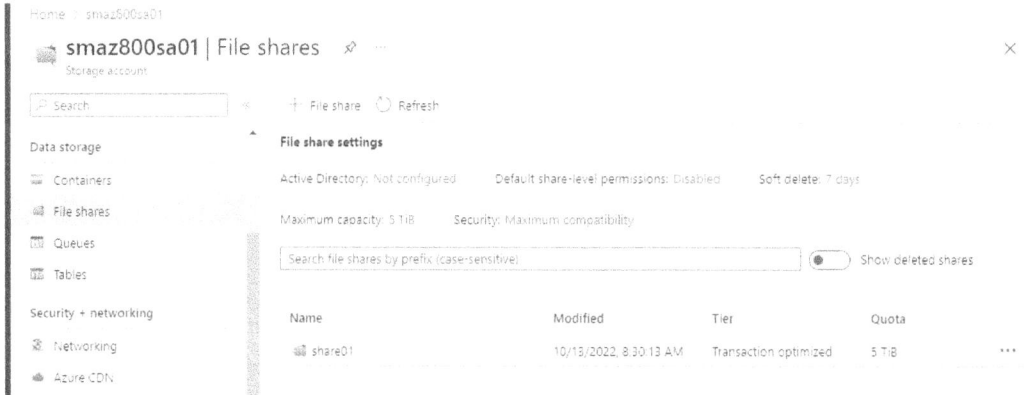

Figure 8.12 – The Files shares blade

11. Click on the *created file share* and review the information and settings on the page that opens:

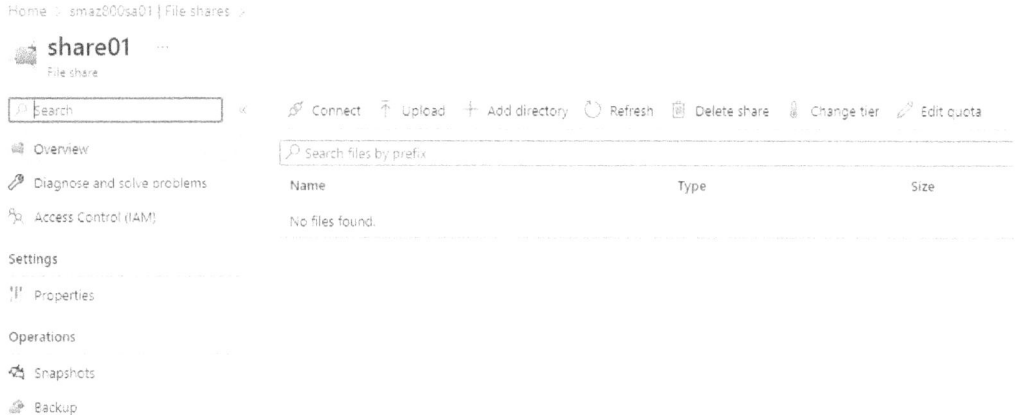

Figure 8.13 – The File shares blade

12. Click on **Connect** to view the connection information:

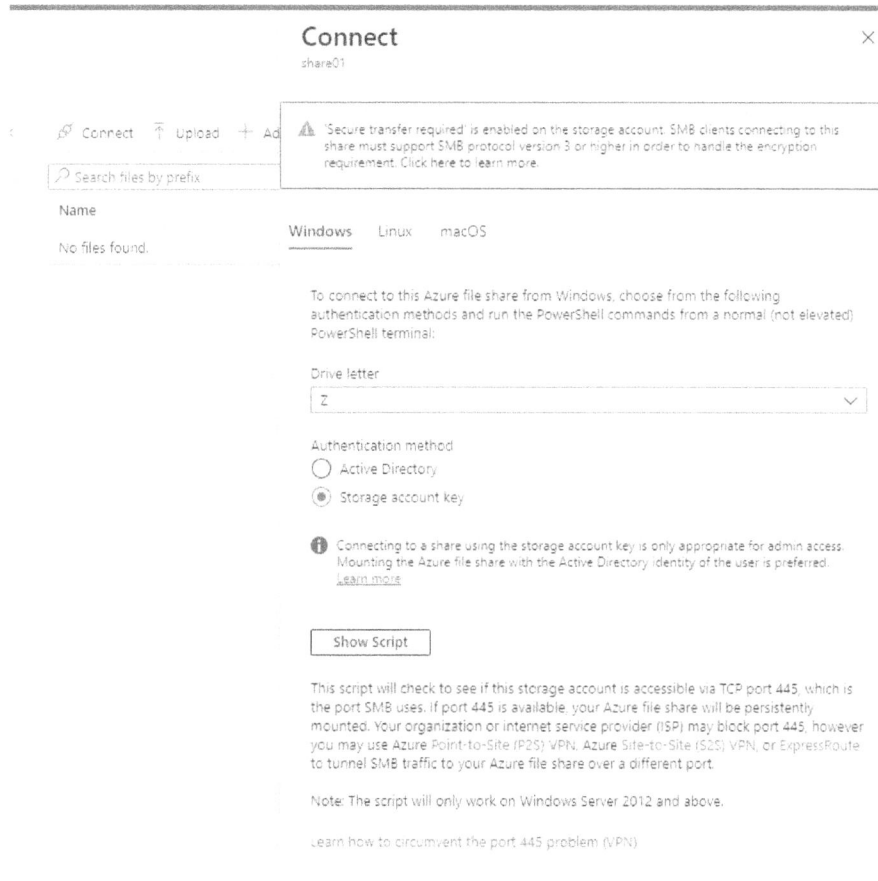

Figure 8.14 – The Connect blade

13. Click on **Access Control (IAM)** to view the information regarding **identity-based authentication**; click the **Learn more** hyperlink for further information:

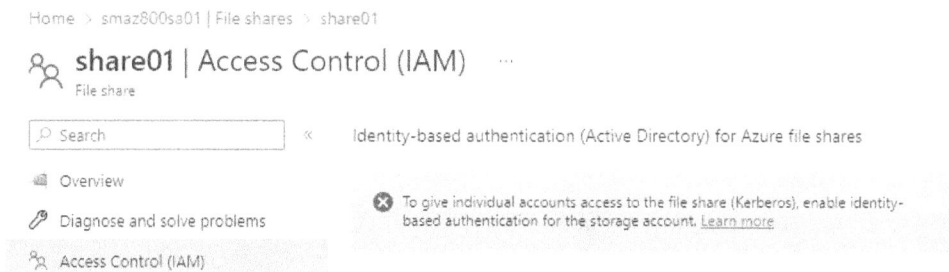

Figure 8.15 – The Access Control (IAM) blade

With that, we have completed this exercise. This exercise taught us the skills to create an Azure file share using standard storage. In the next exercise, we will implement an Azure File share using a premium storage account.

Exercise – creating a premium performance Azure file share

In this exercise, we will implement an Azure file share using a premium storage account.

Follow these steps to get started:

1. Log into the Azure portal at `https://portal.azure.com`. Alternatively, you can use the Azure Desktop app: `https://portal.azure.com/App/Download`.

2. In the search bar, type `storage account`; click on **Storage accounts** from the list of **Services** shown.

3. From the **Storage account** blade, create a new storage account for this exercise. The storage account's **Performance** type must be **Premium**, and the **Premium account** type must be **File Shares**, for this exercise.

4. Click on **File shares** from your **Premium** storage account under the **Data storage** section:

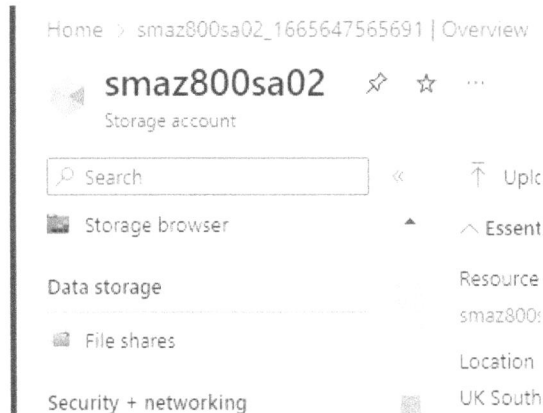

Figure 8.16 – The Storage account screen

5. Click on **+ File share** via the top navigation from the **File shares** blade to create a new file share.

6. Enter a **Name** for the file share.

7. Enter a **Provisioned capacity**.

8. Review the **Performance** information presented.

9. Select the **SMB** or **NFS** protocol as required:

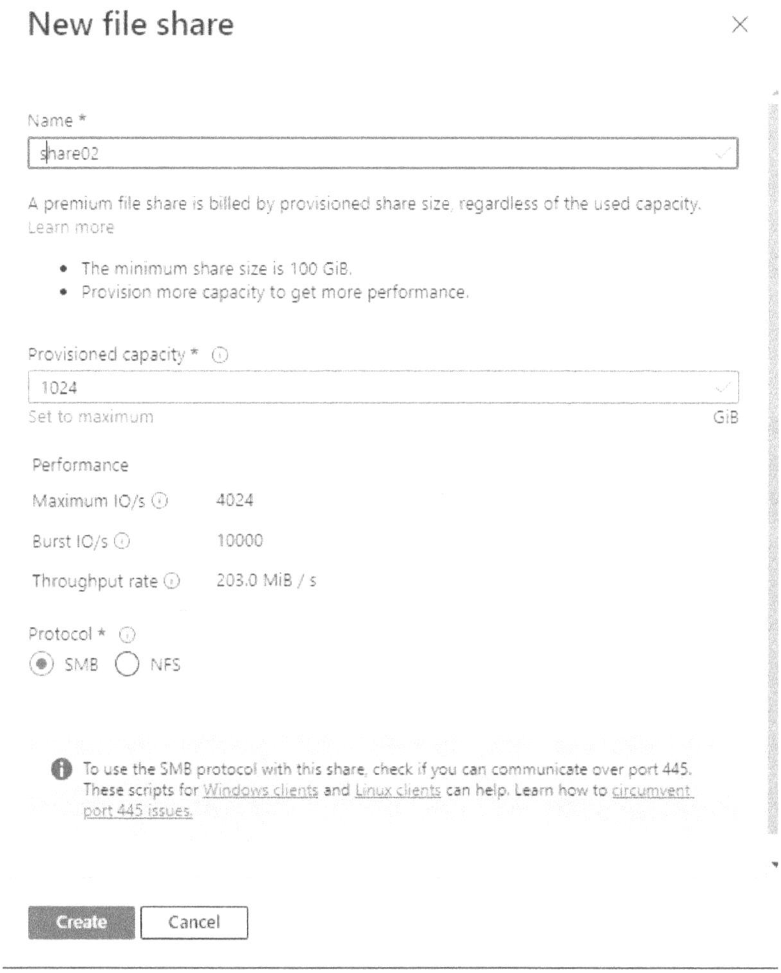

New file share ✕

Name *

share02

A premium file share is billed by provisioned share size, regardless of the used capacity.
Learn more

- The minimum share size is 100 GiB.
- Provision more capacity to get more performance.

Provisioned capacity * ⓘ

1024

Set to maximum   GiB

Performance

Maximum IO/s ⓘ   4024

Burst IO/s ⓘ   10000

Throughput rate ⓘ   203.0 MiB / s

Protocol * ⓘ
◉ SMB   ◯ NFS

ⓘ To use the SMB protocol with this share, check if you can communicate over port 445.
These scripts for Windows clients and Linux clients can help. Learn how to circumvent
port 445 issues.

Create   Cancel

Figure 8.17 – The New file share blade

10. Review the *communication over port 445* information.

11. Click **Create**.

12. You will receive a message that the file share was successfully created.

13. The file share can now be seen in the **File shares** blade:

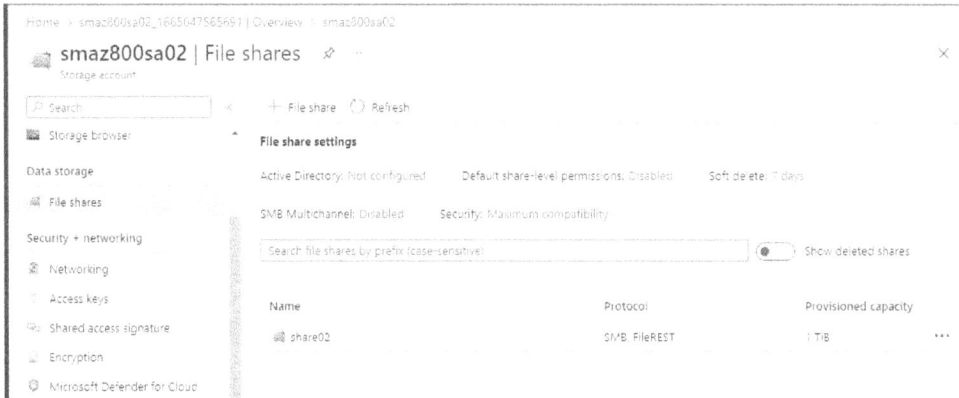

Figure 8.18 – The Files share blade

14. Click on the *created file share* and review the information and settings on the page that opens.

15. Click **Connect** to view the connection information and the port 445 communication problem:

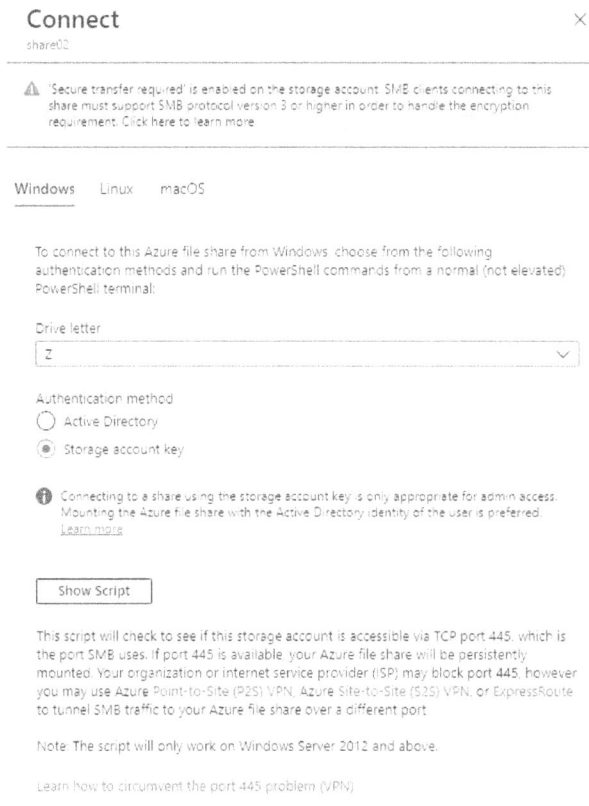

Figure 8.19 – The Connect blade

16. Click on **Access Control (IAM)** to view the information regarding **identity-based authentication**; click the **Learn more** hyperlink for further information:

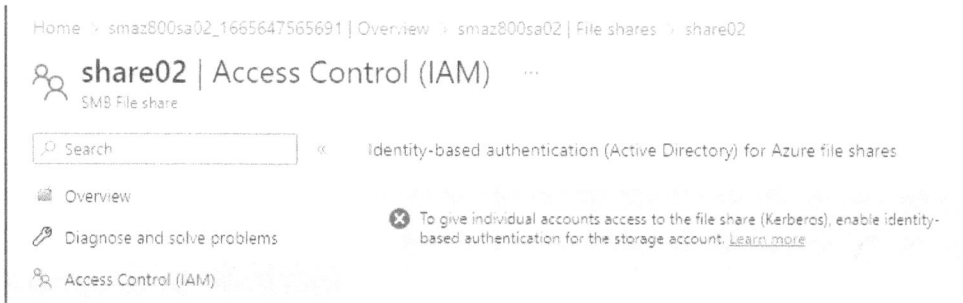

Figure 8.20 – The Access Control (IAM) blade

With that, we have completed this exercise. This exercise taught us the skills to implement an Azure file share using a premium storage account. Now, let's summarize this chapter.

Summary

We started this chapter by introducing the storage services that are available in Azure, their use cases, and where we may select each one for a real-world scenario. Then, we looked at implementing the Azure Files service and Azure File Sync, and how to migrate the service from DFS-R. We finished this chapter with some hands-on exercises to provide you with valuable practical skills.

We ensured that we provided coverage of *AZ-800 Administering Windows Server Hybrid Core Infrastructure: Manage storage and file services*.

This chapter aimed to take your knowledge beyond the exam objectives; we added new skills and learning with the content provided. This will help you develop your knowledge and skills for on-premise network infrastructure services and allow you to be prepared for a real-world, day-to-day hybrid environment-focused role.

In the next chapter, you will learn about implementing Hyper-V on Windows Server.

Further reading

This section provides links to additional study references and additional exam information:

- *Microsoft Certified: Windows Server Hybrid Administrator Associate*: https://docs.microsoft.com/en-us/learn/certifications/windows-server-hybrid-administrator/

- *Exam AZ-800: Administering Windows Server Hybrid Core Infrastructure*: `https://docs.microsoft.com/en-us/learn/certifications/exams/az-800`

- *Exam AZ-800: Study Guide*: `https://query.prod.cms.rt.microsoft.com/cms/api/am/binary/RWKI0r`

- *Microsoft Learn*: `https://docs.microsoft.com/en-us/learn/modules/implement-hybrid-file-server-infrastructure/`

Skills check

Check what you have learned in this chapter by answering the following questions:

1. What four storage types are available in Azure?
2. What is a storage account?
3. What storage account types can be created?
4. Can you change the performance type of a storage account?
5. What is Azure Files?
6. What are some of the use cases for Azure Files?
7. What are some of the considerations when implementing Azure Files?
8. What are the storage tiers?
9. What storage redundancy options are there for Azure Files?
10. What encryption is available for Azure Files?
11. How is storage protected?
12. What storage networking/access options are available?
13. What storage authentication options are available?
14. What are the file share permissions that can be assigned to users?
15. What is Azure File Sync, and what are its use cases?

Part 4:
Hybrid Compute

This part will provide complete coverage of the knowledge and skills required for the skills measured in the *Manage Windows Servers and workloads in a hybrid environment* and *Manage virtual machines and containers* sections of the exam.

This part of the book comprises the following chapters:

9

Implementing and Managing Hyper-V on Windows Server

The previous chapter covered the Storage services available in Azure.

This chapter covers content from *AZ-800 Administering Windows Server Hybrid Core Infrastructure: Manage Hyper-V and Guest Virtual Machines.*

First, we will introduce the concept of virtualization and cover implementing, managing, and securing Microsoft's Hypervisor, Hyper-V. We will conclude with some hands-on exercises to help you develop your skills further.

The following topics are included in this chapter:

- Introduction to virtualization and Hyper-V
- Implementing and managing Hyper-V
- Creating and managing Hyper-V virtual machines
- Securing Hyper-V workloads with Guarded Fabric
- Hands-on exercises

In addition to the topics listed in this chapter, this chapter's goal is to take your knowledge beyond the exam objectives to prepare you for a real-world, day-to-day hybrid environment-focused role.

Introduction to virtualization and Hyper-V

In this section, we will introduce the concept of **virtualization**, specifically in the context of compute – that is, CPU and memory – as opposed to storage or network virtualization.

Computers (*machines*), whether virtual or physical, are the core building blocks of on-premise solutions and Azure IaaS solutions.

The term we use for these virtualized physical machines is **virtual machine** (**VM**). VMs are the virtualization of a computer's physical memory and processor resources; physical computers are emulated in software. This allows a physical machine's resources to be subdivided and have a portion allocated to each virtual machine. Each VM runs its own isolated OS, the same as a physical server. The OS accesses virtual compute resources such as **virtual processors** (**virtual CPU** or **vCPU**) and virtual memory.

The next section will look at the relationship between physical computers and VMs.

Comparing virtual and physical machines

To understand the virtualization model over the physical model, we will briefly explore the physical hardware model of operation.

Historically, an organization operated on a model where, typically, the application and dedicated instance of an OS were mapped one-to-one to physical hardware. This was not very efficient in terms of costs or operations and also not very scalable since each app or new workload requested by the business required a new physical server. This led to scale-out server sprawl with increased rack space and associated costs.

This traditional physical approach model is outlined in the following diagram:

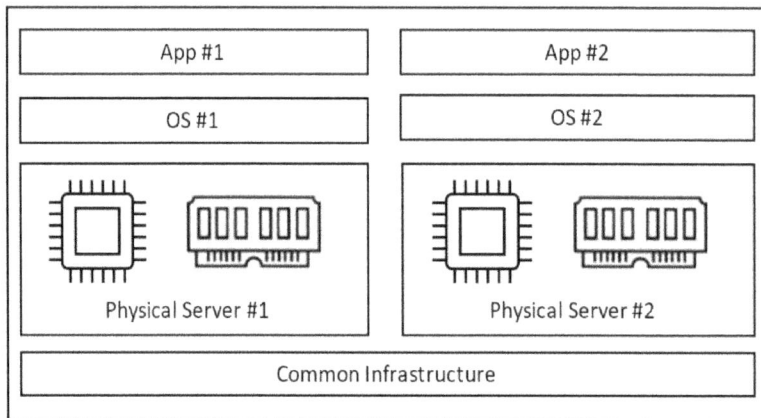

Figure 9.1 – Physical server app deployment model

This traditional app delivery approach with a one-to-one relationship can benefit from the virtualization model. Virtualization addresses the following key challenges of the traditional physical hardware model:

- Utilization
- Scale
- Cost

The concept of virtualization is based on *hardware abstraction*, and for compute virtualization, the *hypervisor* is the technology layer that enables this hardware abstraction.

Abstraction involves removing dependencies and ignoring or filtering out some aspects that are no longer relevant so that what is important can remain the focus. In this hardware abstraction context, the VM is no longer dependent on the hardware layer it runs on when we abstract it (that is, when we remove the hardware from the equation); we no longer need that layer or need to concern ourselves with it.

This means that the value of virtualization is that we can do more with less by utilizing resources at more efficient costs and economies of scale.

We can achieve these efficiencies and optimizations by running multiple VMs from one physical server. This also means less rack space, fewer physical assets to cool, power, and maintain, and possibly less human capital. Technology can be seen as a more tangible value creation center, not a cost center.

The nutshell version is that when we look at the preceding diagram (*Figure 9.1*), we can deliver more apps with fewer physical servers through virtualization, which translates to a lower cost to deliver and operate those apps for the business.

Each VM shares the hardware resources from the underlying physical server (*referred to as the host*) that it is located on. These can be referred to as host resources. The subdivision of the host resources means we can create many virtual processors and memory resources to be assigned to the host VM. The following diagram illustrates this virtualization approach:

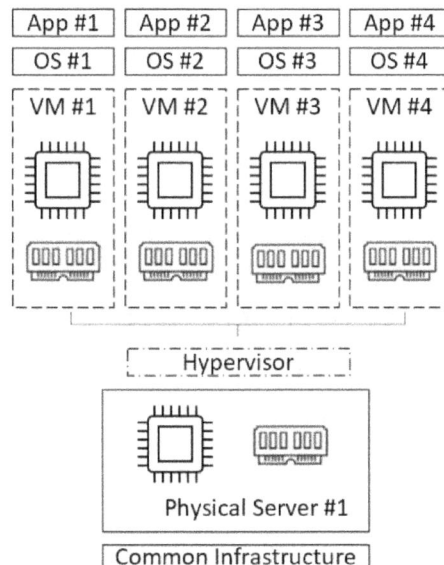

Figure 9.2 – Virtualization app deployment model

The diagram shown in *Figure 9.1* illustrates that we deployed two physical servers to run two applications with virtualization. In the example shown in *Figure 9.2*, we can deploy one physical server to run four applications; this is the benefit and value of virtualization. The underlying physical server host resources will determine how many VMs can be run… *your mileage may vary*.

In this section, we compared physical computers and VMs and learned the value and benefits of virtualization. In the next section, we will introduce Hyper-V.

What is Hyper-V?

Now that we have understood virtualization's value and benefits, we will introduce Hyper-V as the enabler.

Hyper-V is Microsoft's hypervisor, a piece of *virtualization software* and the technology layer that enables this hardware abstraction.

Example use cases for Hyper-V are as follows:

- Physical computer infrastructure consolidation
- **Remote Desktop Virtualization/Virtual Desktop Infrastructure** (VDI) implementations
- Test and development environments
- Private and hybrid cloud infrastructure environments, including **Azure Stack Hub** and **Azure Stack Hyper-Converged Infrastructure** (HCI) implementations

The following diagram outlines the Hyper-V system architecture:

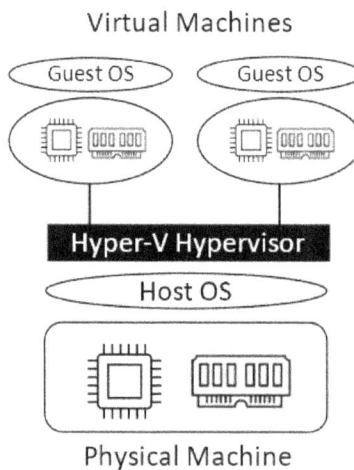

Figure 9.3 – Hyper-V system architecture

The preceding diagram shows that interaction with the physical machine layer is passed through the hypervisor software. The *physical machine OS* is known as the **host OS**, and the *virtual machine OS* is known as the **guest OS**.

In this section, we introduced Hyper-V as the virtualization software layer that allows us to create and run VMs. In the next section, we will learn how to implement and manage Hyper-V.

Implementing and managing Hyper-V

Hyper-V can be installed as a *server role* in Windows Server using Server Manager, PowerShell, or **Windows Admin Center (WAC)**.

The PowerShell `Install-WindowsFeature` cmdlet can be used to install the role locally or on remote hosts. The PowerShell `Get-WindowsFeature` cmdlet can then be run to verify the installation of Hyper-V. We will look at the installation steps in the *Hands-on exercises* section.

When installing the Hyper-V role, it should be the only server role that's installed on the host server; you should *not install* roles such as *Domain Services* or a *File Server* role.

The *Server Core* installation of Hyper-V has the following benefits over the full GUI OS installation option:

- Minimized hardware resources and better utilization for OS and guest VMs
- Minimized OS resources running and better utilization for OS and guest VMs
- Fewer software updates
- Smaller attack/vulnerable surface area; fewer services are running
- Streamlined deployment and remote management

The system hardware requirements for implementing Hyper-V are as follows:

- 64-bit processor
- **Second-Level Address Translation (SLAT)**
- VM Monitor Mode extensions
- **Intel Virtualization Technology (Intel VT)** or **AMD Virtualization (AMD-V)**
- Hardware-enforced **Data Execution Prevention (DEP)**
- Intel XD bit and AMD NX bit

The Hyper-V hosts should have sufficient processors and memory resources to support the number of VMs required to run on each host. Sufficient storage that is fast enough, network speed, and capacity should also be planned.

The OS that runs on a VM is referred to as a guest OS. Hyper-V supports the following guest OS types:

- All supported versions of Windows
- CentOS, Red Hat Enterprise Linux, Debian, Oracle Linux, SUSE, Ubuntu, and FreeBSD

Where possible, *generation 2* VMs should be used for the guests; this provides capabilities such as *secure boot*, *shielded VMs*, and *hot add/remove virtual network adapters*.

Hyper-V networking

For communication on the network, the two primary components are as follows:

- **Virtual network adapter**
- **Virtual network switch**

The VM is configured with a virtual network adapter that connects to a port on the virtual network switch.

Two virtual network adapter types are supported:

- **Legacy network adapter**: These are only in generation 1 VMs and can be used for installation tasks using a network boot.
- **Network adapter**: Also called a synthetic adapter, it can be used with generation 1 and 2 VMs. It does not support the network boot for generation 1 VMs.

Three types of virtual network switches are supported:

- **Private**: This provides communication between VMs on the same Hyper-V host only; there is no communication between VMs and the Hyper-V host.
- **Internal**: This provides communication between VMs on the same Hyper-V host and between VMs and the Hyper-V host.
- **External**: This provides communication outside the host server, such as to the internet or another network.

A **Virtual LAN (VLAN)** can logically separate networks and partition traffic.

Virtual switches can be created and managed using Hyper-V Manager, Windows Admin Center, or PowerShell. The PowerShell cmdlet for this is `New-VMSwitch`.

Nested virtualization

Nested virtualization allows an instance of Hyper-V to be installed on a VM, which then runs VMs created within this *second layer* of virtualization; the host VM runs as a virtualized Hyper-V host. The following diagram visualizes this concept:

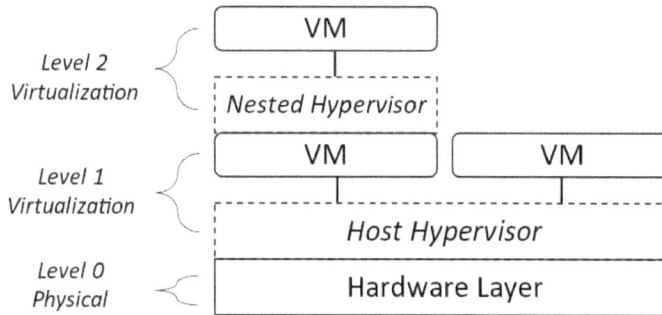

Figure 9.4 – Nested virtualization

The preceding diagram shows that *Level 0* is a physical layer that contains the *virtualization host(s) hardware*, onto which a hypervisor is installed; this host-level hypervisor creates the *Level 1 virtualization* platform.

The virtualized layer running inside a VM as opposed to running on hardware is what we term *Level 2 virtualization*; this means it is running a *nested virtualization layer*.

The following are the requirements for nested virtualization:

- Server 2016 or later

- Enough memory resources

- Version 8.0 or later for guest VMs

- An Intel processor with **Virtual Machine Extensions** (VT-x) and **Extended Page Tables** (EPTs) for the physical host server

- MAC address spoofing enabled

To enable nested virtualization, use the following PowerShell cmdlet from the physical host server:

```
Set-VMProcessor -VMName <VMName>
-ExposeVirtualizationExtensions $true
```

The guest VM must be in the Off state.

Installing the Hyper-V role on the guest VM is the same process as installing the physical host server of the Hyper-V role.

Managing Hyper-V

Hyper-V Manager is a component that provides a GUI for managing *local* and *remote* Hyper-V hosts. Hyper-V Manager is shown in the following diagram:

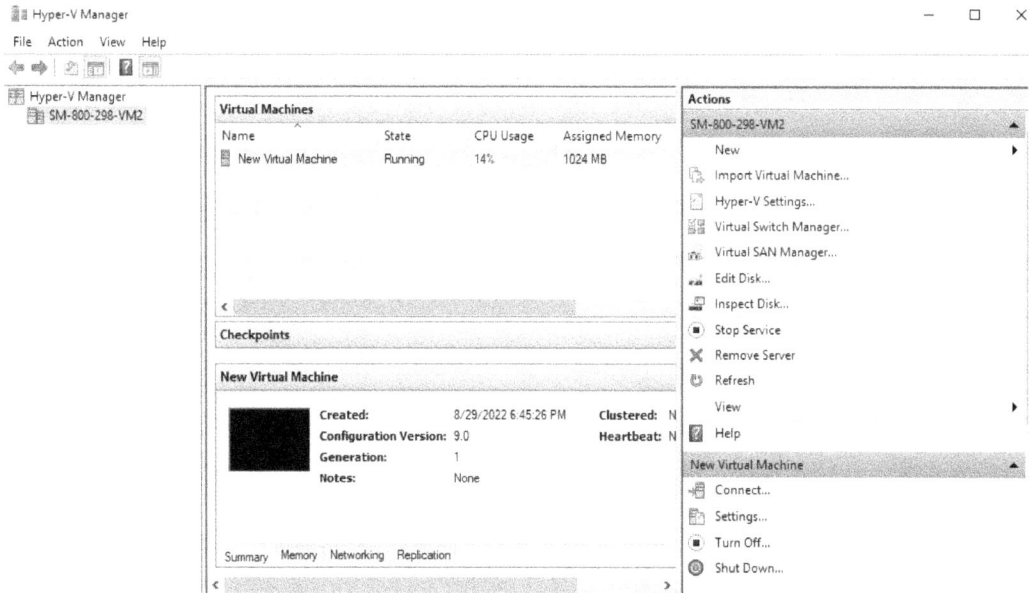

Figure 9.5 – Hyper-V Manager

The following functions are available in Hyper-V Manager:

- Connect to and manage local and remote Hyper-V host servers

- Configure Hyper-V settings

- Manage Hyper-V Integration Services

- Manage Hyper-V guest VMs

- Manage virtual hard disks

- Manage virtual switches – Virtual Switch Manager

- Manage virtual SANs – Virtual SAN Manager

- Manage Hyper-V replication

- Manage checkpoints

- Support for the WS-Management protocol

- Support for previous versions of Hyper-V hosts

In addition to Hyper-V Manager, PowerShell and WAC can also be used to manage Hyper-V hosts.

In this section, we learned how to create and manage Hyper-V host computers. In the next section, we will learn how to create and manage the VMs that will run on the Hyper-V hosts.

Creating and managing Hyper-V virtual machines

We can create and manage Hyper-V using the **New Virtual Machine Wizard** area in the **Hyper-V Manager** console or with PowerShell using the New-VM cmdlet.

The process for implementing a VM consists of the following steps:

1. Create a VM through Hyper-V Manager or PowerShell.
2. Install the guest OS on the VM using Hyper-V Manager.
3. Install or upgrade Integration Services using Hyper-V Manager.

We will learn how to create and manage VMs in more detail in the *Hands-on exercises* section of this chapter.

VM generation versions

The VM generation determines what hardware features and capabilities are available to the VM, such as the *OS type supported*, *size of boot volume*, *boot methods*, and so on.

Hyper-V supports **generation 1** and **generation 2** VMs.

Generation 1 VMs support the following features:

- 32-bit and 64-bit guest OS
- Maximum of 2 TB boot volume with four partitions
- Legacy BIOS
- VHD and VHDX format disks

Generation 2 VMs support the following features:

- 64-bit only guest OS
- Maximum of 64 TB boot volume
- UEFI
- Secure boot
- Shielded VMs
- VHDX-only formatted disks

A VM generation is selected at the time of VM creation and *cannot* be changed after. The following screenshot shows the VM creation wizard screen:

Choose the generation of this virtual machine.

○ Generation 1

This virtual machine generation supports 32-bit and 64-bit guest operating systems and provides virtual hardware which has been available in all previous versions of Hyper-V.

◉ Generation 2

This virtual machine generation provides support for newer virtualization features, has UEFI-based firmware, and requires a supported 64-bit guest operating system.

⚠ Once a virtual machine has been created, you cannot change its generation.

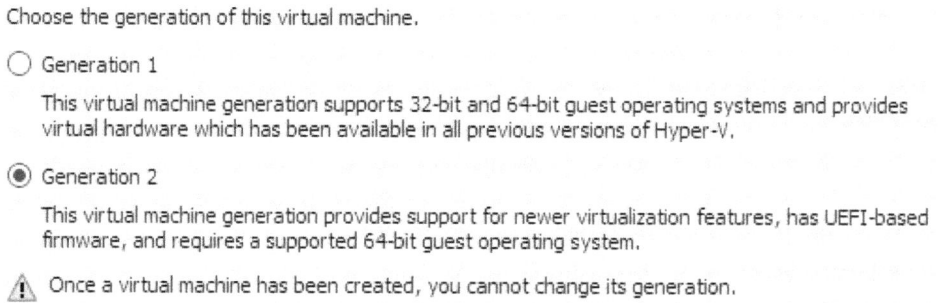

Figure 9.6 – VM generations

There is a negligible difference in performance between *generation 1* and *generation 2* VMs; this should not be the main deciding factor. The main difference is 64-bit OS only in generation 2 and UEFI only.

VM configuration versions

A VM configuration version provides information about the *compatibility* of VM components. The Hyper-V version determines the VM configuration the VM will receive at the time it is created.

For each VM, its configuration version identifies the following information:

- **Configuration**: This provides information about the processor, memory, and storage attached to the VM

- **Checkpoint**: These are files that represent runtime state and configuration files, which are used to create checkpoints

The configuration version of a VM can be found in the **Summary** tab. Here, you can choose the **Configuration Version** value from **Hyper-V Manager**.

The following PowerShell cmdlet can also get the VM version:

```
Get-VM *  |  Format-Table Name, Version.
```

The VM configuration versions that a Hyper-V host supports can be found using the following PowerShell cmdlet:

```
Get-VMHostSupportedVersion
```

The default version that will be used for all VMs that are created on a Hyper-V host can be found using the following PowerShell cmdlet:

```
Get-VMHostSupportedVersion -Default
```

The VM configuration version can be updated with the following PowerShell cmdlet:

```
Update-VMVersion \<vmname\>
```

To update the version, the VM must be *turned off*. It is *not possible* to *downgrade* a VM configuration version once it has been *upgraded*.

The following VM configuration versions are supported for each Hyper-V host Windows version:

*   **Windows Server 2012 R2**, **2016**, **2019**, and **2022**
*   **Windows 10 20H2**, **Enterprise 2016 LTSC**, and **2019 LTSC**
*   **Windows 11**

For additional reference, the following are some examples of the minimum configuration versions for compatibility with certain Hyper-V requirements:

*   **Hibernation support**: Version 9.0
*   **Hot add/remove memory**: Version 6.2
*   **Large memory VMs**: Version 8.0
*   **Nested virtualization**: Version 8.0
*   **PowerShell Direct**: Version 6.2
*   **Secure Boot for Linux VMs**: Version 6.2
*   **Virtual processor count**: Version 8.0
*   **Virtual Trusted Platform Module (vTPM)**: Version 7.0

The configuration version is important to know since VMs created on earlier Hyper-V hosts may not work on newer Hyper-V hosts with a newer OS version. You may need to move a VM between Hyper-V hosts, and you will want to know that the hosts support the VM's configuration. A host that doesn't support the VM configuration version of the VM will *not allow* that VM to start.

VM settings

The VM's configuration and properties settings are grouped into **Hardware** and **Management**; they are represented for each generation type, as shown in the following figure:

Generation 1
Settings

☆ **Hardware**
   📱 Add Hardware
   📟 Firmware
      Boot from File
   🎛 Memory
      512 MB
  ⊞ 📱 Processor
      2 Virtual processors
  ⊟ 📟 SCSI Controller
   ⊞ 💾 Hard Drive
      DirectAccess.vhdx
    💿 DVD Drive
      en_windows_server_2012_r2_...
  ⊟ 🎤 Network Adapter
      hyperv_usingSONIO
     Hardware Acceleration
     Advanced Features
  ⊞ 🎤 Network Adapter
      hyperv_usingSONIO
☆ **Management**
   ℹ Name
      DirectAccess
   📋 Integration Services
      Some services offered
   📁 Checkpoint File Location
      C:\VM_Machines\Virtual Machines\...
   📁 Smart Paging File Location
      C:\VM_Machines\Virtual Machines\...
   ▶ Automatic Start Action
      Always start
   ⏹ Automatic Stop Action
      Save

Generation 2
Settings

☆ **Hardware**
   📱 Add Hardware
   📟 Firmware
      Boot from Network Adapter
   🎛 Memory
      3072 MB
  ⊞ 📱 Processor
      4 Virtual processors
  ⊟ 📟 SCSI Controller
   ⊞ 💾 Hard Drive
      spccvbox1.vhdx
  ⊞ 🎤 Network Adapter
      Team_1
☆ **Management**
   ℹ Name
      SPCCVBOX1
   📋 Integration Services
      Some services offered
   📁 Checkpoint File Location
      C:\Virtual Machines\SPCCVBOX1
   📁 Smart Paging File Location
      C:\Virtual Machines\SPCCVBOX1
   ▶ Automatic Start Action
      None
   ⏹ Automatic Stop Action
      Shut Down

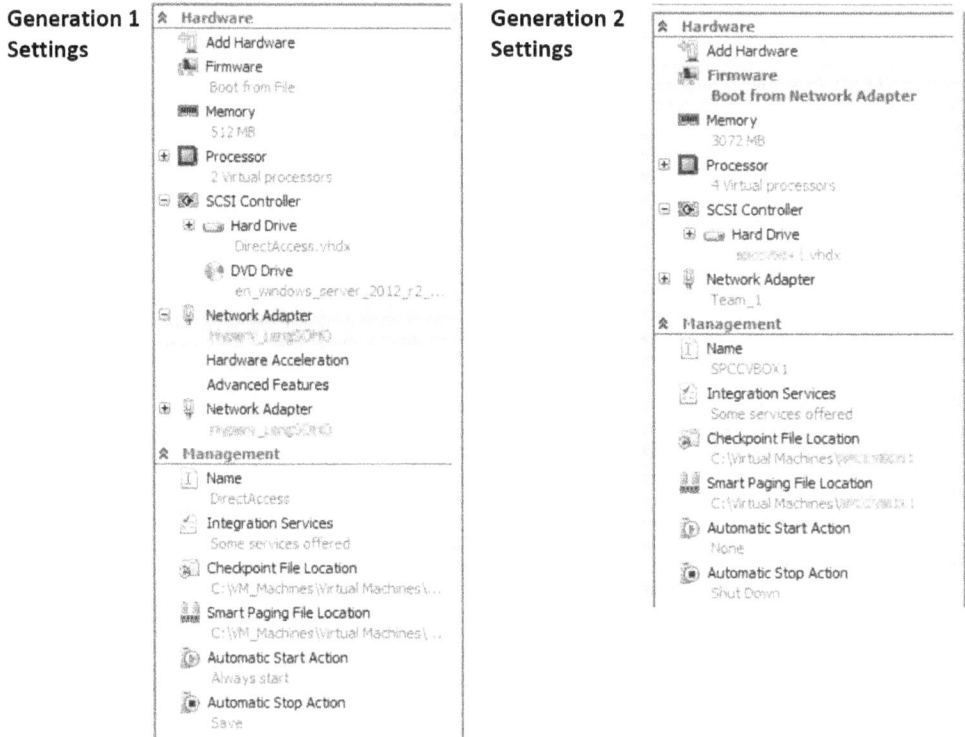

Figure 9.7 – VM generation settings

The available simulated hardware components that mediate access to the real hardware components depend on the generation version of the VM.

Generation 1 VMs have the following hardware that can be accessed:

- BIOS

- Memory

- Processor

- **Non-Uniform Memory Access (NUMA)**

- IDE controller

- SCSI controller

- Network

- COM port

- Diskette drive

Generation 2 VMs have the following hardware that can be accessed:

- Firmware
- Memory
- Processor
- SCSI controller
- Network adapter
- Fibre channel adapter

The management settings provide configuration information on the characteristics of a VM; these are as follows:

- **Name**: Used for the display name of the VM; does not change the VM's hostname
- **Integration Services**: Used to configure which integration settings are enabled for the VM
- **Checkpoints**: Used to *enable*, *disable*, specify the *type* of checkpoints, and specify where they will be stored:
  - A VM can be in one of the following states: **Off**, **Starting**, **Running**, **Paused**, or **Saved**
  - There are two types of checkpoints that are point-in-time snapshots of a VM: **production** and **standard** checkpoints
  - There are a maximum of *50 checkpoints* for a VM
- **Smart Paging File Location**: Used to configure the location where Smart Paging files are stored
- **Automatic Start Action**: Used to define how a VM responds when you turn on a Hyper-V host
- **Automatic Stop Action**: Used to define how a VM responds when you shut down a Hyper-V host gracefully

The **Hardware** and **Management** information is stored in configuration files. The following two file formats are used:

- **.vmcx files**: This format is used for *VM configuration*
- **.vmrs files**: This format is used for *runtime data*

These are binary files that cannot be manually edited; Hyper-V Manager or PowerShell is required to read these files and make changes to them.

VM virtual hard disk formats

A **virtual hard disk** (VHD) is a software representation of a physical hard disk, the same as a VM's processor, and memory is a software representation of a physical computer's processor and memory. You can create partitions, files, and folders, just like you can for a physical disk. You install the VM's OS and any applications and services on these VHDs. Multiple VHDs, such as physical computers, can be attached to meet storage requirement use cases.

VHDs can be created and managed using the following:

- Hyper-V Manager
- Windows Disk Management
- PowerShell, using the New-VHD cmdlet
- The Disk Management console
- The command line, using the **Diskpart tool**

Two VHD formats are available when a new VHD is created for a VM; these are as follows:

- **VHD**: This is the format that was created by older Hyper-V versions and can create disks with a maximum size of 2,040 GB.
- **VHDX**: This is the newer format and can create disks up to 64 TB in size. It supports large block sizes for better performance.

You can convert a disk between formats using the **Edit Virtual Hard Disk Wizard** area of Hyper-V Manager or with PowerShell using the Convert-VHD cmdlet. This creates a *new disk* and copies the contents of the *existing disk* into it.

VM disk types

Hyper-V provides different disk types to suit different needs:

- **Fixed-size VHD**: Immediately allocates all of the disk space.
- **Dynamic size VHD**: A maximum size is defined when created and only allocates space as needed to grow.
- **Differencing VHD**: This is linked to a parent VHD with a base configuration. Changes are only made to the differencing disk, preserving the configuration on the parent. This could be a use case for multiple child disk configuration images based on the same parent configuration image.
- **Pass-through physical disk**: This allows you to directly connect to a physical hard disk through the host computer.

You can convert a disk between types using the **Edit Virtual Hard Disk Wizard** area of Hyper-V Manager; the same disk format should be used.

In this section, we looked at implementing and managing Hyper-V VMs; we covered generation versions, configuration versions, VM settings, disk formats, and types. Now, let's look at VM memory.

VM memory

A VM's memory can be configured as required from the **Hardware** section of the VM settings in Hyper-V Manager, as shown in the following screenshot:

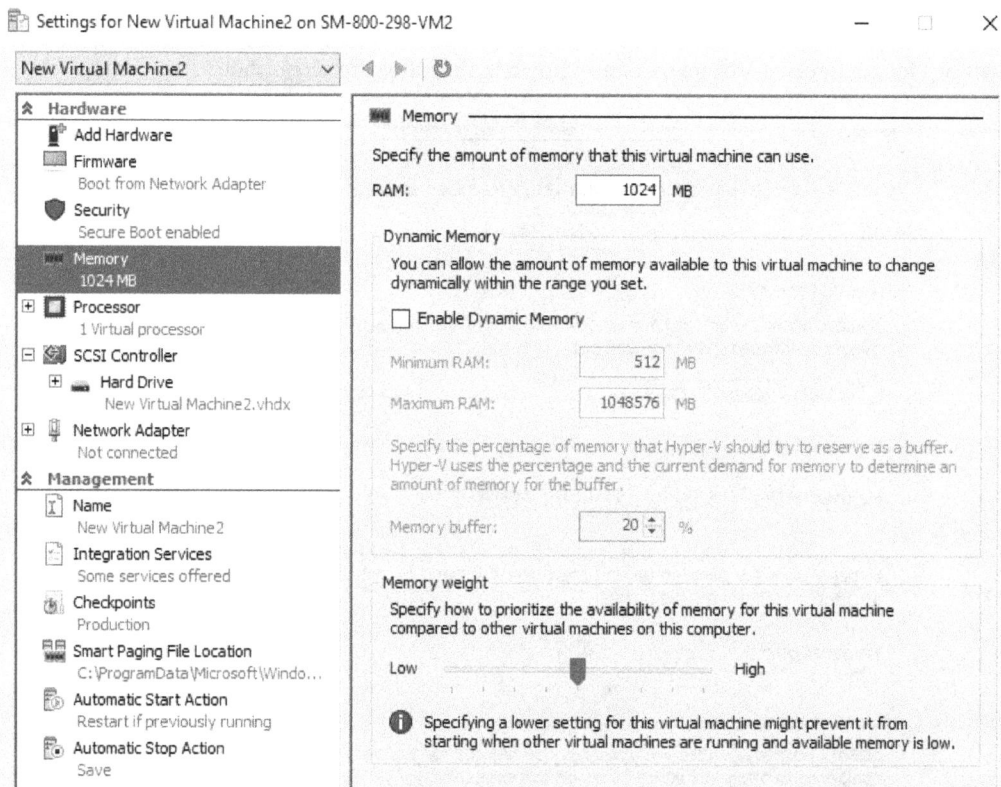

Figure 9.8 – VM memory settings

A VM can be enabled to use **Dynamic Memory**. As its name suggests, this allows you to dynamically size and allocate the available memory to a VM based on demand. A benefit of this is that the VM doesn't need to be powered off, as in the case without Dynamic Memory.

There are four aspects of configuring Dynamic Memory when it is enabled; these are as follows:

- **Startup RAM**: This is the amount of memory a VM has when it starts, irrespective of Dynamic Memory being enabled.

- **Minimum RAM**: This is the minimum level that will be dynamically allocated based on the load.

- **Maximum RAM**: This is the maximum level that will be dynamically allocated based on the load.

- **Memory Buffer**: This determines the size of memory allocation that will grow in chunks of memory and increase to the maximum RAM. It is represented as a percentage of the RAM configured. The default is 20%, meaning it will increase memory in 204 MB chunks if the total memory is 1,024 MB.

The settings for configuring VM memory are shown in the following screenshot:

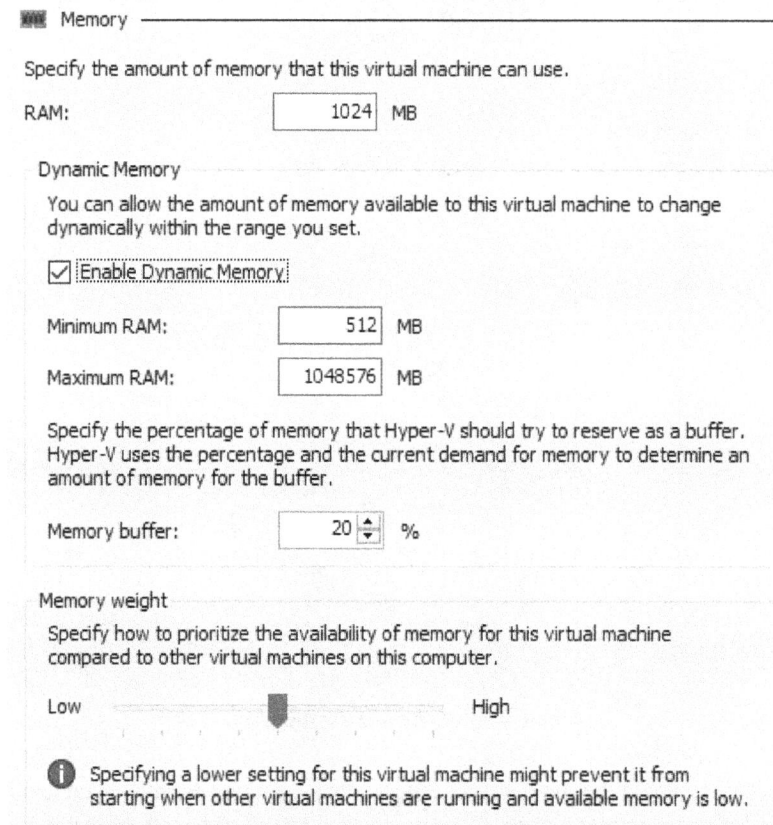

Figure 9.9 – Configuring a VM's memory

Dynamic Memory cannot be used with NUMA. Integration Services must be updated as Dynamic Memory requires these services to control this function.

Hyper-V Integration Services

Hyper-V Integration Services provides operational functionality to VMs, also called **integration components**. These services (*components*) allow the VM to communicate with the Hyper-V host for its operations. Keeping these services (*components*) updated is important to ensure the smooth operation of the VM's functions and capabilities.

You will need to install Integration Services on Hyper-V hosts earlier than Windows Server 2016 and Windows 10. To install it, from Hyper-V Manager, **connect to the VM** and, from the **action** menu, select **Insert Integration Services Setup Disk**. Windows Server 2016 and Windows 10 or newer Hyper-V hosts do not require this.

The following Integration Services are available:

- **Hyper-V Heartbeat Service**: Reports whether a VM is running correctly
- **Hyper-V Guest Shutdown Service**: Permits a host to trigger a VM shutdown
- **Hyper-V Time Synchronization Service**: Synchronizes the VM's clock with the host
- **Hyper-V Data Exchange Service (KVP)**: Provides a metadata exchange between the VM and the host
- **Hyper-V Volume Shadow Copy Requestor**: Provides a VSS backup of the VM without it being shut down
- **Hyper-V Guest Service Interface**: Provides an interface to copy files to or from a VM from the host
- **Hyper-V PowerShell Direct Service**: Provides PowerShell access to a VM without a network connection

Integration Services can be turned *on* and *off* using Hyper-V Manager or PowerShell.

To use Hyper-V Manager, right-click on the VM to access its settings, then select **Integration Services** under the **Management** section.

For PowerShell, you can use the following cmdlets:

```
Enable-VMIntegrationService
Disable-VMIntegrationService
```

To get a list of the Integration Services enabled on a VM, use the following PowerShell cmdlet:

```
Get-VMIntegrationService -VMName "VM1"
```

Discrete Device Assignment

Discrete Device Assignment (DDA) allows a VM to directly access physical PCIe hardware on the host whenever it's available and supported. Two device classes are supported:

- Graphics adapters
- NVMe storage devices

Generation 1 and 2 VMs support DDA. When a device is attached, the following VM features are restricted and not available to the VM:

- Dynamic Memory
- VM save/restore
- Live migration
- High availability cluster member

The following steps should be taken to configure DDA:

1. Turn off the Automatic Stop Action by using the following PowerShell cmdlet:

    ```
    Set-VM -Name VMName -AutomaticStopAction TurnOff.
    ```

2. Disable the device by using Device Manager or the following PowerShell cmdlets:

    ```
    Get-PnpDevice
    Disable-PnpDevice
    ```

3. Dismount the device from the host partition using the following PowerShell cmdlet:

    ```
    Dismount-VMHostAssignableDevice -LocationPath
    $locationPath.
    ```

4. Assign the device to the VM using the following PowerShell cmdlet:

    ```
    Add-VMAssignableDevice -LocationPath $locationPath
    -VMName VMName.
    ```

If you wish to remove the device from the VM and return to the host, you can run the following from PowerShell ISE:

```
#Remove the device from the VM
Remove-VMAssignableDevice -LocationPath $locationPath -VMName
VMName
```

```
#Mount the device back in the host
Mount-VMHostAssignableDevice -LocationPath $locationPath
```

The device can then be re-enabled on the host. Use Device Manager or the following PowerShell cmdlets to do so:

```
Get-PnpDevice
Enable-PnpDevice
```

Enhanced session mode

Enhanced Session Mode allows local computer resources such as *printers, clipboards, local drives, USB drives*, and so on to be redirected for use by the guest VMs. This is enabled through **Virtual Machine Connection (VMConnect)**.

The following are the requirements for using the local resources in a VM:

- Both the Enhanced Session Mode policy and Enhanced Session Mode must be enabled on the Hyper-V Host
- The computer running VMConnect must use Windows 8.1/10 and Windows Server 2012 R2/2016
- The VM must use Windows 8.1/10, Windows Server 2012 R2/2016, and have Remote Desktop Services enabled

If available on the computer where *VMConnect* is run, the following local resources can be redirected:

- Drives
- USB devices
- Supported plug-and-play devices
- Smart cards
- Printers
- Audio
- Display settings
- Clipboards

By default, Windows 10 OS Hyper-V hosts have **Enhanced Session Mode**; Windows 2012 R2 and Windows Server 2016 are now considered out of support. The following steps cover **Enhanced Session Mode**:

1. This can be turned on from **Hyper-V Settings**; select **Enhanced Session Mode Policy** under the **Server** settings, as shown in the following screenshot:

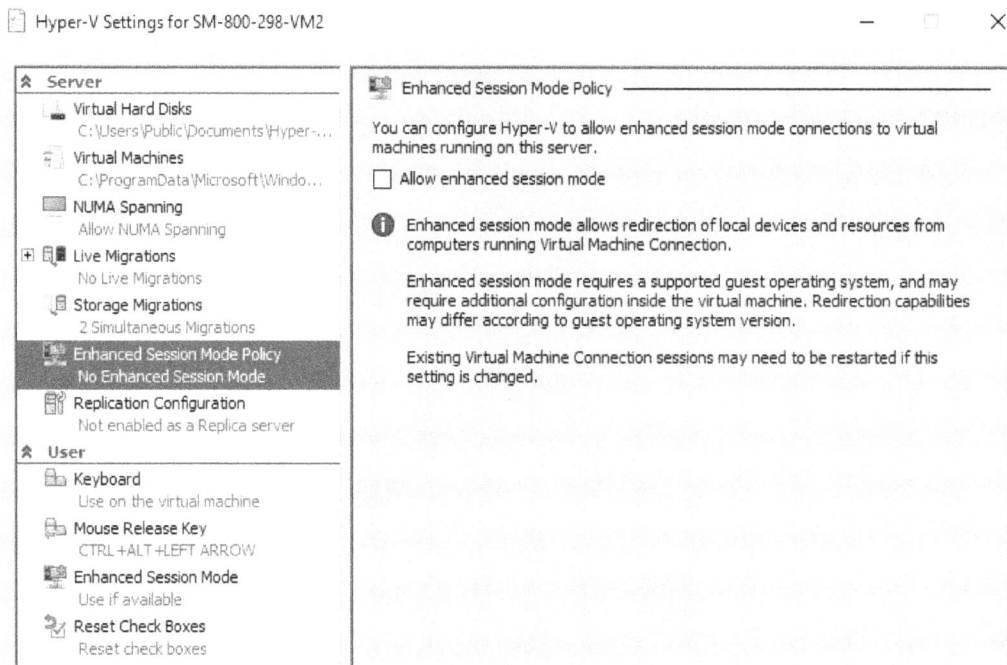

Figure 9.10 – Enabling the Enhanced Session Mode Policy setting

2. Next, under the **User** settings, select **Enhanced Session Mode** as shown in the following screenshot:

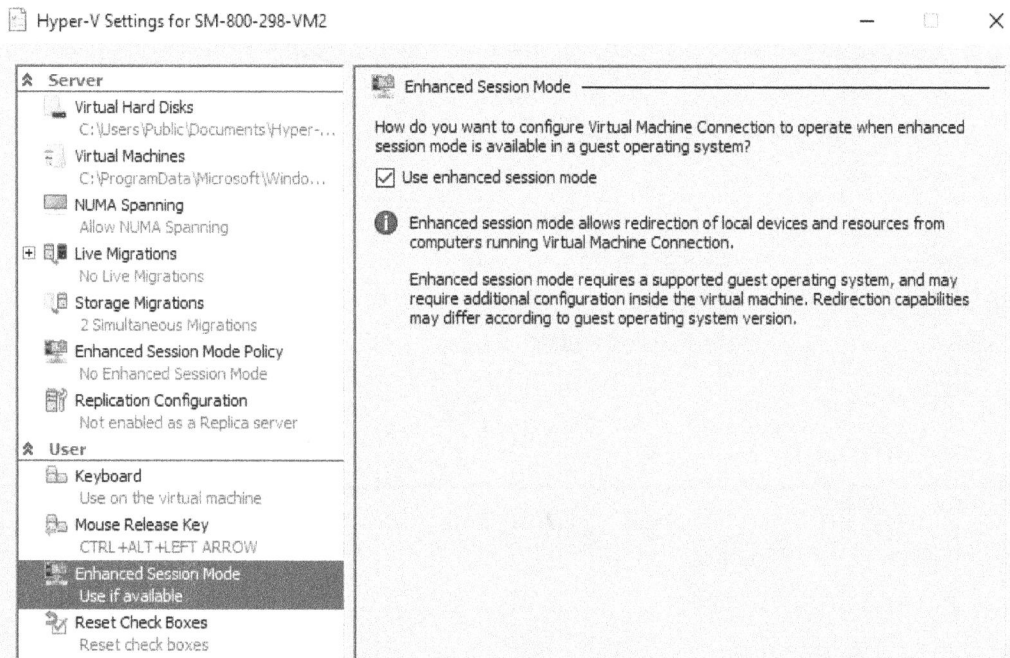

Figure 9.11 – Enable the Enhanced Session Mode setting

When you open **VMConnect**, you will see the VM's **local device resources** that can be accessed and used:

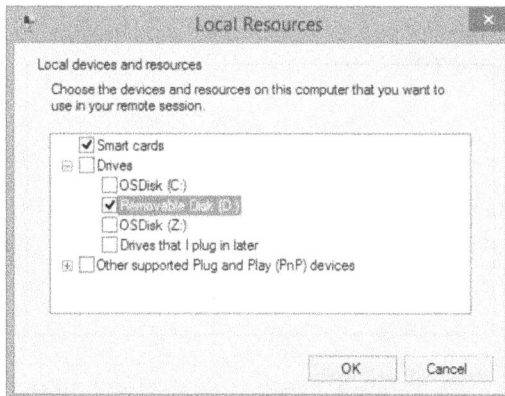

Figure 9.12 – Accessing local resources via VMConnect

The settings for VMConnect can be changed using Command Prompt or PowerShell with the following command:

```
VMConnect.exe <ServerName> <VMName> /edit
```

VM resource groups

You can use VM resource groups to group VMs and perform orchestrated operations at scale on the group rather than individual VMs.

The following are the PowerShell cmdlets for this function:

- `Add-VMGroupMember`
- `Get-VMGroup`
- `New-VMGroup`
- `Remove-VMGroup`
- `Remove-VMGroupMember`
- `Rename-VMGroup`

This can *only* be implemented using PowerShell as there is *no* Hyper-V Manager option.

CPU groups

VM CPU groups allow you to manage and allocate host CPU resources across guest VMs with ease. With CPU groups, you can do the following:

- Create groups of VMs with differing allocations shared across the group for the host's CPU resources. Different classes of services can be created for differing VM types.
- Set specific groups so that they have CPU resource limits; this caps the group's amount of host CPU resources it can consume. This can enforce a class of service that's desired for that group.
- Set a group's constraints to run only on a specific set of the host's CPUs. This can provide VM isolation between different CPU group members.

The Hyper-V **Host Compute Service** (**HCS**) manages CPU groups. Here, the cpugroups.exe command-line utility can be used.

CPU groups can be quite complex and there's more to them than what's been detailed in this section; please refer to the following Microsoft documentation on this functionality: https://docs. microsoft.com/en-us/windows-server/virtualization/hyper-v/manage/ manage-hyper-v-cpugroups.

Hypervisor schedulers

Windows Server 2016 introduced new *virtual processor scheduling logic modes*. How work is managed and allocated across guest virtual processors by the Hyper-V hosts is defined by these *modes* or *scheduler logic types*.

The scheduler types are as follows:

- Classic scheduler
- Core scheduler
- Root scheduler

More details on each of these scheduler types can be found in the following Microsoft article on the topic: `https://docs.microsoft.com/en-gb/windows-server/virtualization/hyper-v/manage/about-hyper-v-scheduler-type-selection`.

Beginning with Windows Server 2019, the *core scheduler* is the default for the maximum-security posture to ensure security is by default. For these security reasons, Microsoft recommends changing the default *classic scheduler* in Windows Server 2016 to enable the *core scheduler* to protect against potentially malicious VMs.

When the *core scheduler* is configured, the guest VMs can optionally use **simultaneous multithreading** (**SMT**), which allows separate, independent execution threads to share the same processor resources.

SMT can be enabled on a guest VM with the following PowerShell command:

```
Set-VMProcessor -VMName <VMName> -HwThreadCountPerCore <n>
```

The scheduler is configured on the host via the entry in the `hypervisorschedulertype` BCD. You can change the scheduler type from Command Prompt with administrator privileges using the following command:

```
bcdedit /set hypervisorschedulertype "type"
```

Here, `"type"` is either the *classic*, *core*, or *root* scheduler that can be entered into the command.

> **Note**
> Hyper-V does not support, at the time of writing, the *root scheduler* and should *not* be configured for *virtualization scenarios*.

VM checkpoints

Each guest VM in Hyper-V can have one of the following checkpoints configured:

- **Production checkpoint**: This is a point-in-time image of a VM that can be restored. This uses a backup function within the VM to create the checkpoint instead of the saved state function.

- **Standard checkpoint**: This is a running VM's state, data, and hardware configuration capture. It is intended for test and development use cases.

Checkpoints for each VM can be enabled or disabled. Follow these steps to enable checkpoints:

1. From **Hyper-V Manager**, right-click the VM and click **Settings**.
2. Select **Checkpoints** under the **Management** section.
3. Select **Enable checkpoints**; to *disable* it, clear the checkbox.
4. Click **Apply** to save your changes, then click **OK** to finish.

To change the checkpoint type, follow these steps:

1. From **Hyper-V Manager**, right-click the VM and click **Settings**.
2. Select **Checkpoints** under the **Management** section.
3. Select **Production** or **Standard** checkpoints.
4. Optionally, you can choose to store the checkpoints in a different location; use the **Checkpoint File Location** section and set it as *required*.
5. Click **Apply** to save your changes, then click **OK** to finish.

The default for new machines is *production*.

NIC teaming

Network Interface Card (**NIC**) teaming is a technology provided by third-party hardware vendors and uses multiple network adapters to aggregate bandwidth. If a network adapter fails, traffic failover is provided to avoid connectivity and traffic flow loss.

However, Microsoft provides no support channel for this hardware vendor's technology and no configuration information for the individual hardware vendors implementing this technology; your hardware vendor should provide configuration assistance for your scenario.

Please refer to the following official Microsoft support policy for NIC teaming with Hyper-V: https://docs.microsoft.com/en-us/troubleshoot/windows-server/virtualization/support-policy-nic-teaming.

Hyper-V Replica

Hyper-V Replica is a built-in capability in the Hyper-V role. It allows you to replicate VMs between Hyper-V hosts to support high availability goals; it creates a live copy of a VM that's replicated to an offline VM. The data can be configured to synchronize every *30 seconds, 5 minutes*, or *30 minutes* to meet your **Recovery Point Objective** (**RPO**) target.

The primary and secondary host servers can be in the same physical location or replicated over a WAN link. Hyper-V Replica supports standalone and clustered Hyper-V hosts. There is no dependency on Active Directory for the Hyper-V hosts.

Follow these steps to configure Hyper-V Replica:

1. Configure the Hyper-V hosts:

    I.   A minimum of two hosts are required, with one or more VMs on each.

    II. From the *Hyper-V settings* of the host, select **Enable this computer as a Replica server** in the **Replication Configuration** section.

    III. Select **Use Kerberos for HTTP** or use **Certificate-based authentication** for HTTPS.

    IV. To allow the replica server to allow VM replication traffic from any authenticated primary server, select **Allow replication from any authenticated server**. To accept VM replication traffic only from the specifically selected primary servers, select **Allow replication from the specified servers**.

    V.   Click **OK** to complete.

2. Configure the firewall(s). VM replication traffic must be permitted across the network; you should configure the Windows firewall and any third-party firewalls.

3. Enable VM replication. For each VM to replicate, do the following:

    I.   From **Hyper-V Manager**, right-click a VM and open the **Enable Replication Wizard** area by clicking on **Enable replication**. Complete the wizard's steps as per your requirements. Detailed steps can be found in the following Microsoft documentation: `https://docs.microsoft.com/en-gb/windows-server/virtualization/hyper-v/manage/set-up-hyper-v-replica`.

4. Carry out a failover. You can run a *test failover*, a *planned failover*, or an *unplanned failover*.

Only a single recovery point with the latest replication from the primary server is the default configuration for Hyper-V Replica. Additional recovery points can be stored up to a maximum of *24-hourly recovery points*. Both *crash-consistent* and *app-consistent* recovery points can be created.

Managing a VM remotely

The following tools are all ways to connect remotely to a VM running on a Hyper-V Host; this may be useful for automated configuration and operations, as well as troubleshooting and support tasks where console access may have been lost:

- **PowerShell Remoting**: This method allows a local user's PowerShell session to send commands to be executed locally on a remote VM running on the Hyper-V host. This can be enabled with the PowerShell `Enable-PSRemoting` cmdlet.

- **PowerShell Direct**: This method allows you to directly connect inside a VM running on a Hyper-V host without the need for a network connection. This uses the *Hyper-V VMBus* to connect to the VM. You can use the PowerShell `PSSession` or `Invoke-Command` cmdlet.

- **HVC.exe**: This method allows a direct **secure shell protocol** (**SSH**) connection to a Linux VM on a Hyper-V host.

In this section, we learned how to create and manage Hyper-V guest VMs. In the next section, we will look at securing the workloads running on the Hyper-V hosts.

Securing Hyper-V workloads with Guarded Fabric

We will introduce this section by looking at the Guarded Fabric security solution at an overview level and then explore each component in the following subsections.

Guarded Fabric

Guarded Fabric is a collection of component services and capabilities that allows a security solution to protect VMs against inspection, theft, and tampering, and malicious actors, humans, or malware from compromised VMs or hosts.

Guarded Fabric is comprised of the following components at its core:

- **Shielded VM**: You specify templates and images that a VM must use to be a Shielded VM.

- **Guarded Host**: You specify which hosts are secure for running Shielded VMs.

- **Host Guardian Service** (**HGS**): This ensures that only authorized and secure Guarded Fabric Hosts can run Shielded VMs. It provides the key service to the Guarded Hosts.

The Guarded Fabric security solution is shown in the following diagram:

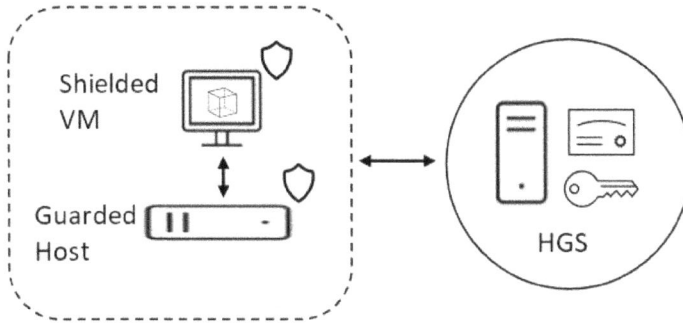

Figure 9.13 – Guarded Fabric components

The following VM types can run in Guarded Fabric with different security levels:

- **Shielded VM**: Provides the greatest protection with the highest security measures

- **Encrypted Standard VMs**: Not as protected as Shielded VMs

- **Standard VMs**: No protection and provides the least security measures

In this section, we introduced the Guarded Fabric security solution. The following subsection will look at the Host Guardian Service's components and operation.

Host Guardian Service

The **Host Guardian Service** (**HGS**) comprises two core components:

- **HGS Attestation service**: Only authorized and safeguarded hosts can run a Shielded VM through Attestation. Two Attestation modes are available to ensure Guarded Hosts only run approved code; these are as follows:

  - **TPM-trusted Attestation**: The Guarded Host approval in this mode is based on three types of information: Trusted Platform Module identity, Measured Boot sequence, and the code integrity policy.

  - **Host key Attestation**: The Guarded Host approval in this model is based on a key possession.

- **Key Protection Service** (**KPS**): This allows a Guarded Host to power on Shielded VMs through provided keys. A Shielded VM can also be migrated to another Guarded Host with these security keys.

The **Admin-trusted Attestation** mode was included in Windows Server 2016 but was *deprecated* in Windows Server 2019.

The following steps can be used to start a Shielded VM in Guarded Fabric:

1. The Shielded VM is requested to be started by a user.

2. The Guarded Host requests Attestation for the HGS before the Shielded VM can start.

3. The HGS Attestation service validates the credentials of the Guarded Host.

4. A certificate is sent to the Guarded Host by the HGS Attestation service.

5. The Attestation certificate is submitted to the KPS by the Guarded Host. Using an encrypted secret, a key is requested to unlock the Shielded VM.

6. The KPS determines the certificate's validity from the Guarded Host, retrieving the key to unlock the Shielded VM by decrypting the secret.

7. The Guarded Host receives the key from KPS.

8. The Shielded VM is started by it being unlocked with the key.

The HGS role can be implemented using the **Add Roles and Features Wizard** area of Server Manager or with the following PowerShell cmdlet:

```
Install-WindowsFeature HostGuardianServiceRole
-IncludeManagementTools -Restart
```

In this section, we looked at securing Hyper-V workloads. In the next section, we will complete some hands-on exercises to reinforce some of the concepts covered in this chapter.

Hands-on exercises

To support your learning with some practical skills, we will utilize the concepts and understanding we gained from this chapter and put them to practical use.

We will look at the following exercises:

*   Exercise – implementing the Hyper-V role using Server Manager

*   Exercise – implementing the Hyper-V role using PowerShell

*   Exercise – creating a guest VM

Getting started

To get started with this section, you will need access to a Windows server; this can be physical, virtual, or an Azure IaaS VM. You must ensure the physical computer or VM meets the installation requirements for Hyper-V, as per the following article: `https://docs.microsoft.com/en-us/windows-server/virtualization/hyper-v/system-requirements-for-hyper-v-on-windows`.

Let's move on to the exercises for this chapter.

Exercise – Implementing the Hyper-V role using Server Manager

In this exercise, we will implement the Hyper-V role using Server Manager. For this exercise, we will be using a Windows Server 2019 Gen2 VM.

Follow these steps to get started:

1. Log into your Windows Server for this exercise using an admin account, click **Add Roles and Features** from the **Manage** screen of **Server Manager**, and click **Next** on the **Before You Begin** screen.

2. Select **Role-based or feature-based installation** on the **Installation Type** screen and click **Next**.

3. From the **Server Selection** screen, accept the default selection of the current server and click **Next**.

4. Select **Hyper-V** from the **Roles** list on the **Server Roles** screen.

5. Click **Add Features** on the pop-up screen. Then, click **Next**:

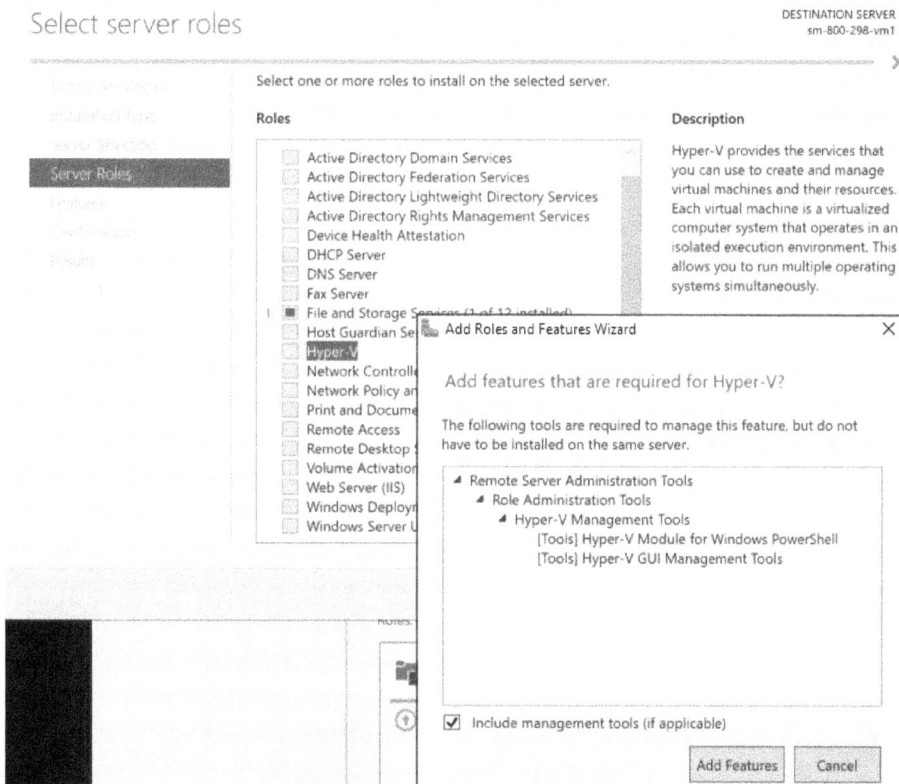

Figure 9.14 – Adding a Hyper-V server role

6. On the **Server Roles** screen, click **Next**.

7. On the **Features** screen, click **Next**.

8. On the **Hyper-V** screen, click **Next**:

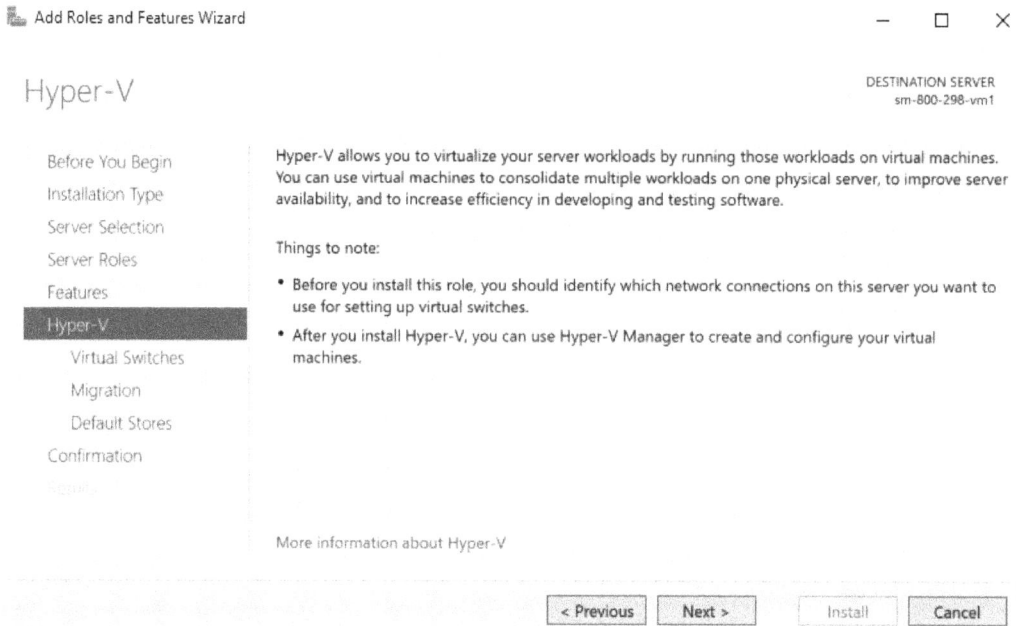

Figure 9.15 – Adding a Hyper-V server role

9. On the **Create Virtual Switches** screen, select a network adapter and click **Next**:

Figure 9.16 – The Create Virtual Switches screen

10. On the **Virtual Machine Migration** screen, accept the defaults and click **Next**:

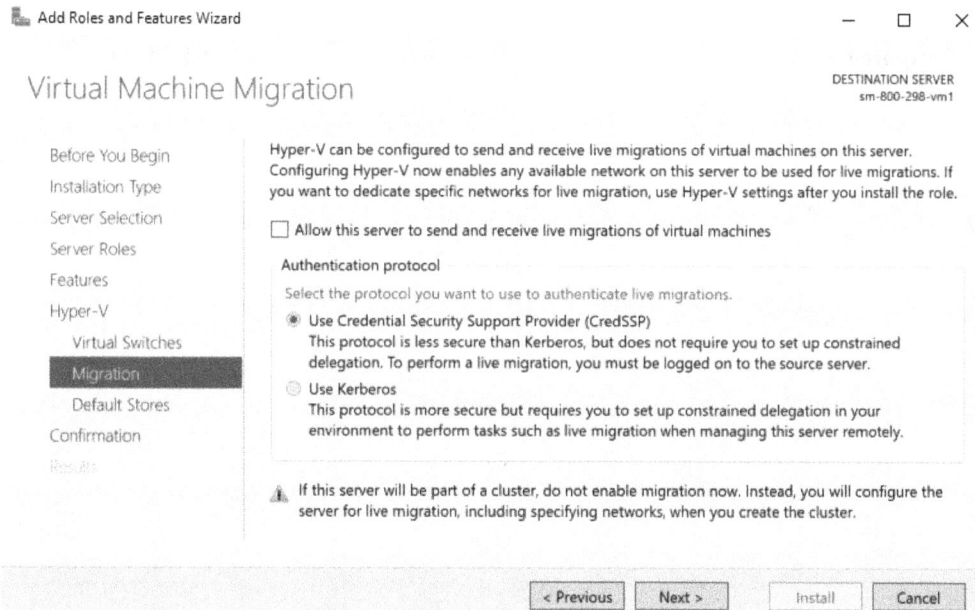

Figure 9.17 – Enabling migrations

11. On the **Default Stores** screen, accept the defaults and click **Next**:

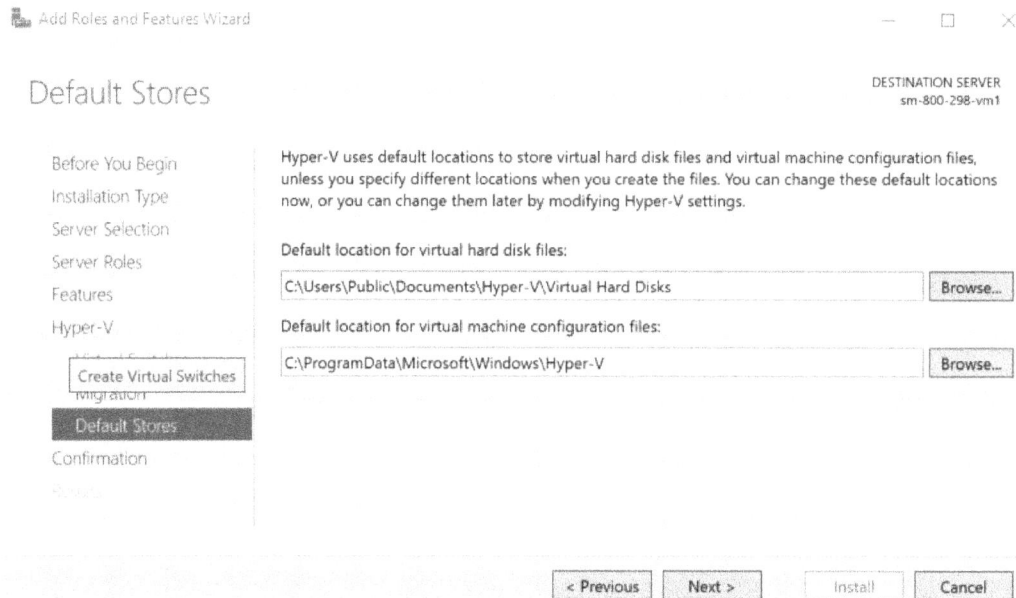

Figure 9.18 – Setting the default stores

12. On the **Confirm installation selections** screen, select **Restart the destination server automatically if required**, click **Yes** on the *server restart popup* that appears, and then click **Install**:

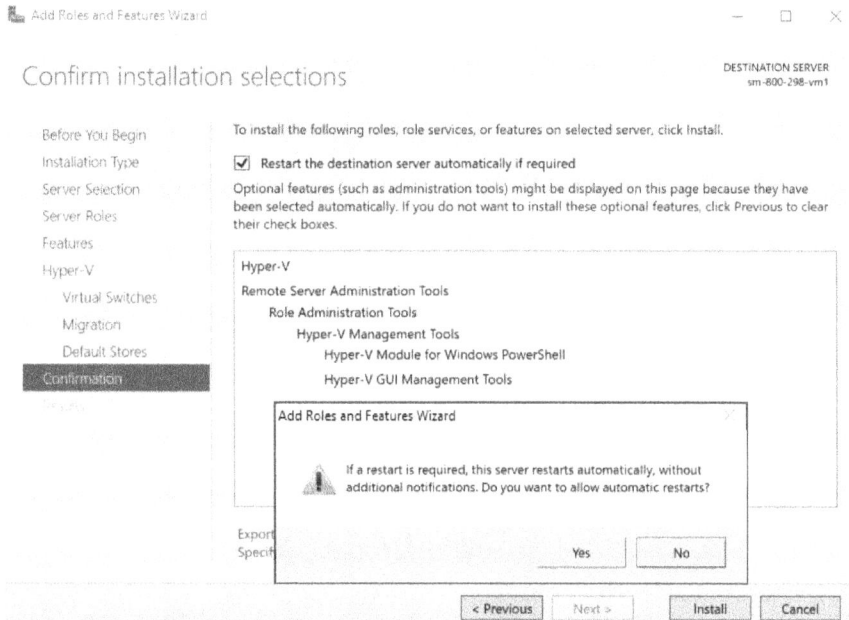

Figure 9.19 – The Confirm installation selections screen

13. The installation will start:

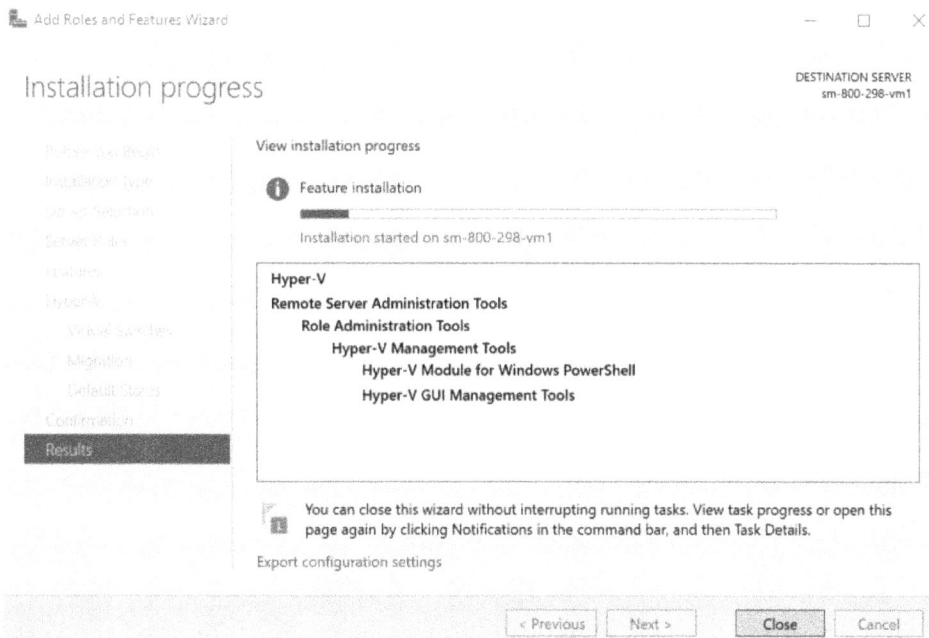

Figure 9.20 – The Installation progress screen

14. When the installation has succeeded, click **Close** to finish installing the wizard. The Hyper-V role will now appear under **Server Manager** > **Dashboard**:

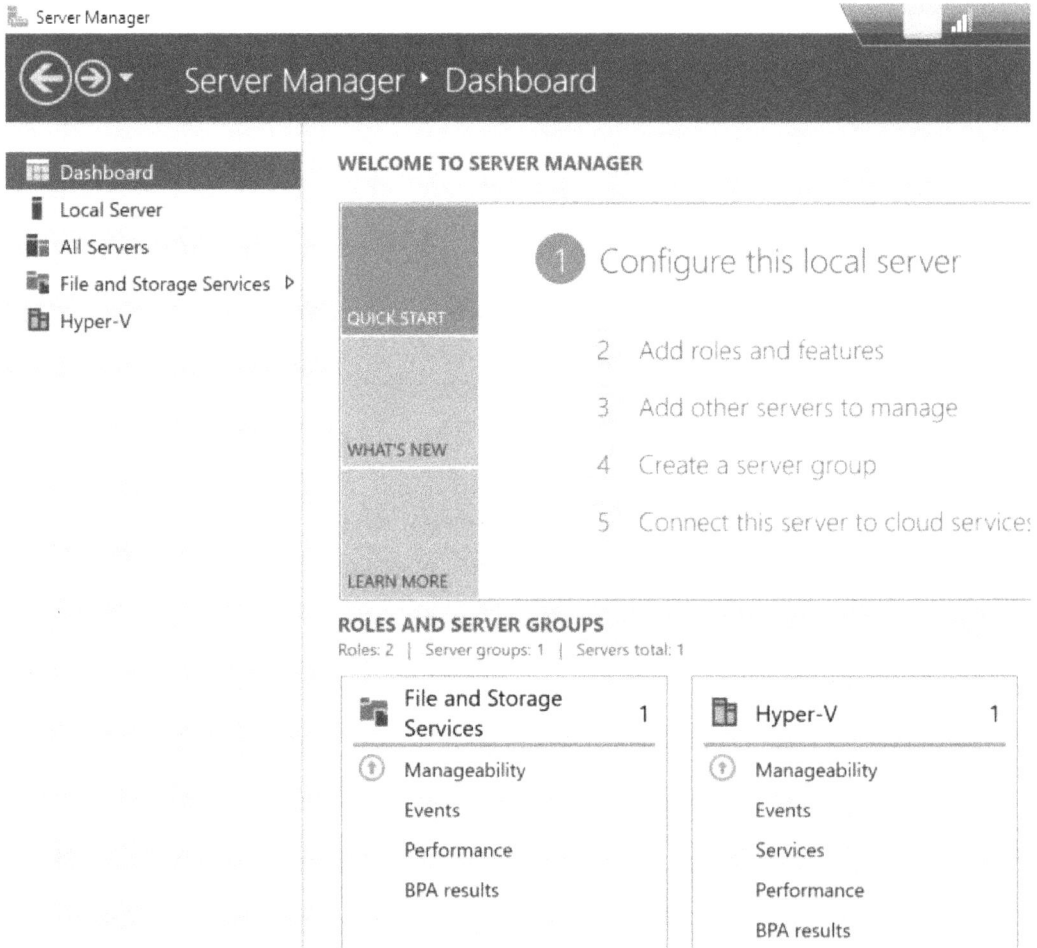

Figure 9.21 – Hyper-V server role installed

With that, we have completed this exercise. This exercise taught us the skills to implement the Hyper-V role on Windows Server using Server Manager. In the next exercise, we will implement the Hyper-V role on Windows Server using PowerShell.

Exercise – implementing the Hyper-V role using PowerShell

In this exercise, we will implement the Hyper-V role using PowerShell. For this exercise, we will use a Windows Server 2019 Gen2 VM and install it locally on our server.

Follow these steps to get started:

1. Log into your Windows Server for this exercise using an admin account and launch a PowerShell session *running as an Administrator*.

2. Run the following command to install Hyper-V locally onto this server:

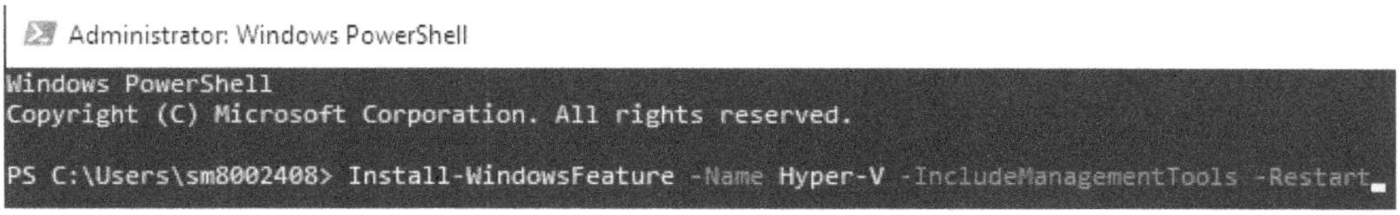

```
Administrator: Windows PowerShell
Windows PowerShell
Copyright (C) Microsoft Corporation. All rights reserved.

PS C:\Users\sm8002408> Install-WindowsFeature -Name Hyper-V -IncludeManagementTools -Restart
```

Figure 9.22 – PowerShell installation

3. The installation will start:

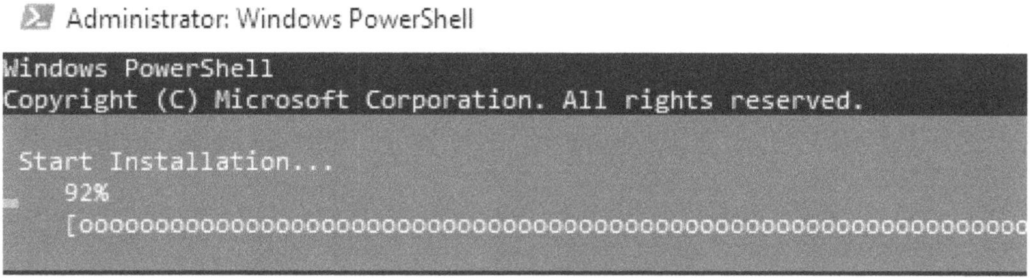

```
Administrator: Windows PowerShell
Windows PowerShell
Copyright (C) Microsoft Corporation. All rights reserved.

Start Installation...
    92%
    [oooooooooooooooooooooooooooooooooooooooooooooooooooooooooooooooooooo
```

Figure 9.23 – Installation in progress

4. The server will restart; the Hyper-V role will now appear under **Server Manager** > **Dashboard**:

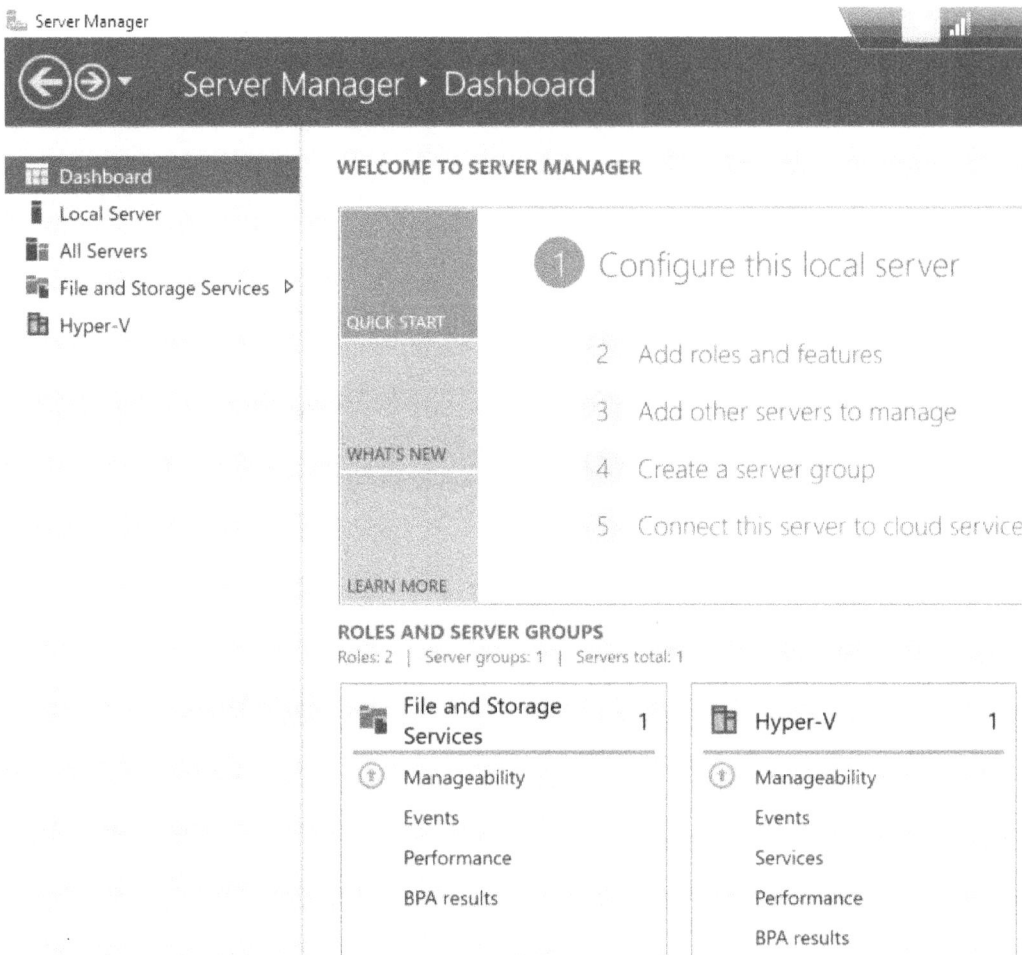

Figure 9.24 – Hyper-V server role installed

With that, we have completed this exercise. This exercise taught us the skills to implement the Hyper-V role on Windows Server using PowerShell. In the next exercise, we will look at creating a guest VM.

Exercise – creating a guest VM

In this exercise, we will create a guest VM. Follow these steps to get started:

1. Log into your Windows Server for this exercise using an admin account and open **Hyper-V Manager**.

2. Click **New** from the **Action** pane and click **Virtual Machine…**:

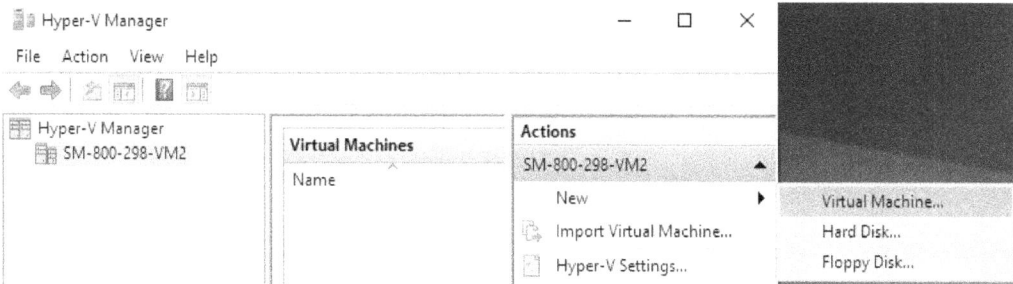

Figure 9.25 – Creating a VM

3. From the **New Virtual Machine Wizard** area, click **Next**:

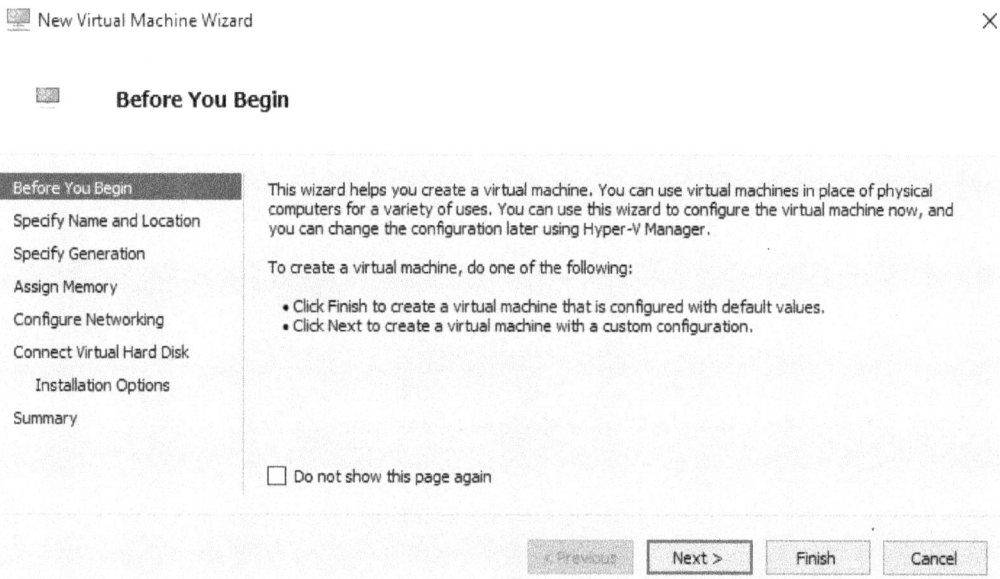

Figure 9.26 – New Virtual Machine Wizard – Before You Begin

4. For this exercise, I have selected all the *defaults* presented. You could select each option as required. You can review each of the options and the defaults by reading the following Microsoft documentation: `https://docs.microsoft.com/en-us/windows-server/virtualization/hyper-v/get-started/create-a-virtual-machine-in-hyper-v#options-in-hyper-v-manager-new-virtual-machine-wizard`.

5.   On the **Summary** screen, click **Finish** once you have confirmed your choices:

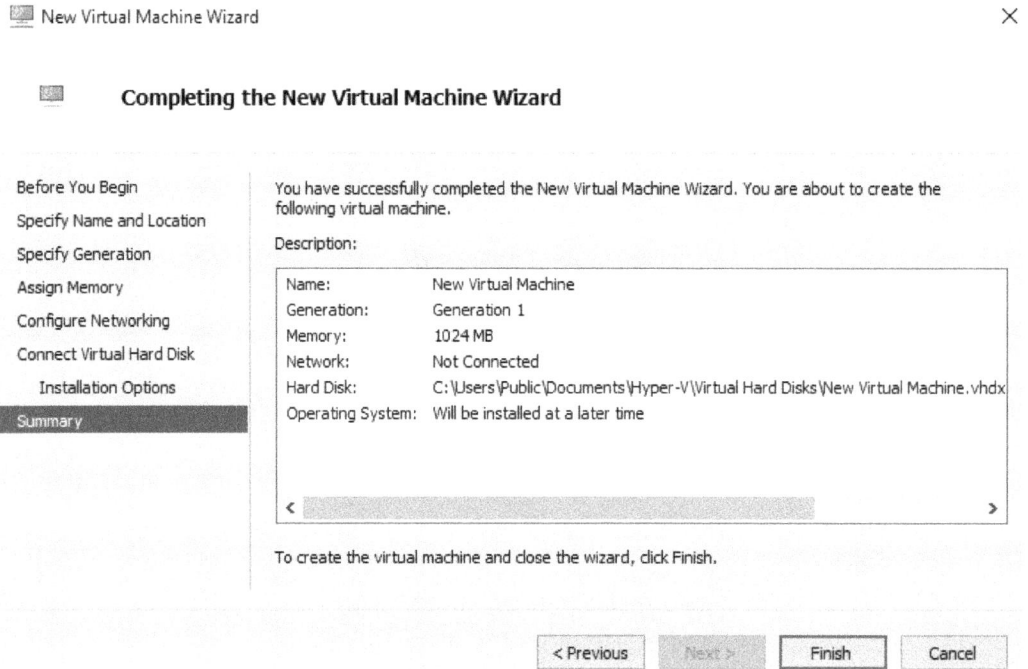

Figure 9.27 – New Virtual Machine Wizard – Summary

6.   Select **Actions** | **Start** from the **Virtual Machine Connection** window:

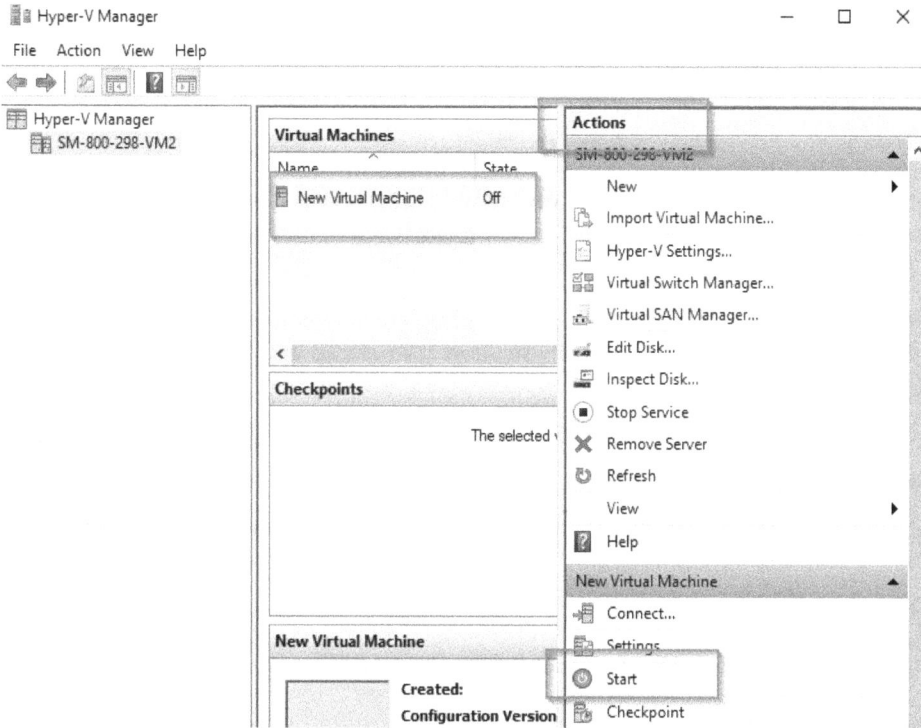

Figure 9.28 – Starting a VM

7. Select **Actions | Connect…** from the **Virtual Machine Connection** window:

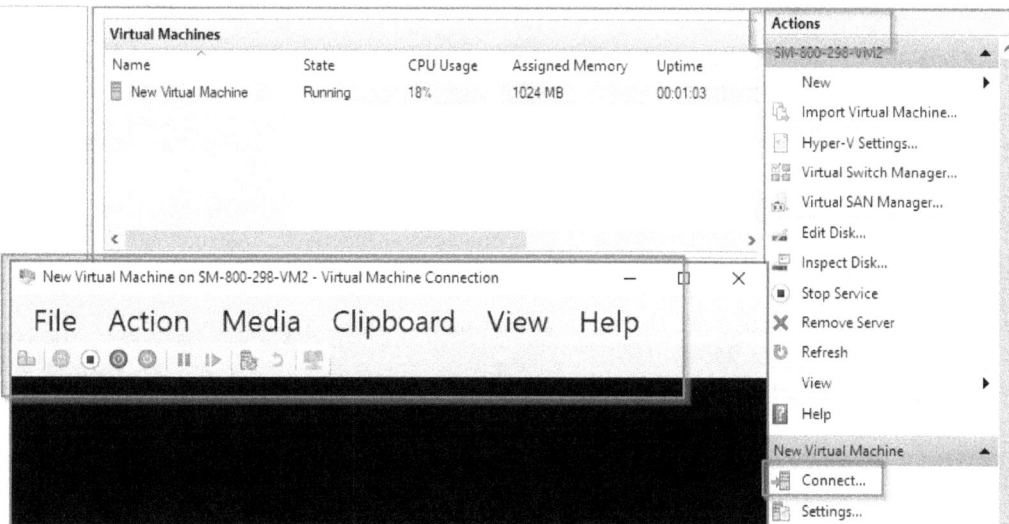

Figure 9.29 – Connecting to a VM

With that, we have completed this exercise. This exercise taught us the skills to create a guest VM. Now, let's summarize this chapter.

Summary

This chapter's content has helped develop your knowledge and skills regarding virtualization with Hyper-V and provided coverage for *AZ-800 Administering Windows Server Hybrid Core Infrastructure: Manage Hyper-V and Guest Virtual Machines*.

First, you learned about virtualization and how to implement and manage Hyper-V and secure workloads using the Guarded Fabric security solution. We finished with hands-on exercises to help you develop your skills further.

You have gained new skills through the information provided and taken your knowledge beyond the exam objectives to prepare you for a real-world, day-to-day hybrid environment-focused role.

In the next chapter, you will learn how to implement and manage Windows Server containers.

Further reading

This section provides links to additional study references and additional exam information:

- *Microsoft Certified: Windows Server Hybrid Administrator Associate*: `https://docs.microsoft.com/en-us/learn/certifications/windows-server-hybrid-administrator/`

- *Exam AZ-800: Administering Windows Server Hybrid Core Infrastructure*: `https://docs.microsoft.com/en-us/learn/certifications/exams/az-800`

- *Exam AZ-800: Study Guide*: `https://query.prod.cms.rt.microsoft.com/cms/api/am/binary/RWKI0r`

- *Microsoft Learn*: `https://docs.microsoft.com/en-us/learn/paths/manage-virtualization-containers-hybrid-environment/`

Skills check

Check what you have learned in this chapter by answering the following questions:

1. What are the benefits and values of the virtualization computing model over the traditional physical computing model?

2. What is Hyper-V?

3. What are some of the hardware requirements for implementing Hyper-V?

4. What are some of the benefits of implementing Hyper-V on Server Core instead of a full GUI OS installation?

5. What guest OS types are supported?

6. What are the two primary network components?

7. What are the three virtual network switch types, and how do they differ in functionality?

8. What is nested virtualization?

9. Name some of the requirements for nested virtualization.

10. Name at least six of the functions available in Hyper-V Manager.

11. In what ways can we manage Hyper-V hosts and guest VMs?

12. What are VM generation versions?

13. What are VM configuration versions?

14. What is the difference between generation 1 and generation 2 VM settings?

15. What VHD disk formats are supported, and how do they differ?

16. What are the different VHD disk types, and what are the use cases for these different types?

17. What is Dynamic Memory, and how is it configured?

18. What is Hyper-V Integration Services?

19. What is Discrete Device Assignment?

20. What functionality does Enhanced Session Mode provide?

21. What is the benefit of VM resource groups and CPU groups?

22. What are the three scheduler types?

23. Explain the two VM checkpoint types.

24. How is high availability provided?

25. Explain the components and operation of Guarded Fabric.

10

Implementing and Managing Windows Server Containers

In the previous chapter, we added skills for implementing and managing Hyper-V on Windows Server.

This chapter covers content from *AZ-800 Administering Windows Server Hybrid Core Infrastructure: Create and manage containers*.

This content does not aim to cover all aspects of containers and their creation, operation, and orchestration; the content is limited in scope to only the exam skills outline.

In this chapter, we will look at the concept of containerization, introduce Docker, and look at Windows Server containers. We will conclude with some hands-on exercise to help you develop your skills further.

The following topics are included in this chapter:

- Introduction to containers
- Introduction to Docker
- Securing Windows Server containers
- Networking for Windows Server containers
- Creating Windows Server container images
- Managing Windows Server container images
- Hands-on exercises

In addition to the topics listed in this chapter, this chapter's goal is to take your knowledge beyond the exam objectives to prepare you for a real-world, day-to-day hybrid environment-focused role.

Introduction to containers

In this section, we will introduce the concept of **containers** and look at their relationship with **virtual machines (VMs)**.

As a concept, containers enable standardized and isolated deployable compute resource units that are lightweight and portable.

Containers are built around encapsulation, which involves packaging the code of an application and its dependencies so that it can be deployed into development or production environments seamlessly with repeatable, predictable, and consistent results.

The benefit and value of containers are that they are *self-contained computing units* that are portable and lightweight.

In the next section, we will look at the relationship between containers and VMs.

Comparing containers and VMs

To understand the containers model, we will briefly look at it alongside virtualization (VMs).

Like a VM, a container is a compute resource unit; they have the same aim, which is to host and execute code. However, VMs carry a lot of overhead; they are non-standard, large in size, resource-intensive, monolithic, fragile, hard to move, and anti-agile. In contrast, containers could be considered *anti-VM*; their characteristics are the opposite in that they are standard, lightweight, small in size, utilize fewer resources, are portable, more agile, responsive, and have quicker boot times. They are well-positioned for digital transformation scenarios and modernizing your data center workloads.

The approach of using containers to deliver code execution is almost thought of as being closer to an application delivery model than a virtualization model.

This leads us to the next area of thought, where containers consume virtualization to deliver code execution by virtualizing the software resources – that is, the OS. This is a stark contrast to virtualization, where the VMs virtualize the hardware resources.

In the following diagram, we can visualize the concept of containers as compute units proportionate VMs, and the physical servers as the compute units:

```
          VM #2              Container
        ┌─────────┐            #2
        │ App #2  │         ┌─────────┐
        ├─────────┤         │ App #2  │
        │ OS #2   │         └─────────┘
        └─────────┘

          VM #1              Container
        ┌─────────┐            #1
        │ App #1  │         ┌─────────┐
        ├─────────┤         │ App #1  │
        │ OS #1   │         └─────────┘
        └─────────┘
```

App #1	App #2		Hypervisor		Docker
OS #1	OS #2		OS		OS

| Sever #1 | Sever #2 | | Sever #1 | | Sever #1 |

| Physical | Virtualization | Containerization |

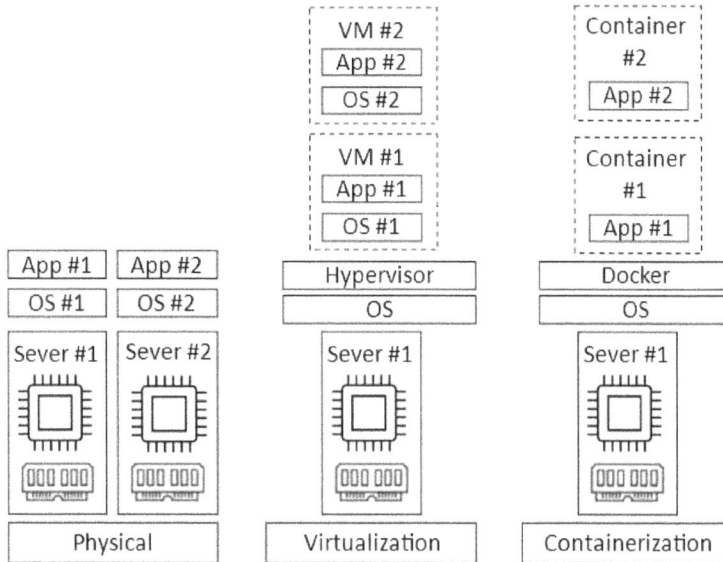

Figure 10.1 – Physical versus VM versus containers

The preceding diagram shows that, in a nutshell, fewer physical servers are required to deliver the same number of apps.

Looking closely, we can see that the traditional approach of having single apps per server meant we needed two physical servers to deliver the two apps. This diagram shows that the same two apps now only need one physical server with virtualization. Still, we needed to purchase, maintain, secure, and support the OS on each VM, which can be quite a cost and operations burden. With containers, however, we deliver the same number of apps, still on the same number of physical servers as virtualization, but we only need one copy of the OS.

To summarize, with containers, we *virtualize the software*; this allows many of these compute instances (containers) to share the underlying host software resources of a single-host OS. By doing so, we achieve greater efficiency, scale, and agility.

This section introduced containers and compared them to VMs and the traditional physical server model to help you understand the value and benefits of running containers to host and execute your application's code. In the next section, we will introduce Docker.

Introduction to Docker

Docker (`https://www.docker.com`) is an open source containerization platform that packages an app's code into the standardized unit of compute that we looked at in the previous sections of this chapter. Docker, being an open standard, runs on all major OS types, any infrastructure, and any cloud platform.

The two core components of the Docker platform are as follows:

- **Docker Engine**: This is a runtime environment that is lightweight, secure, and can run on Windows-based OSs, Linux, and macOS
- **Docker Client**: This is the **command-line interface** (**CLI**) for the engine and runs on the host computer

Docker can be installed on Windows Server using package management. The PowerShell `PackageManagement` module is available in Windows Server 2019 and later.

Docker can also be installed using the PowerShell `DockerMicrosoftProvider` module.

We will learn how to install Docker in the *Hands-on exercises* section of this chapter.

In this section, we introduced Docker. In the next section, we will look at securing Windows Server containers.

Securing Windows Server containers

This section looks at security in the context of isolation modes.

Windows Server containers provide two isolation modes: **process** and **Hyper-V isolation**. The degree of isolation that's created between each container and the host OS differentiates these isolation modes. Let's outline each mode:

- **Process isolation** (this is the traditional isolation mode):
  - The same kernel is shared between the containers and the host
  - There is a user mode per container
  - When starting a Docker container, the `-isolation=process` command switch is used
- **Hyper-V isolation**:
  - With this isolation mode, there is a kernel per container
  - Each container runs inside a VM
  - Hardware-level isolation is provided between each container and the host
  - When starting a Docker container, the `-isolation=hyperv` command switch is used

These two isolation modes are shown in the following diagram:

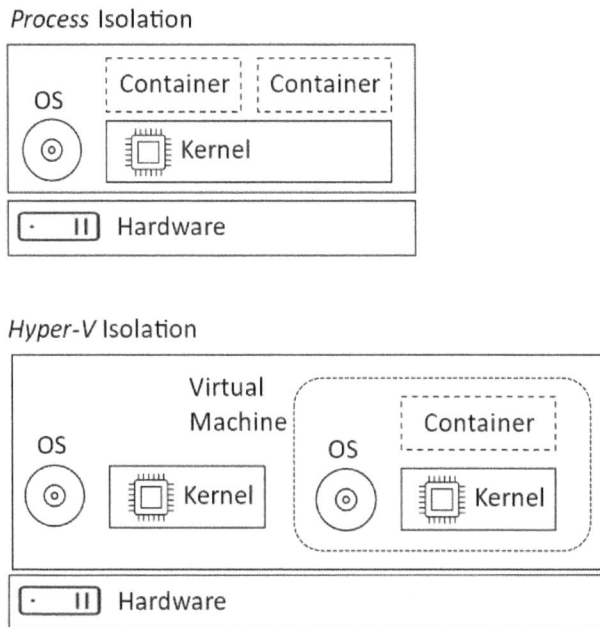

Process Isolation

Hyper-V Isolation

Figure 10.2 – Isolation modes

Process isolation is the default mode for Windows containers on Windows Server hosts. Windows containers running on Windows 10 hosts use *Hyper-V isolation*.

In this section, we looked at security in the context of isolation in Windows containers. In the next section, we will look at networking in Windows containers.

Networking for Windows Server containers

The concept of networking with Windows Server containers is similar in function to VMs. A **virtual network adapter** (**vNIC**) is attached to a container, which connects to a Hyper-V **virtual switch** (**vSwitch**). There are five networking drivers, or modes, that can be created through Docker to support Windows; these are as follows:

- L2bridge network driver
- L2tunnel network driver
- NAT network driver
- Overlay network driver
- Transparent network driver

Further details on each of these can be found in the following Microsoft documentation article: `https://docs.microsoft.com/en-gb/virtualization/windowscontainers/container-networking/network-drivers-topologies`.

Depending on the physical network and host networking requirements, the one that best suits your needs should be chosen.

You can run the `docker network ls` command to list the available networks.

NAT is the default network type that's created the first time Docker Engine runs. It uses the **WinNAT** Windows component and an internal vSwitch.

Two vSwitches are available; these are as follows:

- **Internal vSwitch**: This is not connected directly to a network adapter on the container host
- **External vSwitch**: This is connected directly to a network adapter on the container host

In this section, we looked at networking in Windows Server containers. In the next section, we will look at the process of creating Windows Server container images.

Creating Windows Server container images

A base image provides a foundation layer of the OS for a container; all containers will be based on that image. The base image contains the user mode OS files, runtimes files, required app dependencies, and any required miscellaneous configuration files.

Microsoft provides the following base images for Windows Server containers:

- **Windows**: This includes the full Windows APIs and system services but no server roles. It is not available on Windows Server 2022. This is the largest image (3.4 GB).
- **Windows Server**: This includes the full Windows APIs and more server features than the Windows image, has GPU support, and has no IIS limits. It requires Windows Server 2022 and is slightly smaller than the Windows image (3.1 GB).
- **Server Core**: This and Nano Server are the most common images. It includes a subset of the Windows APIs, including the .NET Framework and most server roles, and a medium-sized image. It is best for **lift and shift** scenarios.
- **Nano Server**: This includes the .NET Core APIs and some server roles; it is an ultralight image. The key image difference is a significantly smaller API surface; PowerShell, WMI, and the Windows servicing stack are not included.

The following information summarizes the appropriate base image to use for a given scenario:

- *Is a full .NET Framework required for the application?* You should use the **Windows Server Core** base image.

- *Is the app based on .NET Core?* You should use the **Nano Server** base image.

- *Does your app require a dependency that's missing from Windows Server Core?* You should use the **Windows** base image.

- *Is GPU acceleration support required?* You should use the **Windows Server** base image.

These base images can be discovered on Docker Hub: `https://hub.docker.com/_/microsoft-windows-base-os-images`.

A list of all the Microsoft container base images available can be found by running the following PowerShell command:

```
docker search Microsoft
```

These images can be downloaded from the **Microsoft Container Registry** (**MCR**).

To download a base image, you can use the `docker pull` command. The following is an example:

```
docker pull mcr.microsoft.com/windows/servercore:ltsc2022
```

Once the image(s) have been downloaded, you can run the following command to verify which images are now available locally and display metadata:

```
docker image ls
```

In this section, we outlined Microsoft-based container creation using Docker. In the next section, we will look at managing Windows Server container images.

Managing Windows Server container images

A Windows container is created, run, and managed through Docker. **Docker Engine** provides tools that help automate the creation process. The Docker components that enable this are as follows:

- **The Docker text file**, `Dockerfile`: This file includes instructions for creating a container image. This includes identifying the base image that will be used and the commands that will be run when creating the image and identifying when new instances of the image are deployed.

- **The Docker command**, `docker build`: This command calls the *Dockerfile*, which then triggers the image creation process.

In addition to automation, Docker commands can be run manually. The following are some common commands for reference:

- `docker images`: This command lists installed images on the container host

- `docker run`: This command creates a container using a container image

- `docker commit`: This command commits the changes that have been made to a container to a new container image

- `docker stop`: This command stops a running container

- `docker rm`: This command removes an existing container

The Docker approach allows you to store images as code, perform continuous integration with a development life cycle, and recreate quick and repeatable images for upgrade and maintenance purposes with consistent results.

Containers can also be managed by the browser-based GUI known as **Windows Admin Center (WAC)**. **Container extensions** must be installed so that container management in WAC is seen. This can be seen in the following screenshot:

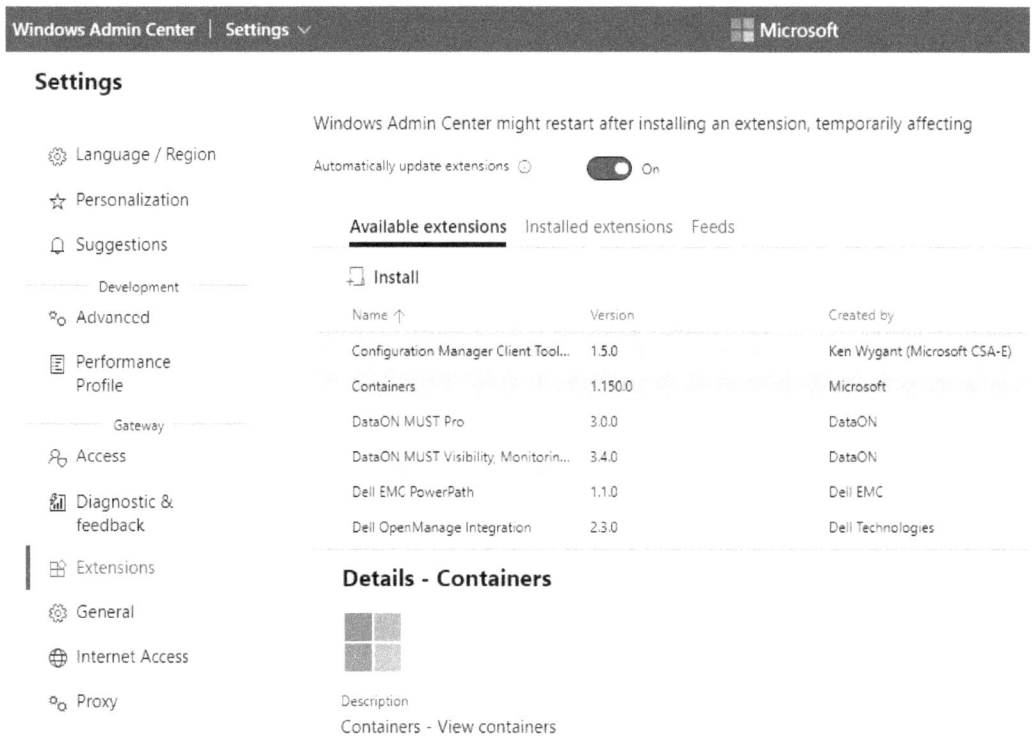

Figure 10.3 – Installing container extensions for WAC

Once the extensions have been installed, you will be prompted to install Docker if it does not exist on the host, as shown in the following screenshot:

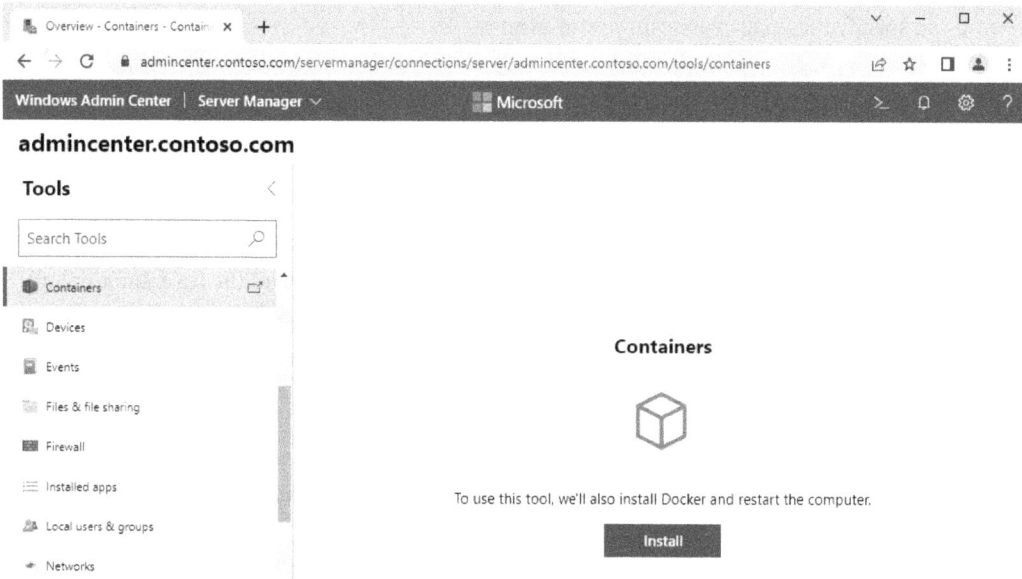

Figure 10.4 – Installing Docker via WAC

The WAC container management experience is shown in the following screenshot:

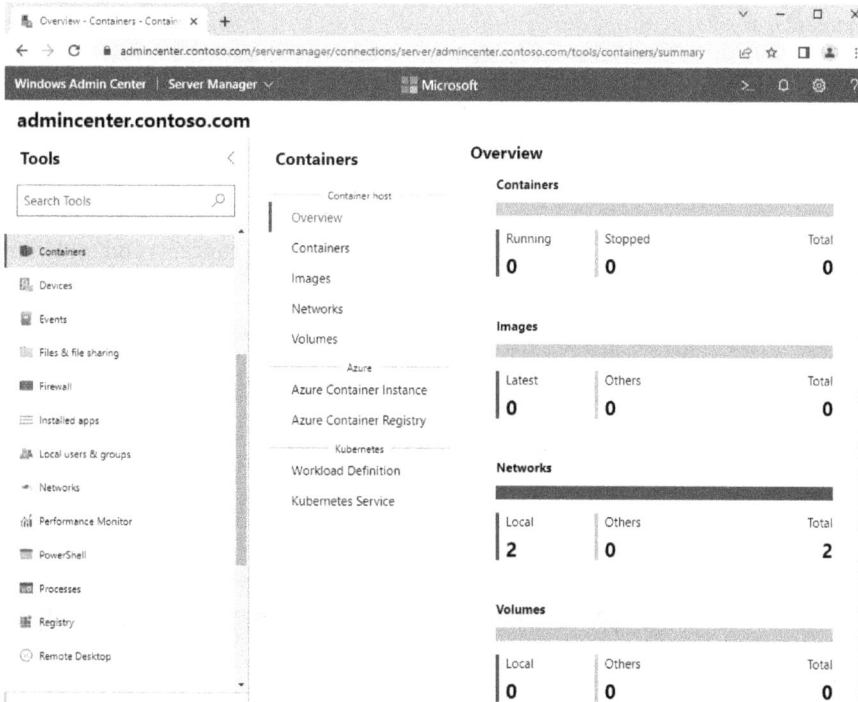

Figure 10.5 – Container management with WAC

Now, you can specify the image you want to run from MCR.

Maintaining container images

Container images should be regularly updated to ensure they receive changes that are made via code and patching.

The **base OS** layer contains OS elements that the host OS does not share. It comprises two images: the **base layer** and the **update layer**. The base layer is the larger layer but changes the least amount and needs the least amount of maintenance; the update layer is smaller. The following diagram represents these layers:

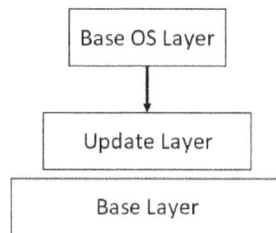

```
┌──────────────────────┐
│    Base OS Layer     │
└──────────┬───────────┘
           │
           ▼
┌──────────────────────┐
│    Update Layer      │
└──────────────────────┘
┌──────────────────────────┐
│      Base Layer          │
└──────────────────────────┘
```

Figure 10.6 – Container image update layers

In the update layer, a new image is created for each code change. The base OS layer gets its updates from the *updates* layer.

In this section, we looked at managing Windows containers. In the next section, we will complete a hands-on exercise to reinforce some of the concepts that were covered in this chapter.

Hands-on exercises

To support your learning with some practical skills, we will utilize the concepts and understanding we gained from this chapter and put them to practical use.

We will look at the following exercises:

- Exercise – installing Docker on Windows Server using PowerShell
- Exercise – installing Docker on Windows Server using WAC

Getting started

To get started with this section, you will need access to a Windows server; this can be physical, virtual, or an Azure IaaS VM. You will need access to an **elevated PowerShell session** and have **WAC** installed with admin access.

Let's move on to the exercises for this chapter.

Exercise – installing Docker on Windows Server using PowerShell

In this exercise, we will install Docker on a Windows Server using PowerShell. To do so, we will use a Windows Server 2019 Gen2 VM.

Follow these steps to get started:

1. Log into your Windows Server for this exercise using an admin account and open an *elevated PowerShell session*.

2. Run the following command to install the *Docker-Microsoft PackageManagement Provider* from the Windows PowerShell Gallery:

   ```
   Install-Module -Name DockerMsftProvider -Repository
   PSGallery -Force
   ```

3. Install the latest version of Docker using the `PackageManagement Windows` PowerShell module:

   ```
   Install-Package -Name docker -ProviderName
   DockerMsftProvider`
   ```

   The following is the output from the command:

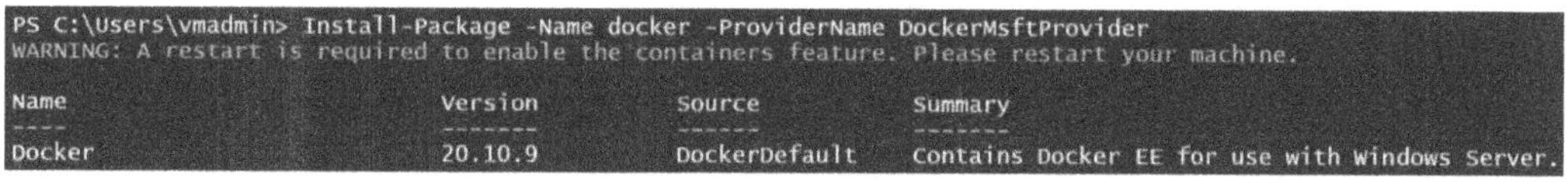

Figure 10.7 – Installing Docker

4. Restart the host when the Docker installation is complete using the following command:

   ```
   Restart-Computer
   ```

5. Log back into the server and re-open an elevated PowerShell session.

6. Run the following command to ensure the Docker service is running on the host:

   ```
   Start-Service docker
   ```

7. Run the following command to check the Docker version that's been installed:

   ```
   docker version
   ```

The following is the output from the command:

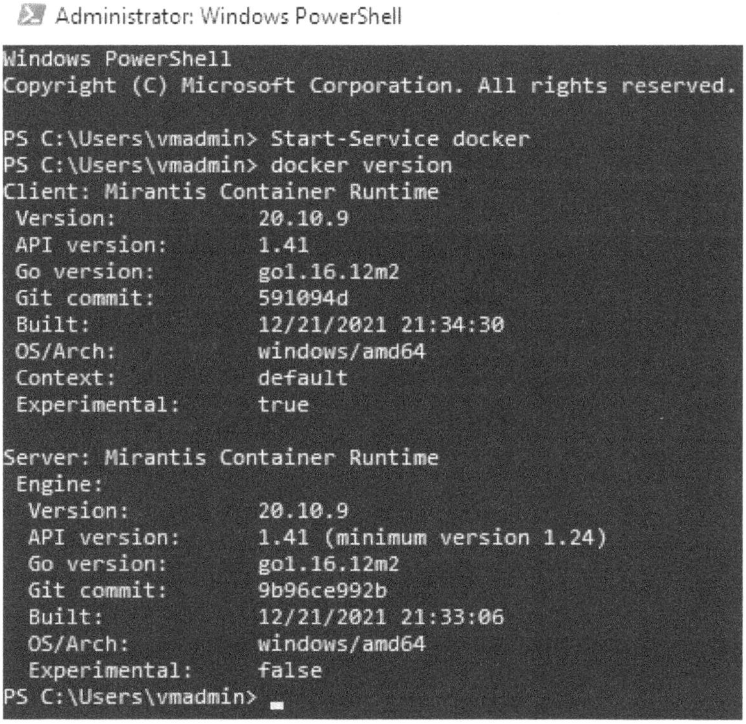

Figure 10.8 – Checking the Docker version

With that, we have completed this exercise. This exercise taught us how to install Docker on Windows Server using PowerShell.

Exercise – installing Docker on Windows Server using WAC

In this exercise, we will install Docker on a Windows Server using the **Windows Admin Center** (**WAC**). To do so, we will use a Windows Server 2019 Gen2 VM.

Follow these steps to get started:

1. Log into your Windows Server where WAC is installed and, using an admin account, launch WAC.

2. Once you've logged into WAC, click the gear icon to access the **Settings** screen from the top-right toolbar and click **Extensions**:

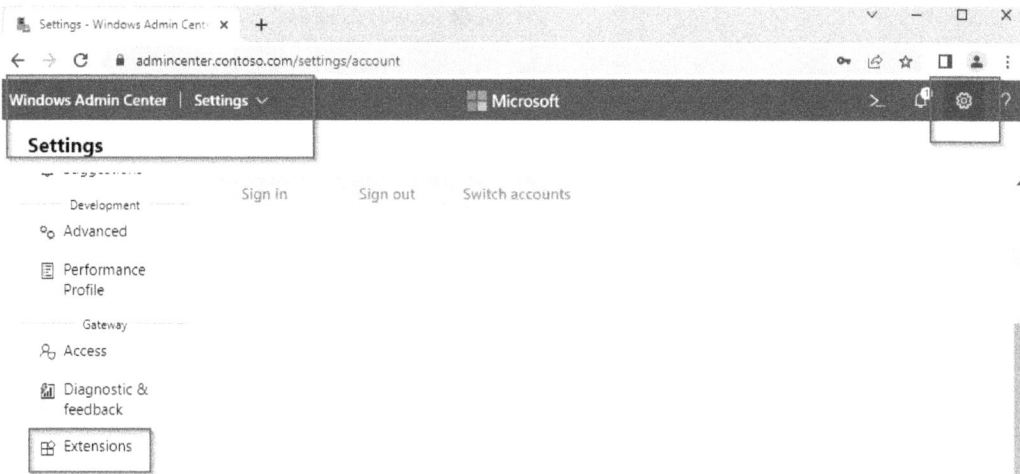

Figure 10.9 – WAC Settings screen

3. From the **Extensions** screen, search for the **Containers** extension:

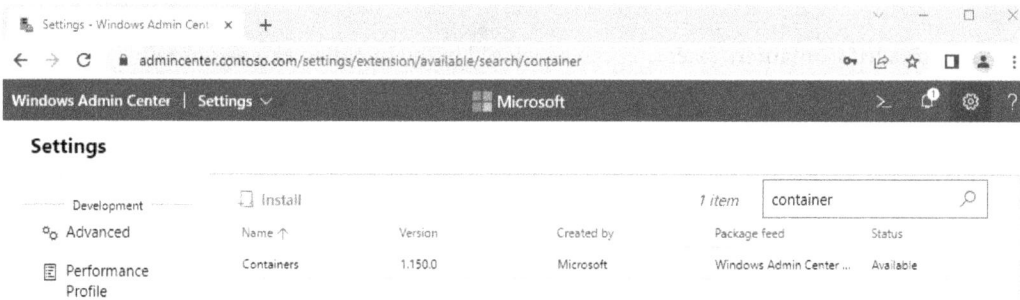

Figure 10.10 – WAC Containers extension

4. Click on the **Containers** extension, then click **Install**:

Install

Name ↑	Version	Created by	Package feed	Status
Containers	1.150.0	Microsoft	Windows Admin Center ...	Available

Installing the extension
Installing the extension: 'Containers'.
localhost   9:01:01 AM

Installing...

Please wait while we install 'Containers'. This page will reload after installation is complete.

OK

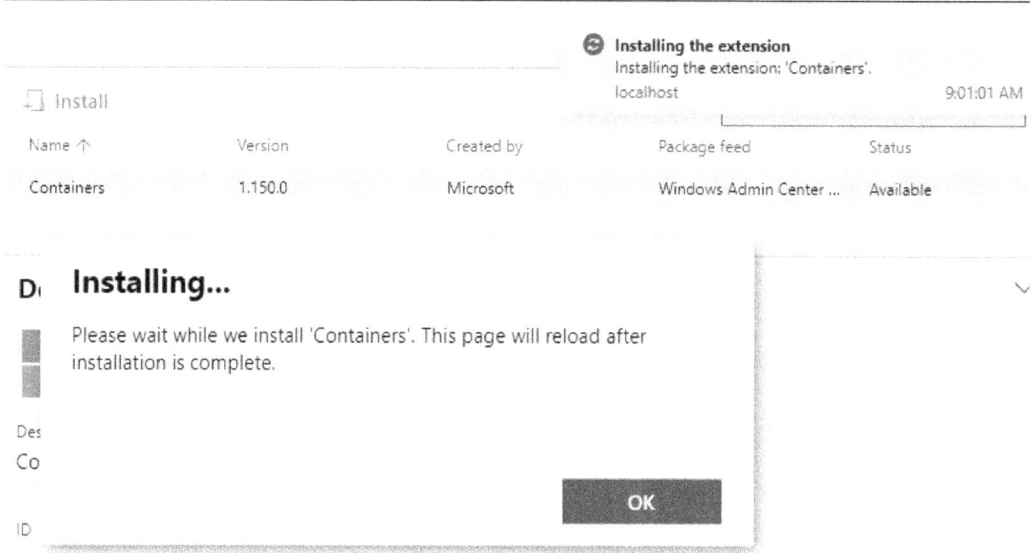

Figure 10.11 – WAC installing the extension

5. The installed **Containers** extension can be viewed in the **Installed extensions** tab:

Extensions

Windows Admin Center might restart after installing an extension, temporarily affecting anyone using this instance of Windows Admin Center.

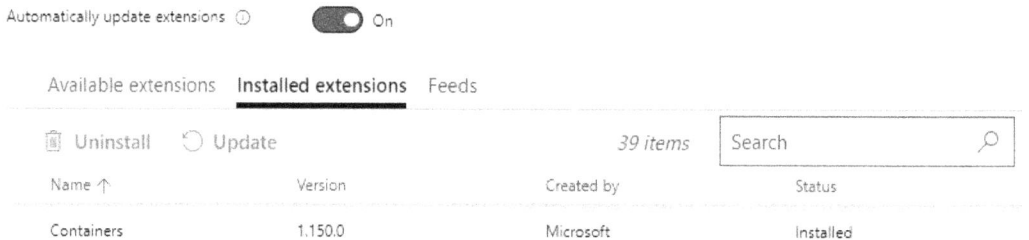

Automatically update extensions ⓘ   ⬤ On

Available extensions   **Installed extensions**   Feeds

🗑 Uninstall   ↻ Update   *39 items*   Search 🔍

Name ↑	Version	Created by	Status
Containers	1.150.0	Microsoft	Installed

Figure 10.12 – WAC extension installed

6. From the top-left menu navigation, select **All connections**. Then, select the **Gateway** entry to make a connection:

Windows Admin Center | All connections ∨   ■■ Microsoft

+ Add   ▭ Connect   ▭ Manage as   🗑 Remove   ✎ Edit Tags   *1 item*   *1 selected* ✕

✓	Name ↑	Type	Last connected	Managing as
✓	■ admincenter.contoso.com [Gateway]	Servers	Never	ADMINCENTER\Ad

Connecting...

• •• •

Cancel

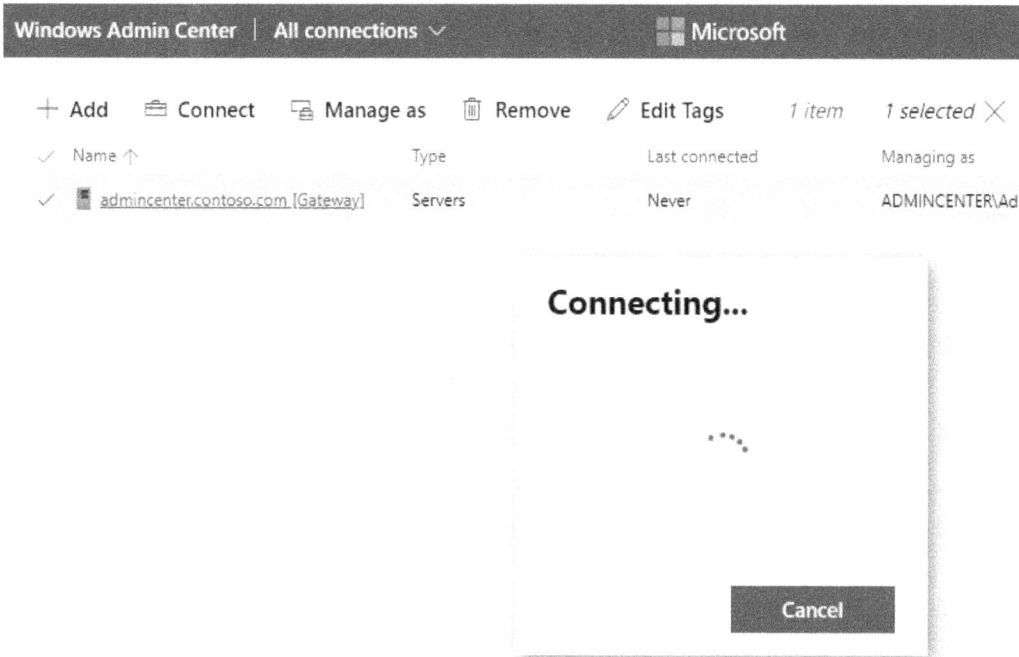

Figure 10.13 – WAC making a connection

7. From the left tools menu, click **Containers**:

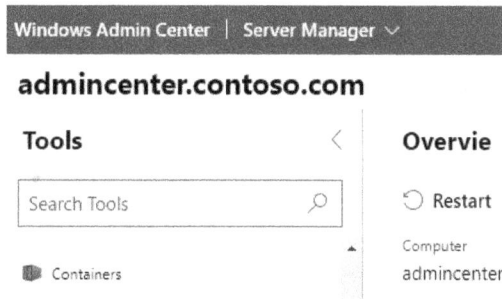

Windows Admin Center | Server Manager ∨

admincenter.contoso.com

Tools   ‹   **Overvie**

[Search Tools   ⌕]   ↻ Restart

                              ▲   Computer
🟦 Containers   admincenter

Figure 10.14 – Opening the Containers screen

8. When the page loads, click **Install**:

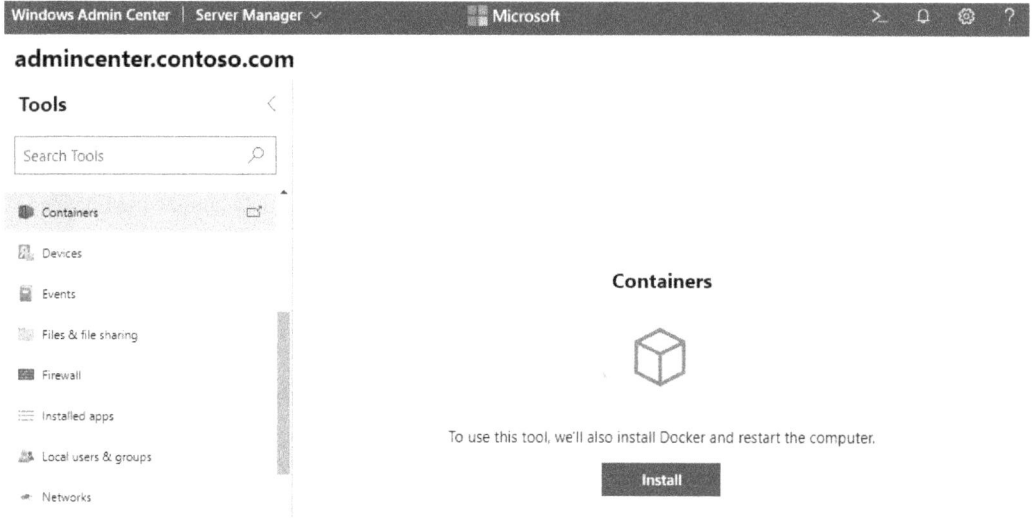

Figure 10.15 – WAC installing Docker

9. Wait a few moments while Docker is installed. Your computer will reboot once it's been installed:

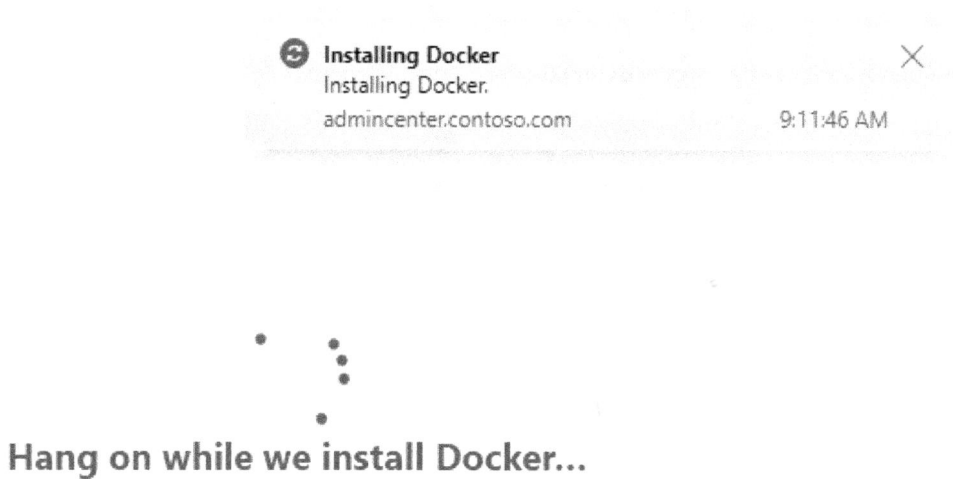

Figure 10.16 – Docker has been installed

10. Once your computer has restarted, launch WAC. You will see that Docker has been installed and that you are ready to create and manage containers:

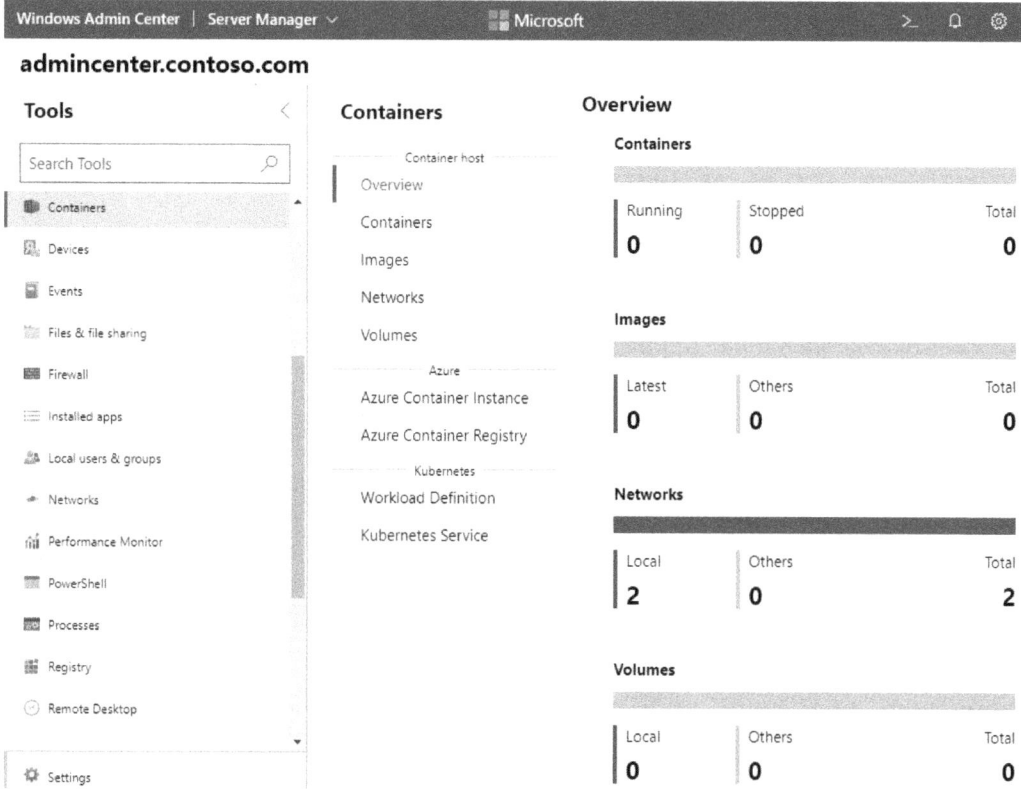

Figure 10.17 – Accessing Docker from WAC

With that, we have completed this exercise. This exercise taught us how to install Docker on Windows Server using WAC. Now, let's summarize this chapter.

Summary

This chapter has helped further your knowledge and skills regarding container technologies by providing coverage for *AZ-800 Administering Windows Server Hybrid Core Infrastructure: Create and manage containers*.

First, you learned about containers and compared them against VMs. Then, you learned how to create and manage Windows Server container images and perform container networking.

These new skills will take your knowledge beyond the exam objectives to prepare you for a real-world, day-to-day hybrid environment-focused role.

In the next chapter, you will learn how to manage Windows Server VMs.

Further reading

This section provides links to additional study references and additional exam information:

- *Microsoft Certified: Windows Server Hybrid Administrator Associate*: `https://docs.microsoft.com/en-us/learn/certifications/windows-server-hybrid-administrator/`

- *Exam AZ-800: Administering Windows Server Hybrid Core Infrastructure*: `https://docs.microsoft.com/en-us/learn/certifications/exams/az-800`

- *Exam AZ-800: Study Guide*: `https://query.prod.cms.rt.microsoft.com/cms/api/am/binary/RWKI0r`

- *Microsoft Learn*: `https://docs.microsoft.com/en-us/learn/modules/run-containers-windows-server/`

Skills check

Check what you have learned in this chapter by answering the following questions:

1. What are containers, and how do they differ from VMs?

2. What is Docker?

3. What arc thc two core components of the Docker platform?

4. What are the isolation modes, and how do they differ?

5. What are container images?

6. What are the four base Windows Server container images?

7. How do these four base Windows Server container images differ?

8. Where can the container base images be downloaded from?

9. In what ways can a container be managed?

10. Explain networking for Windows Server containers.

11

Managing Windows Server Azure Virtual Machines

In the previous chapter, we learned about containerization as a concept and saw how to implement and manage Windows Server containers.

This chapter covers content from *AZ-800 Administering Windows Server Hybrid Core Infrastructure: Manage Azure Virtual Machines that run Windows Server*.

This content does not aim to cover all aspects of the management of Windows Server Azure **Virtual Machines** (**VMs**); the content is limited in scope to only the exam skills outline. Some content from the exam objectives has already been covered in other chapters and will not be duplicated.

In addition to covering the following topics, this chapter's goal is to take your knowledge beyond the exam objectives to prepare you for a real-world, day-to-day, hybrid-environment-focused role.

The following topics are covered in this chapter:

- Managing Azure VM data disks
- Resizing Azure VMs
- Configuring continuous delivery for an Azure VM
- Configuring connections to VMs

We will then conclude with a hands-on exercise to develop your skills further.

Managing Azure VM data disks

When a VM is created, it must use a system disk that holds the operating system; this will be the C:\ drive. In addition, for some VM types, you can specify a **temp disk** that is automatically provisioned at the time of VM creation; this will appear as the D:\ drive.

We will not focus or cover content on the system disk types as this exam objective section discusses the data disk type.

The **data disk** can be attached to a VM in addition to the system and temp disk and is used if additional disk storage or layout is needed. This may be for installing software such as SQL Server, where we want different disk layouts for data, logs, and so on, or you may select data disks for performance reasons due to the effects of disk caching on the OS disk.

For applications and operations that are performance-sensitive, you should ensure you use a data disk for the installation path and storage of data rather than the OS disk. The reason for this is that by default, the OS disk has *read/write caching* enabled, which can impact performance on certain applications, such as a database; this includes a domain controller's database.

Microsoft advises domain controller IaaS VMs to use a data disk with caching disabled (by default, it's read caching) and *not* to use the OS disk to prevent write caching issues with operations. The recommendation is that the AD DS database (`ntds.dit`), logs, and **System Volume** (**SYSVOL**) use a volume on a data disk with caching disabled. Disk caching can be set using the Azure portal or with PowerShell. We can use the following PowerShell cmdlet:

```
AzVMDataDisk
```

More information can be found on caching performance and configuration in the following Microsoft Learn modules:

- `https://docs.microsoft.com/en-us/learn/modules/caching-and-performance-azure-storage-and-disks/`

- `https://docs.microsoft.com/en-us/learn/modules/caching-and-performance-azure-storage-and-disks/6-exercise-manage-cache-settings-with-powershell`

The data disk will be detached and reattached when changing the caching, but the VM does not need to be restarted, so there is no downtime with this operation; however, changing the system disk caching will cause a reboot of the VM.

Adding a new disk and **detaching** a disk will not require downtime of the VM; no VM restart is required. If the disk is not detached and you wish to resize the disk, the VM the disk is attached to will need to be deallocated; this will cause downtime. One feature, **LiveResize**, allows a drive to be expanded without downtime; information can be found in the following Microsoft documentation: `https://docs.microsoft.com/en-us/azure/virtual-machines/windows/expand-os-disk#expand-without-downtime`.

The following PowerShell cmdlets can be used for managing data disks:

- `New-AzDiskConfig`: This cmdlet creates a configurable disk object

- `Add-AzVMDataDisk`: This cmdlet attaches a data disk to a VM
- `$myVM.StorageProfile.DataDisks`: This provides info on the data disks, such as *name*, *disk size*, *caching*, **logical unit number** (**lun id**), and so on

The following are the different disk types that can be attached to an Azure IaaS VM:

- **Standard HDD**: Provides the lowest cost storage, lowest performance, and lowest durability
- **Standard SSD**: Provides consistent performance and low latency
- **Premium SSD**: Provides high performance and low latency
- **Ultra Disk**: Provides sub-millisecond latency

The disk **stock-keeping units** (**SKUs**) available to be selected for a data disk are represented in the following screenshot:

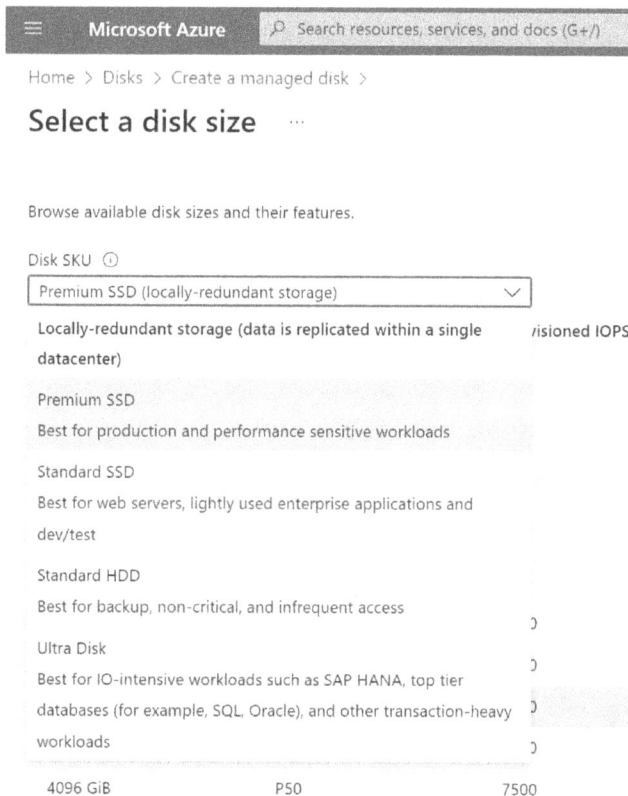

Figure 11.1 – Azure VM disk SKUs

There are several limitations of when the **Ultra Disk** SKU can be used; these are outlined in the following Microsoft documentation: `https://docs.microsoft.com/en-us/azure/virtual-machines/disks-types#ultra-disk-limitations`.

To use a **Premium** disk SKU as a data disk to attach to your VM, the VM type must support Premium disks; this can be seen in the Azure portal as represented in the following figure:

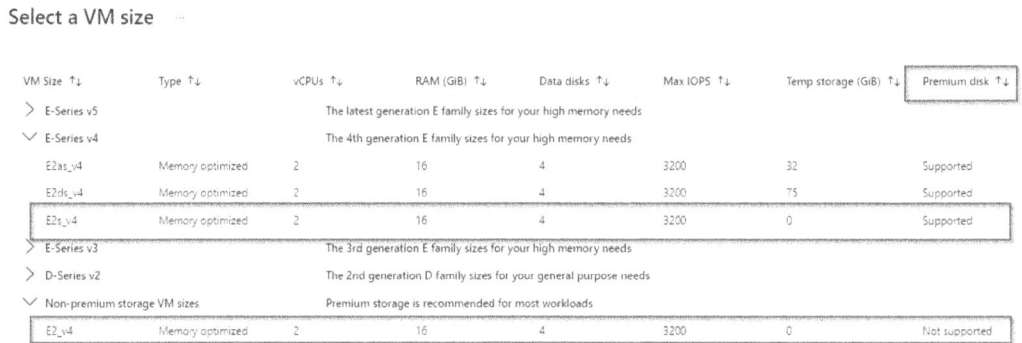

Select a VM size

VM Size ↑↓	Type ↑↓	vCPUs ↑↓	RAM (GiB) ↑↓	Data disks ↑↓	Max IOPS ↑↓	Temp storage (GiB) ↑↓	Premium disk ↑↓
> E-Series v5			The latest generation E family sizes for your high memory needs				
∨ E-Series v4			The 4th generation E family sizes for your high memory needs				
E2as_v4	Memory optimized	2	16	4	3200	32	Supported
E2ds_v4	Memory optimized	2	16	4	3200	75	Supported
E2s_v4	Memory optimized	2	16	4	3200	0	Supported
> E-Series v3			The 3rd generation E family sizes for your high memory needs				
> D-Series v2			The 2nd generation D family sizes for your general purpose needs				
∨ Non-premium storage VM sizes			Premium storage is recommended for most workloads				
E2_v4	Memory optimized	2	16	4	3200	0	Not supported

Figure 11.2 – Azure VM Premium disk support

The preceding screenshot shows that the **E2s_v4** VM size is showing as **Supported** for attaching a Premium disk, but the **E2_v4** VM size shows as **Not supported**. The "s" in the naming of the VM denotes that these VM sizes can support a Premium disk SKU.

To simplify selecting a VM size that meets your requirements, such as needing a Premium disk SKU for the data disk, you can add a filter for VMs that support Premium disks. This is represented in the following figure:

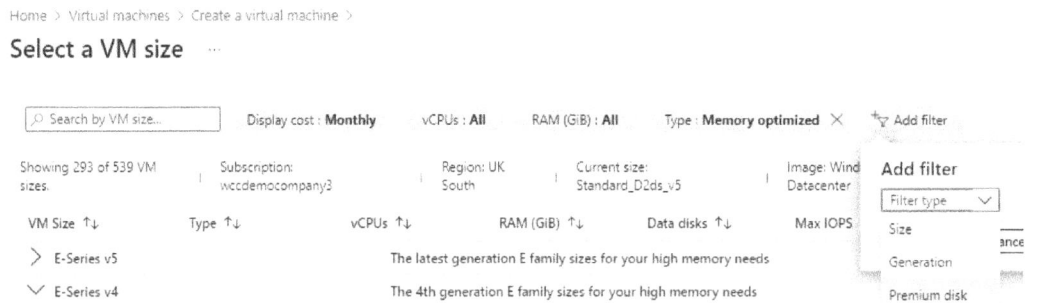

Home > Virtual machines > Create a virtual machine >

Select a VM size

	Display cost : **Monthly**	vCPUs : **All**	RAM (GiB) : **All**	Type : **Memory optimized** ✕	Add filter

Showing 293 of 539 VM sizes. | Subscription: wccdemocompany3 | Region: UK South | Current size: Standard_D2ds_v5 | Image: Wind Datacenter

Add filter

Filter type ∨

VM Size ↑↓	Type ↑↓	vCPUs ↑↓	RAM (GiB) ↑↓	Data disks ↑↓	Max IOPS
> E-Series v5			The latest generation E family sizes for your high memory needs		
∨ E-Series v4			The 4th generation E family sizes for your high memory needs		

Size
Generation
Premium disk

Figure 11.3 – Azure VM size Premium disk filter

In this section, we looked at managing IaaS VM data disks. In the next section, we will look at how to resize VMs.

Resizing Azure VMs

A VM can be scaled up or down to meet your requirements to any other VM size available in that region. This may be needed to meet performance demands for a VM running at peak capacity or where a VM is oversized and has underutilized resources that can be scaled back to optimize costs.

When you wish to resize a VM, your options will depend upon whether the VM is running or in the stopped/deallocated state.

Not all VM sizes may be available when a VM is running, as the underlying VM hardware cluster has to support the size you need.

It is also worth noting that not all regions hold all VM sizes to choose from. It might be considered to try and standardize a VM size(s) that is available across all your regions.

In some cases, you may need to ensure your VM is in the stopped/deallocated state, giving you the broadest choice of all sizes available in that region for your VM.

This is represented in the following figure:

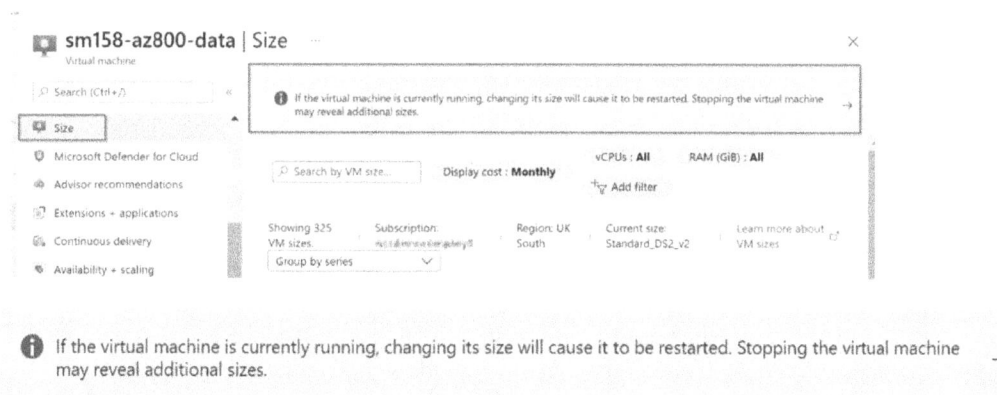

Figure 11.4 – Azure VM resize

The next aspect that may also impact your resize is if there is insufficient capacity in the region to support your needs. This is uncommon; however, scenarios such as the global pandemic saw a peak in demand, which meant some regions had no available capacity for a period. This can be addressed through the purchase of guaranteed reserved capacity, but there is a cost impact to be considered.

Another impact of scaling up in size is that subscriptions will have a quota limit on cores or VM types, meaning you won't be able to resize the VM. The following figure shows a core quota limit being reached.

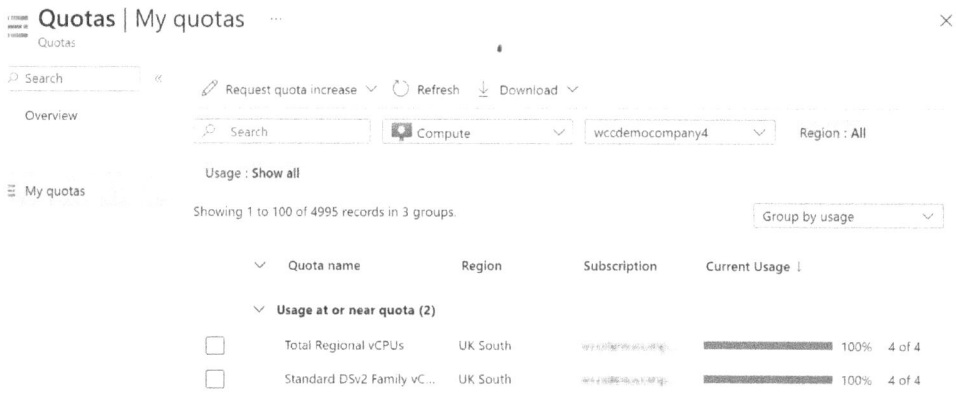

Figure 11.5 – Azure core quota limit at 100%

The following shows the message that is displayed when the quota is exceeded for a new VM:

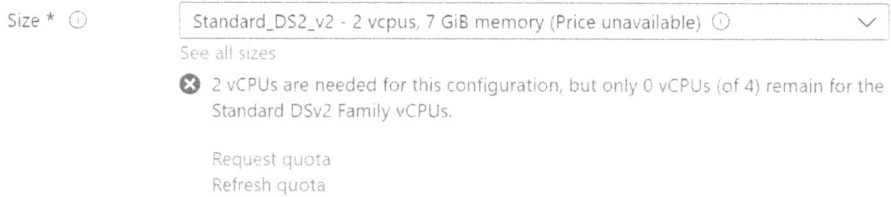

Figure 11.6 – Azure core quota exceeded message

However, this can be resolved with a service request to Microsoft, who can raise the quota, with the caveat that your purchase plan type must support an increase in quota; sandbox environments don't allow a core quota increase. The request quota increase is represented in the following figure:

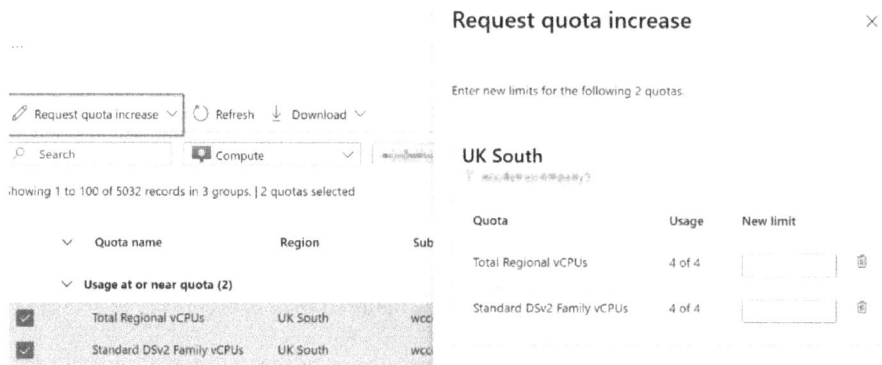

Figure 11.7 – Quota limit increase request

You must also ensure you resize to the same VM type; for example, if you have a VM size of Dds4, this means you have a D series with four vCPUs that provide a local temp disk and support SSD disks. You want to ensure you resize to keep those VM characteristics. That is, resize from Dds4 to Dds8, not D8; D8 would not support your SSD disks or provide the local temp disk.

Changing the VM size will cause it to be restarted (at the time of writing). This should be considered if you are using dynamic IP address allocation, as this will be lost when the VM is restarted; you should configure static IP allocation to resolve this.

This chapter will look at changing the VM size in the *Hands-on exercise* section.

In this section, we looked at the operations of resizing VMs. In the next section, we will look at configuring the continuous delivery of an Azure IaaS VM.

Configuring continuous delivery for an Azure VM

Continuous delivery in the context of Azure first-party solutions for Azure IaaS VMs relates to Azure Pipelines.

Azure Pipelines is better known as a software development tool but can be used to implement an **Infrastructure-as-Code (IaC)** approach to automate the deployment of *artifacts* to IaaS VMs in Azure as the compute target. The following figure visualizes where continuous delivery fits into the larger DevOps methodology.

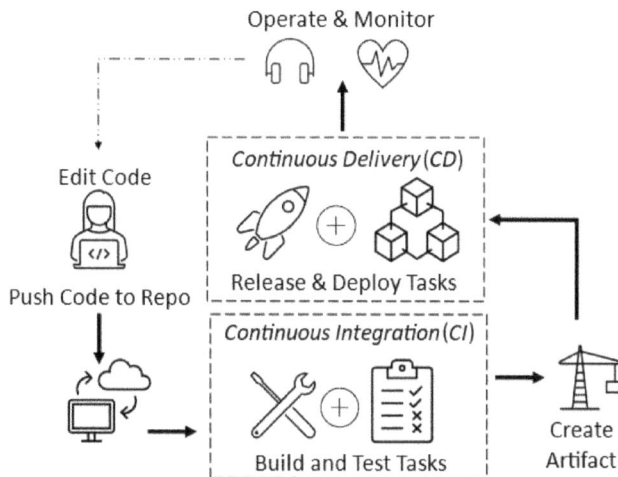

Figure 11.8 – CD within DevOps methodology

How this continuous delivery approach relates specifically to the IaC approach for deployment is illustrated in the following figure, which outlines a simplified view of the flow and components; for clarity, not all flow components are shown:

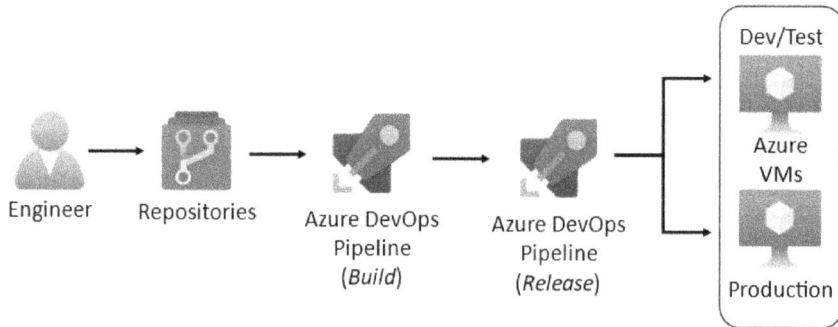

Figure 11.9 – Continuous delivery using Azure DevOps

The Azure portal continuous delivery solution for VMs is based on Azure Pipelines; this is shown in the following screenshot:

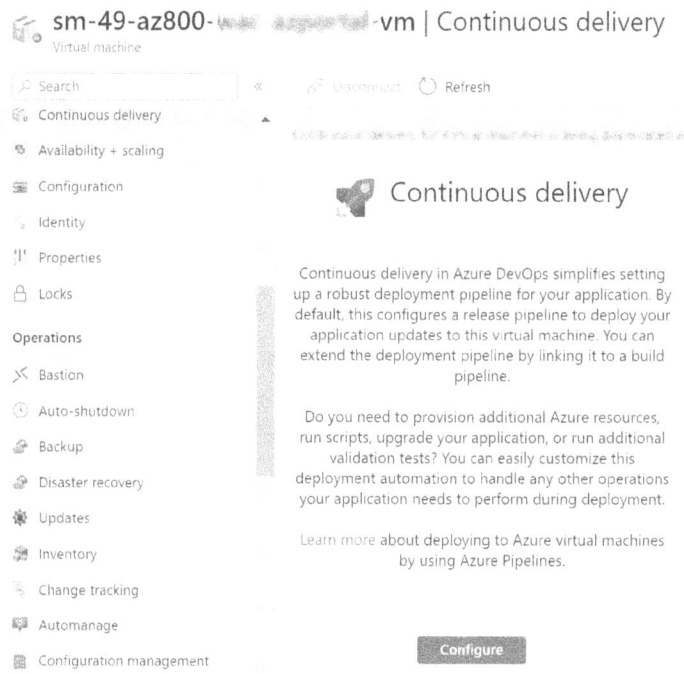

Figure 11.10 – Azure VM continuous delivery

To configure continuous delivery with a deployment pipeline, you will need to have an **Azure DevOps organization** that you can select; if you do not have an Azure DevOps organization to select, you will be prompted to create one. This is shown in the following figure:

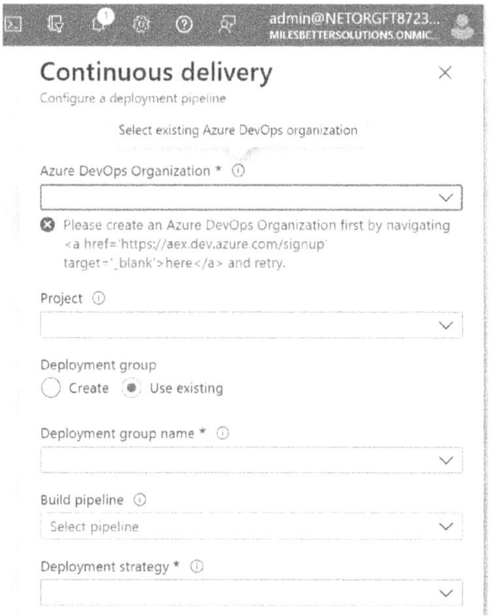

Figure 11.11 – Configuring a deployment pipeline

Full details on how to configure a deployment pipeline for an Azure VM using Azure DevOps can be found in the following Microsoft documentation: `https://docs.microsoft.com/en-us/azure/virtual-machines/linux/tutorial-devops-azure-pipelines-classic`.

This section looked at configuring continuous delivery for an Azure VM. In the next section, we will look at configuring connections to VMs.

Configuring connections to VMs

Once an Azure VM has been created, we must consider how to allow administrative access. We will look at the access methods of **Windows Remote Management** (**WinRM**) and **Remote Desktop Services Protocol** (**RDP**) in the following sections.

WinRM access

We can connect via PowerShell Remoting to run PowerShell commands on remote computers such as Azure IaaS VMs; this uses the WinRM protocol and allows us to establish a remote connection to the computer.

To enable PowerShell Remoting, we need to use the following PowerShell cmdlet: `Enable-PSRemoting`. The WinRM service will be started so that commands can be operated on the remote computer. The ports used are `HTTP:5985` and `HTTPS:5986`; all PowerShell remote communications are always encrypted by WinRM regardless of the protocol used (whether HTTP or HTTPS).

The outline of the process for setting up WinRM access to an Azure VM is as follows:

1. An Azure key vault is created.

2. A self-signed certificate is created.

3. The self-signed certificate is stored in Azure Key Vault.

4. The URL for the self-signed certificate stored in Azure Key Vault is retrieved.

5. When creating a VM, the URL for the self-signed certificate is referenced for access.

6. Ensure the VM is enabled for WinRM by enabling PSRemoting with the following PowerShell cmdlet:

   ```
   Enable-PSRemoting -Force
   ```

7. A connection can then be made using the following:

   ```
   Enter-PSSession -ConnectionUri https://<public-ip-
   dns-of-the-vm>:5986 -Credential $cred -SessionOption
   (New-PSSessionOption -SkipCACheck -SkipCNCheck
   -SkipRevocationCheck) -Authentication Negotiate
   ```

Further information on setting up can be found in the following Microsoft documentation: `https://docs.microsoft.com/en-gb/azure/virtual-machines/windows/connect-winrm`.

RDP access

As an alternative remote connection method, we can connect to an Azure VM using a full interactive desktop session using RDP for Windows VMs and **Secure Socket Shell** (**SSH**) for Linux servers. For the scope of the exam objective, we will cover Windows Server access using RDP.

Connectivity to the Windows Server Azure VM requires using TCP port `3389` by default. If you are using a firewall or a **Network Security Group** (**NSG**), you will need to allow inbound access to this port and protocol; for security purposes, you limit the IP address range you will accept connections from. Further details on configuring the NSG for RDP access can be found in the following Microsoft documentation: `https://docs.microsoft.com/en-us/troubleshoot/azure/virtual-machines/troubleshoot-rdp-nsg-problem`.

If you are not using a VPN or ExpressRoute to connect to resources in the Azure environment, you must also have a public IP associated with the VM. Further details on configuring this can be found in the following Microsoft documentation: `https://docs.microsoft.com/en-gb/azure/virtual-network/ip-services/associate-public-ip-address-vm`.

Chapter 6 included a hands-on exercise for remotely accessing an Azure VM using RDP and configuring an NSG to allow inbound RDP traffic.

In this section, we looked at configuring connections to VMs. In the next section, we will complete a hands-on exercise section to reinforce some of the concepts covered in this chapter.

Hands-on exercise

To support your learning with some practical skills, we will utilize the concepts and understanding gained from the earlier sections of this chapter and put them into practical application.

We will look at the following exercises:

* Exercise – managing a data disk using the Azure portal
* Exercise – resizing an Azure VM

Getting started

To get started with this section, if you do not already have an Azure subscription, you can create a free Azure account at `https://Azure.microsoft.com/free`. This free Azure account provides the following:

* 12 months of free services
* $200 credit to explore Azure for 30 days
* 25+ services that are always free

Let's move on to the exercise for this chapter.

Exercise – managing a data disk using the Azure portal

This exercise will look at managing data disks using the Azure portal.

For this exercise, you will need an Azure VM created to attach and resize data disks; once you have created a VM, return here to complete the exercise or continue if you have an existing VM to use.

In the following sub-sections, you can see the procedure to complete the exercise, segregated into tasks for a better understanding.

Task 1 – accessing the Azure portal

Log in to the Azure portal: `https://portal.azure.com`. You can alternatively use the desktop Azure portal app: `https://portal.azure.com/App/Download`.

Task 2 – attaching a data disk to a VM

1. In the search bar, type in `virtual machines`; click **Virtual machines** from the list of services.

2. Locate your VM for this exercise from the **Virtual machines** blade and click on the name to open the **Virtual machines** blade.

Home >

Virtual machines ✈ ...

milesbettersolutions.onmicrosoft.com (milesbettersolutions.com)

| + Create ∨ | ⮂ Switch to classic | ⏱ Reservations ∨ | ⚙ Manage view ∨ | ↻ Refresh | ↓ Export to CSV | ⚬ Open query | ⊘ Assign tags |

| Filter for any field... | Subscription equals **all** | Type equals **all** | Resource group equals **all** ✕ | Location equals **all** ✕ | ⊹ Add fil |

No grouping

☐ Name ↑	Type ↑↓	Subscription ↑↓	Resource group ↑↓	Location ↑↓	Status ↑↓
☐ 💻 sm029-az800-vm1	Virtual machine	Azure subscription 1	SM029-RG	UK South	Running

Figure 11.12 – VM for the exercise

3. From the **Virtual machines** blade, click **Disks** under **Settings**.

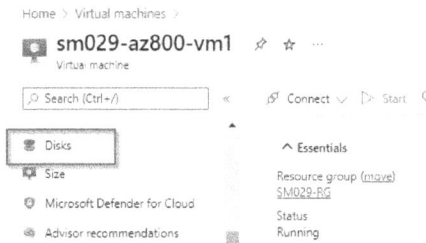

Home > Virtual machines >

💻 sm029-az800-vm1 ✈ ☆ ...
Virtual machine

| 🔍 Search (Ctrl+/) | « | 🔗 Connect ∨ | ▷ Start | ↻ |

🖴 Disks		∧ Essentials
🖴 Size		Resource group (move)
🛡 Microsoft Defender for Cloud		SM029-RG
⚙ Advisor recommendations		Status
		Running

Figure 11.13 – VM disk settings

4. From the **Disks** blade for the VM, click + **Create and attach a new disk**.

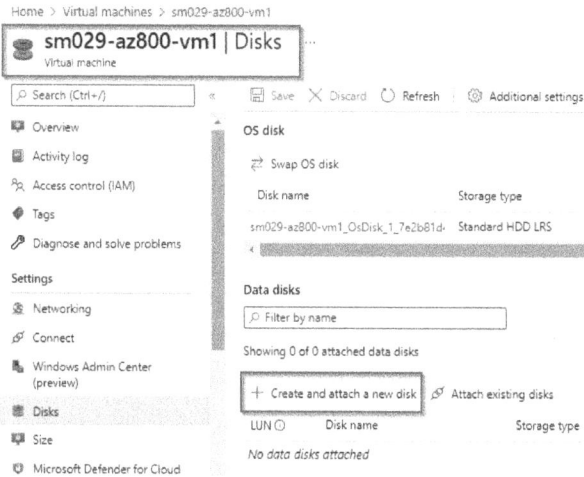

Figure 11.14 – Data disks

5. For the **LUN** number, leave it at the default of **0**, set **Disk name** as **Data**, set **Storage type** as **Standard HDD**, and leave **Size** as the default, then click **Save** from the top menu bar.

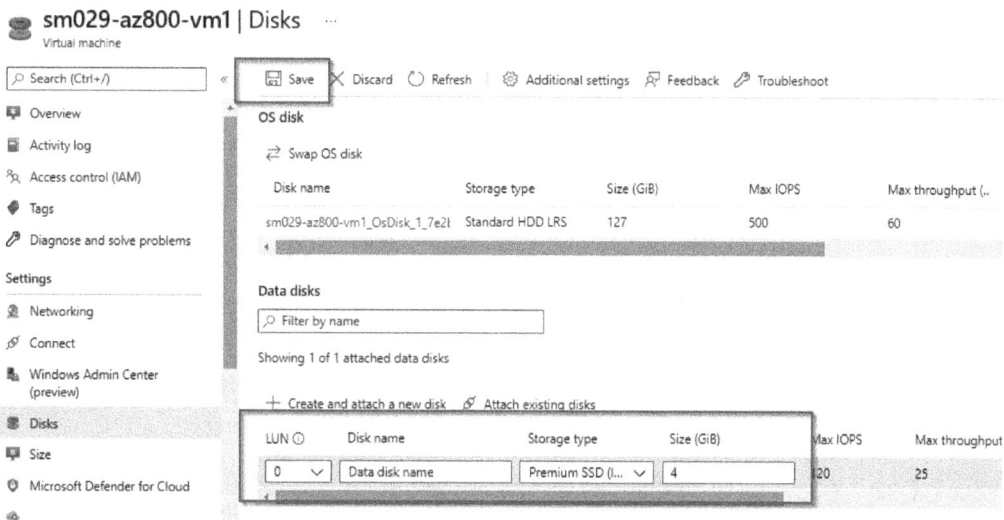

Figure 11.15 – Adding a data disk

6.   You will receive a message to say that the disk was created successfully and will be displayed in the **Disks** blade for the VM.

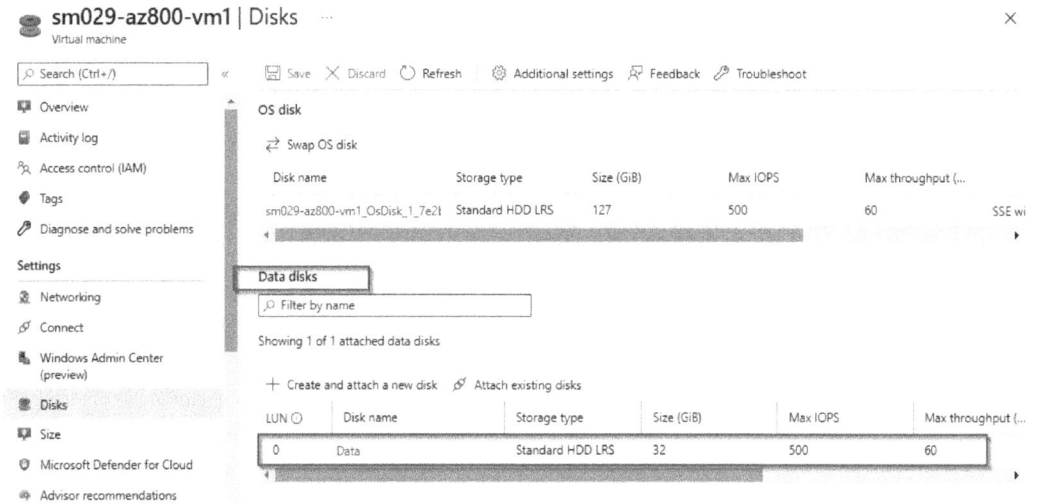

Figure 11.16 – Data disks

7.   Repeat the process to add another data disk; this time, call the disk **Logs** and create a Premium SSD; we will now see this second disk in the **Disk** blade.

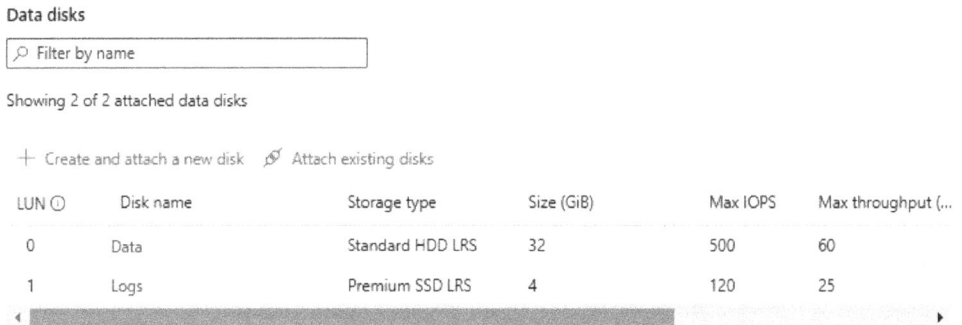

Figure 11.17 – Adding a data disk

8. Navigate to the **Disks** blade, and you will see both the data disks and the OS disk for the VM.

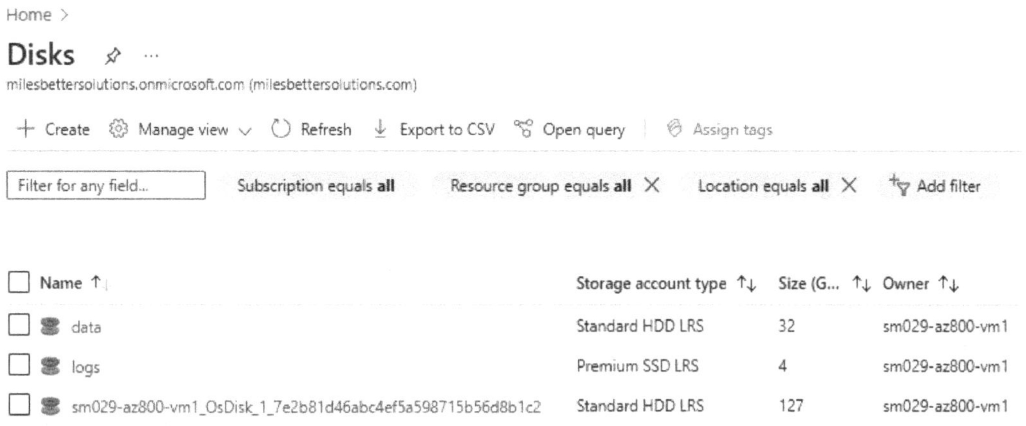

Home >

Disks ✗ ···
milesbettersolutions.onmicrosoft.com (milesbettersolutions.com)

+ Create   ⚙ Manage view ∨   ↻ Refresh   ↓ Export to CSV   ⁗ Open query | ⊘ Assign tags

| Filter for any field... | Subscription equals **all** | Resource group equals **all** ✕ | Location equals **all** ✕ | ⁺ᵧ Add filter |

☐ Name ↑	Storage account type ↑↓	Size (G... ↑↓	Owner ↑↓
☐ 🗄 data	Standard HDD LRS	32	sm029-az800-vm1
☐ 🗄 logs	Premium SSD LRS	4	sm029-az800-vm1
☐ 🗄 sm029-az800-vm1_OsDisk_1_7e2b81d46abc4ef5a598715b56d8b1c2	Standard HDD LRS	127	sm029-az800-vm1

Figure 11.18 – Data disks

In this task, we attached two data disks to a VM. In the next tasks, we will resize the created data disks.

Task 3 – resizing a data disk for a VM

We will resize a disk for this exercise by deallocating the VM instead of detaching and reattaching:

1. Stop/deallocate your VM by clicking the **Stop** option.

Figure 11.19 – Stopping the VM

2. Once the VM has been stopped, click on **Disks** under **Settings** to open the **Disks** blade for the VM.

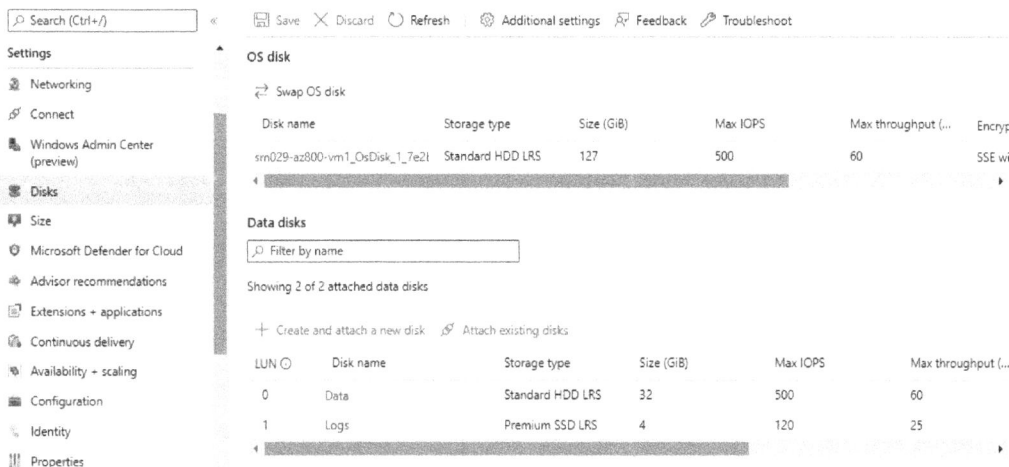

Figure 11.20 – Disks blade

3. Click the data disk named **Data** (LUN 0), open the blade for that disk, and click **Size + performance**.

Figure 11.21 – Disk size

4. From the **Size + performance** blade for the disk, select a new size of **64 GiB** and click **Resize**.

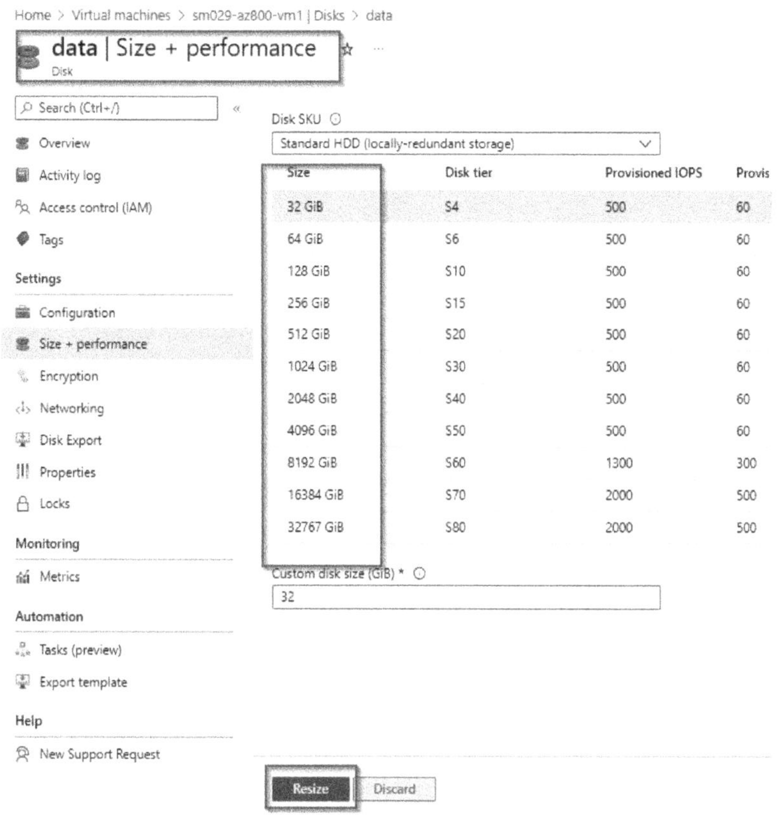

Home > Virtual machines > sm029-az800-vm1 | Disks > data

data | Size + performance
Disk

Search (Ctrl+/)	«	Disk SKU ○

Standard HDD (locally-redundant storage)

Size	Disk tier	Provisioned IOPS	Provis
32 GiB	S4	500	60
64 GiB	S6	500	60
128 GiB	S10	500	60
256 GiB	S15	500	60
512 GiB	S20	500	60
1024 GiB	S30	500	60
2048 GiB	S40	500	60
4096 GiB	S50	500	60
8192 GiB	S60	1300	300
16384 GiB	S70	2000	500
32767 GiB	S80	2000	500

Left navigation:
- Overview
- Activity log
- Access control (IAM)
- Tags

Settings
- Configuration
- Size + performance
- Encryption
- Networking
- Disk Export
- Properties
- Locks

Monitoring
- Metrics

Automation
- Tasks (preview)
- Export template

Help
- New Support Request

Custom disk size (GiB) * ○
32

Resize   Discard

Figure 11.22 – Select disk size

5. You will see a notification that the disk was updated successfully.

6. Return to the **Disks** blade, and the new disk size will be shown as **64 GiB**; you may need to click **Refresh** on the blade.

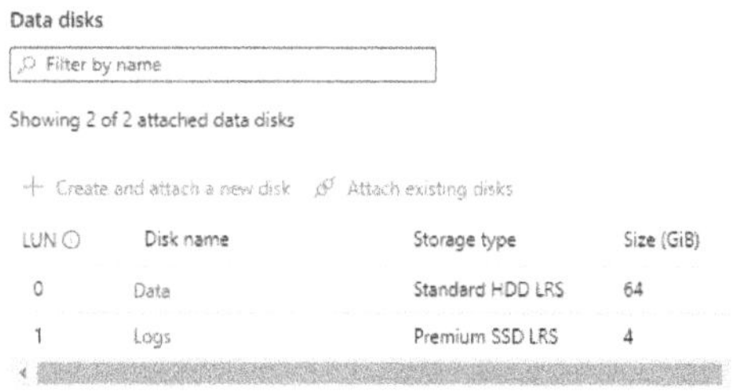

Data disks

Filter by name

Showing 2 of 2 attached data disks

+ Create and attach a new disk   Attach existing disks

LUN ○	Disk name	Storage type	Size (GiB)
0	Data	Standard HDD LRS	64
1	Logs	Premium SSD LRS	4

Figure 11.23 – Disk size changed

7. You should now start the VM to return it to operation.

With this, we have completed this exercise. This exercise taught us the skills needed to add and resize data disks. In the next exercise, we look at resizing a VM.

Exercise – resizing an Azure VM

In this exercise, we will resize a VM; you will need to create a VM or use an existing VM for this exercise.

Follow these steps to get started:

1. In the search bar, type in `virtual machines`; click **Virtual machines** from the list of services.

2. Locate your VM for this exercise from the **Virtual machines** blade and click on the name to open the **Virtual machines** blade.

3. From the **Virtual machines** blade, click **Size** under **Settings**.

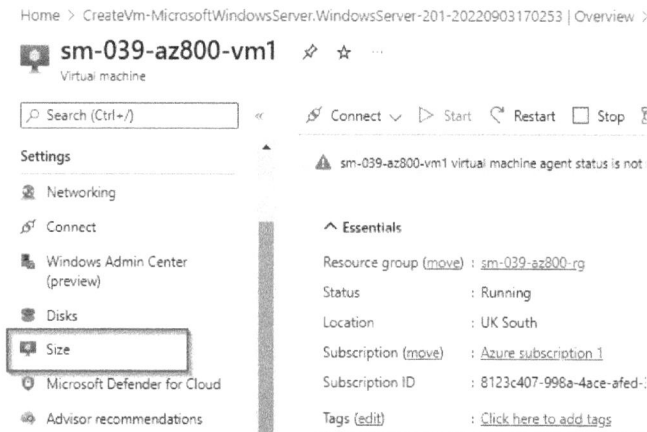

Figure 11.24 – VM size settings

4. The **Size** blade will show the VM's current size, that is, **Standard_ D2ds_v5**, in the example in the figure.

Figure 11.25 – Current VM size

5. Select the **D4ds_v5** VM size, which keeps the same VM type and characteristics of local temp disk and Premium disk support, and then click **Resize**. Note that this will cause the VM to be restarted.

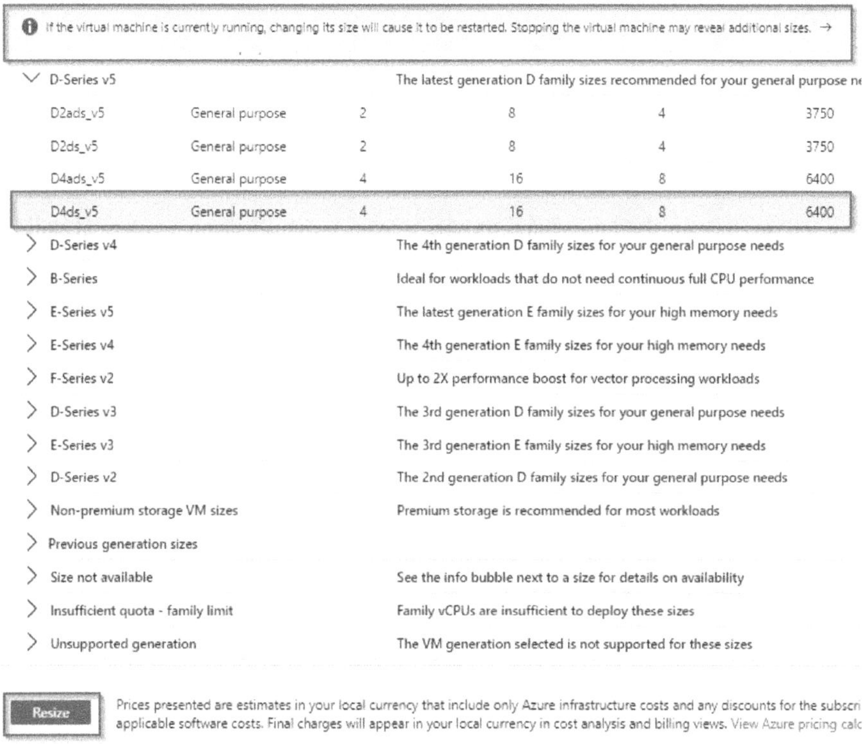

Figure 11.26 – Selecting the new VM size

6. You will receive a notification that the machine has been resized.

7. Return to the **Virtual machines** blade's **Overview** pane, and the new size will be shown as **Standard D4ds v5 (4 vcpus, 16 GiB memory)**.

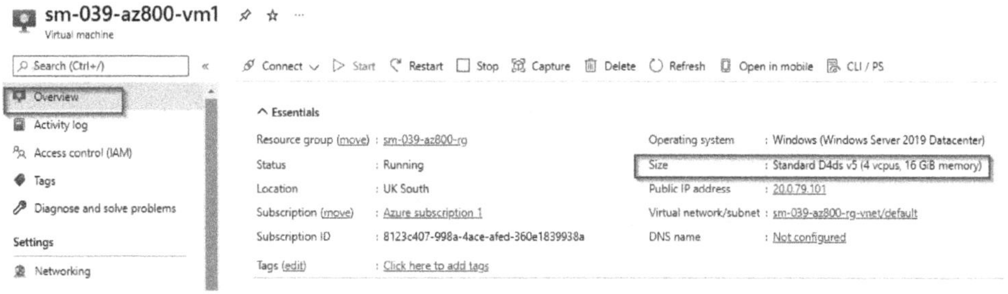

Figure 11.27 – Changed VM size

With this, we have completed this exercise. This exercise taught us the skills needed to resize an Azure VM. Now, let's summarize this chapter.

Summary

This chapter has provided coverage for *AZ-800 Administering Windows Server Hybrid Core Infrastructure: Manage Azure Virtual Machines that run Windows Server.*

We covered content that gave you the skills to manage Windows Server VMs on the Azure platform. We looked at aspects such as managing disks, resizing VMs, and configuring connections to VMs.

A hands-on exercise then finished the chapter to provide you with additional practical skills.

You gained new skills through the information provided; this chapter's goal was to take your knowledge beyond the exam objectives to prepare you for a real-world, day-to-day, hybrid-environment-focused role.

The next chapter will teach you about managing Windows Server in a hybrid environment.

Further reading

This section provides links to additional study references and additional exam information:

- *Microsoft Certified: Windows Server Hybrid Administrator Associate*: https://docs.microsoft.com/en-us/learn/certifications/windows-server-hybrid-administrator/

- *Exam AZ-800: Administering Windows Server Hybrid Core Infrastructure*: https://docs.microsoft.com/en-us/learn/certifications/exams/az-800

- *Exam AZ-800: skills outline*: https://query.prod.cms.rt.microsoft.com/cms/api/am/binary/RWKI0r

- *Microsoft Learn*: https://docs.microsoft.com/en-us/learn/paths/manage-virtualization-containers-hybrid-environment/

Skills check

Challenge yourself with what you have learned in this chapter:

1. When is a data disk used?

2. What is the effect of caching on VM disks?

3. What are the different disk SKUs that can be attached to a VM?

4. What are the limitations of an Ultra Disk SKU?

5. What VM type must be used to support Premium disks?

6. Will changing the size of a disk or attaching a new disk require downtime for the VM?

7. Name some scenarios where not all VM sizes are available when needing to resize a VM.

8. Will changing a VM's size require downtime?

9. What is continuous delivery in the context of Azure VMs?

10. What are the different ways we can configure a remote connection to a VM?

12

Managing Windows Server in a Hybrid Environment

In the previous chapter, we added skills for managing Windows Server **virtual machines** (**VMs**) in Azure.

This chapter covers content from *AZ-800 Administering Windows Server Hybrid Core Infrastructure: Manage Windows Servers and workloads in a hybrid environment.*

We will cover implementing **Windows Admin Center** (**WAC**) to simplify remote server management, covering the different deployment methods of the gateway and Azure portal approaches. We will also cover PowerShell remoting and how we can enhance security through the use of the **Just Enough Administration** (**JEA**) capability that can be used in conjunction. We will then conclude with a hands-on exercise to develop your skills further.

This chapter does not aim to cover all aspects; the content is limited in scope to only the exam objectives and will cover the following topics:

- Deploying a WAC gateway server
- Deploying WAC within the Azure portal
- PowerShell remoting
- Hands-on exercise

In addition to the topics listed in this chapter, this chapter's goal is to take your knowledge beyond the exam objectives to prepare you for a real-world, day-to-day, hybrid environment-focused role.

Deploying a WAC gateway server

This section will look at managing multiple Azure VMs by manually deploying WAC to an Azure VM. Before we jump into the configuration, we should introduce how WAC works.

WAC uses an installed gateway component with a web server to manage servers through **PowerShell remoting** and **Windows Management Instrumentation** (**WMI**) over **Windows Remote Management** (**WinRM**). These components are outlined in the following diagram:

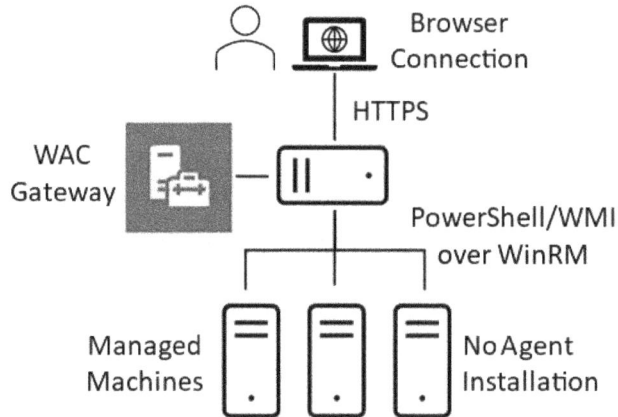

Figure 12.1 – WAC components

The WAC gateway can be deployed to a new or existing Azure VM using a script or manually via the Azure portal.

Script deployment

To deploy the WAG gateway to an Azure VM using the Azure portal, you will first need to download the `Deploy-WACAzVM.ps1` script (`https://aka.ms/deploy-wacazvm`), which runs a **Microsoft installer** (**MSI**) to create all the resources required for the environment, such as a resource group, VM, and **virtual network** (**VNet**). Alternatively, you can deploy to an existing VM and use a local MSI that has been downloaded and reference the file path in the script.

In both cases, you will require a certificate; this should be created and stored in Azure Key Vault, or a self-signed certificate from the MSI could be used.

The script is run from Azure Cloud Shell; the following is an example script.

First, set the variables:

```
$ResourceGroupName = "az800-wac-rg"
$VirtualNetworkName = "az800-wac-vnet"
$SecurityGroupName = "az800-wac-nsg"
$SubnetName = "az800-wac-subnet"
$VaultName = "az800-wac-keyvault"
$CertName = "az800-wac-cert"
```

```
$Location = "uksouth"
$PublicIpAddressName = "az800-wac-publicip"
$Size = "Standard_B2ms_v3"
$Image = "Win2019Datacenter"
$Credential = Get-Credential
```

The following script deploys WAC to a new VM, resource group, and VNet, as set in the preceding variables; it utilizes the MSI from aka.ms/WACDownload and the MSI-generated self-signed certificate:

```
$scriptParams = @{
    ResourceGroupName = $ResourceGroupName
    Name = "az800-wac-vm1"
    Credential = $Credential
    VirtualNetworkName = $VirtualNetworkName
    SubnetName = $SubnetName
    GenerateSslCert = $true
}
./Deploy-WACAzVM.ps1 @scriptParams
```

This deployment method is covered in more detail in the following Microsoft documentation article: https://docs.microsoft.com/en-gb/windows-server/manage/windows-admin-center/azure/deploy-wac-in-azure#deploy-using-script.

Manual deployment

To manually deploy the WAG gateway to an Azure VM, you first create an Azure VM for use as the WAG gateway. You then need to download the MSI installer directly to the VM or a local machine and copy the installer onto the VM; the installer can be downloaded from https://www.microsoft.com/en-gb/evalcenter/download-windows-admin-center.

Launch the *MSI installer* on the WAC gateway VM and follow the wizard instructions. You will need a certificate that you can provide or let the MSI generate a self-signed certificate.

For communication, port 443 (**HTTPS**) is used by default by the installer; you should allow this communication through any network filter, such as a **network security group** (**NSG**) or firewall.

Accessing WAC

WAC is accessed via a browser using the gateway VM's DNS name; that is, `https://<DNS name of your VM>`.

If you set a port other than the default `443`, you will need to get access using the following URL:

```
https://<DNS name of your VM>:<custom port>
```

The browser will prompt for credentials to authenticate the session on the local VM where WAC is installed. These credentials should be for an account that is a member of the VM's *local users* or *administrators* group.

The following screenshot shows the WAC browser-based interface:

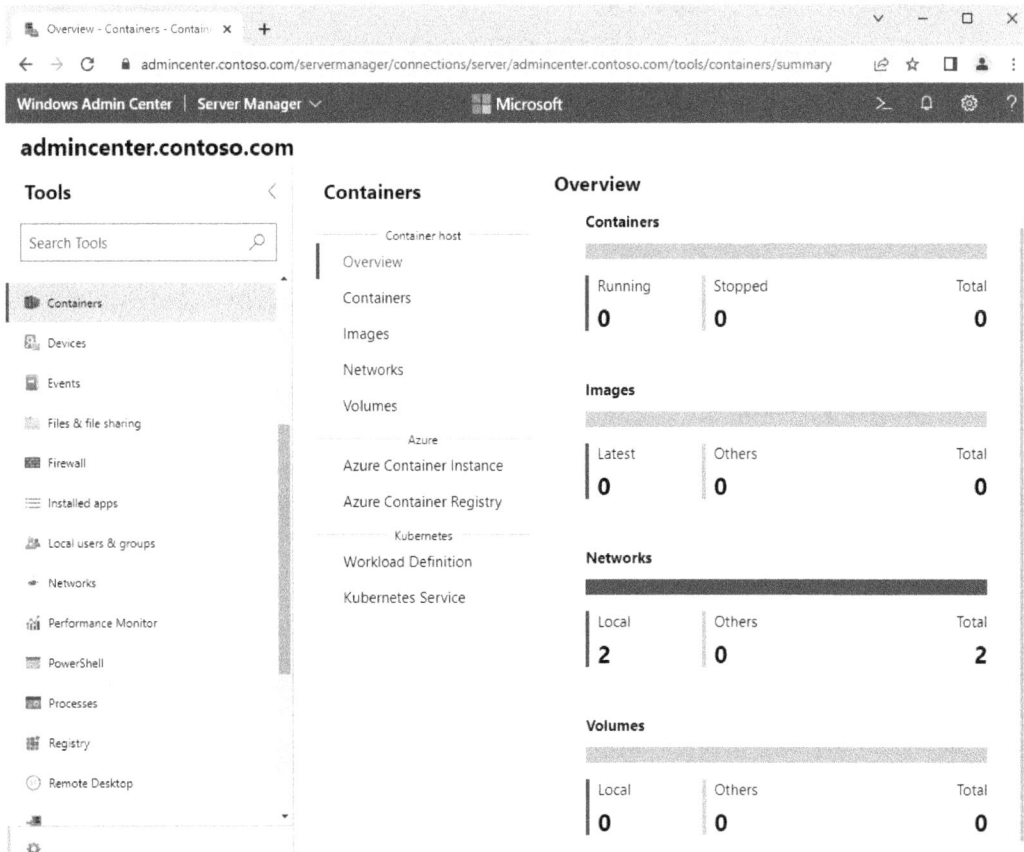

Figure 12.2 – WAC browser interface

In this section, we looked at deploying a WAC gateway VM and covered the manual installation, as well as an automated installation of WAC with a script. In the next section, we will look at installing WAC on a VM so that it can be managed via WAC within the Azure portal.

Deploying WAC within the Azure portal

We can target the management of Azure and hybrid resources using WAC and its integration with Azure services and the Azure portal. The following are the three resources that can be currently targeted for management by WAC using the Azure portal:

- Azure **infrastructure-as-a-service (IaaS)** VMs

- Azure Stack HCI

- Azure Arc-enabled servers

For the exam objectives, we will focus the content of this section on how to target Azure IaaS VMs.

Requirements for WAC

To use the capability of WAC in the Azure portal, WAC is installed on each Azure VM to be managed by it. The Azure VMs to be managed by WAC in the Azure portal must meet the following requirements:

- Windows Server 2016, Windows Server 2019, Windows Server 2022.

- Minimum of 3 GB memory.

- Must be in an Azure public cloud region; Azure China, Azure Government, and other non-public clouds are not supported.

- Outbound internet access allowing HTTPS traffic to the `WindowsAdminCenter` and `AzureActiveDirectory` service tags.

- Inbound access allowing HTTPS traffic.

- Inbound access must be directed to the IP address of a VM; it cannot be redirected from another service, such as a firewall.

In addition to the VM requirements, there are requirements for the management machine that is used to connect to the Azure portal; these are as follows:

- A web browser such as **Microsoft Edge** or **Google Chrome**

Access is recommended via a **virtual private network (VPN)** or ExpressRoute to the same VNet as the VMs that have WAC installed that you need to manage using WAC. This access method is more secure than exposing the VMs to public IP addresses. If you access via a public IP address, you should create an access rule to restrict access from known IP addresses only and allow an inbound rule for ports `443` HTTPS and `6516` TCP.

Once all requirements are met, you can install WAC onto each Azure VM to be managed.

We will continue with this topic as a hands-on installation task in the *Hands-on exercise* section of this chapter.

In this section, we looked at targeting the management of Azure and hybrid resources using WAC and its integration with Azure services and the Azure portal. We walked through implementing WAC inside the Azure portal and accessing an Azure VM to manage it from WAC within the Azure portal.

This concluded the sections on deploying WAC. In the next section, we will look at PowerShell remoting.

PowerShell remoting

PowerShell remoting allows a single command to be run on hundreds of machines or just one remote machine. PowerShell remoting uses the **WS-Management** (**WS-Man**) protocol.

PowerShell remoting is enabled by default in Windows Server 2012/R2, 2016.

PowerShell remoting is enabled with the following elevated command:

```
Enable-PSRemoting
```

The following are the requirements for PowerShell remoting:

- PowerShell installed on the local machine where commands will be run from and a remote machine
- PowerShell remoting must be enabled
- WinRM enabled each machine to connect and run a command

We will look at some example cmdlets that can be run against remote computers.

PowerShell remoting examples

The following cmdlet starts an interactive session with a single remote computer:

```
Enter-PSSession Computer-1
```

Command Prompt will show the remote computer's display name; all commands entered will display on the local computer but are executed on the remote computer.

The interactive session can be ended with the following cmdlet:

```
Exit-PSSession
```

The `Invoke-Command` cmdlet can be used to run scripts on one or more remote computers; an example would be as follows:

```
Invoke-Command -ComputerName Computer-1, Computer-2 -FilePath
C:\az800demo\az800script.ps1
```

The script would be at the path on your local machine and execute on the remote computers; the results would display on your local screen.

The `New-PSSession` cmdlet can be used for a persistent session; an example would be the following:

```
$s = New-PSSession -ComputerName Computer-1,Computer-2
```

Second-hop PowerShell remoting

A challenge arises when you establish a session to a first remote computer as the first hop, but that computer requires access to resources on a second computer.

Access to the second server is a second hop, which will be denied access. This is because the session was established with the first server; the session credentials are not passed from the first server to the second server where the resources are.

The following are the methods of configuration to support and resolve this scenario:

- **Pass credentials inside an Invoke-Command script block**: Provides the simplest configuration but lesser security

- **Configure resource-based Kerberos-constrained delegation**: Provides higher security with a simpler configuration

- **Kerberos-constrained delegation**: Provides the highest security but requires the domain administrator

- **JEA**: Provides the best security but has detailed configuration

- **PSSessionConfiguration using RunAs**: Provides simpler configuration but more complex credential management

- **Configure Credential Security Support Provider (CredSSP)**: The best balance of security and ease of configuration

- **Kerberos delegation (unconstrained)**: Not recommended

CredSSP is disabled by default and should only be used in the most trusted environments. CredSSP caches the credentials on the first-hop remote server and uses those credentials to pass to the second-hop server from where we are directly opening a session. This makes us open to credential theft attacks if the remote server is compromised.

JEA

JEA provides delegated administration through **role-based access control** (**RBAC**) and is built on the foundation of PowerShell remoting.

JEA limits the amount of access and privileges in terms of what users can do and which cmdlets they can run and provides auditing to understand better who did what during a session. It's a key defense against lateral movement attacks through *compromised privileged accounts* and works based on *least privilege* access and rights.

JEA is included in **PowerShell 5.0** and later. To check which version of PowerShell is installed, you can run the following PowerShell cmdlet:

```
$PSVersionTable.PSVersion
```

To start using JEA, you would want to see a result that shows as follows, with the **Major release** being **5** or higher:

```
Major   Minor   Build   Revision
-----   -----   -----   --------
5       1       14393   1000
```

Having the latest version of PowerShell available for your machine is always recommended.

The following list details the JEA availability via OS type:

- Pre-installed: Windows Server 2016+, Windows 10 1607+
- Pre-installed with reduced functionality: Windows 10 1603, 1511
- Full functionality with **Windows Management Framework** (**WMF**) 5.1: Windows Server 2012/R2, Windows 8, 8.1
- Reduced functionality with WMF 5.1: Windows Server 2008/R2, Windows 7
- Not available: Windows 10 1507

PowerShell remoting is a required foundation for JEA and must be enabled before JEA can be used. PowerShell remoting is enabled by default in Windows Server 2012/R2, 2016; the `Enable-PSRemoting` command is used to enable it.

JEA creates configured endpoints that users connect to; there are files associated with a configured endpoint, and these are as follows:

- Role capabilities file
- Session configuration file

The role capabilities file can be identified with a `.prsc` file extension. The session configuration file can be identified with a `.pssc` file extension.

As its name suggests, the role capabilities file describes what can be done in a session using the basis of RBAC; the cmdlets, providers, functions, and external programs are listed.

Who can use the endpoints and which role capabilities they have access to are defined in the session configuration file.

A template for a blank configuration file can be created using the following command:

```
New-PSSessionConfigurationFile -SessionType
RestrictedRemoteServer -Path .\MyJEAEndpoint.pssc
```

The final piece after the role capability file and session configuration file have been created is the registration of the endpoint; the following PowerShell command is used for this:

```
Register-PSSessionConfiguration -Path .\AZ800JEAConfig.pssc
-Name 'JEAServiveDEsk' -Force
```

Further information on configurations can be found in the following Microsoft documentation article: `https://docs.microsoft.com/en-gb/powershell/scripting/learn/remoting/jea/session-configurations`.

This section looked at configuring PowerShell remoting and some challenges and risks. In the next section, we will complete a hands-on exercise section to re-enforce some of the concepts covered in this chapter.

Hands-on exercise

To support your learning with some practical skills, we will utilize the concepts and understanding gained from the earlier sections of this chapter and put them into practical application.

We will look at the following exercise:

- Exercise – implementing WAC within the Azure portal

Getting started

To get started with this section, if you do not already have an Azure subscription, you can create a free Azure account at `https://Azure.microsoft.com/free`. This free Azure account provides the following:

- 12 months of free services
- USD $200 credit to explore Azure for 30 days
- 25+ services that are always free

Let's move on to the exercise for this chapter.

Exercise – implementing WAC within the Azure portal

This exercise will look at implementing WAC within the Azure portal to manage an Azure VM.

You will need the following in place for this exercise:

- An Azure VM to act as the management target.
- Access to the VM with a local admin account.
- Access to the same VNet as the VM from a management machine used to connect to the Azure portal; this could be via a VPN, public IP address, or another Azure VM on the same VNet used as a jump host. If connecting via a public IP address, you should create an inbound port rule for HTTP 443 and TCP 6516.

Follow these steps to get started:

1. You should click on **Windows Admin Center** (preview) from the VM to be managed under the **Settings** section:

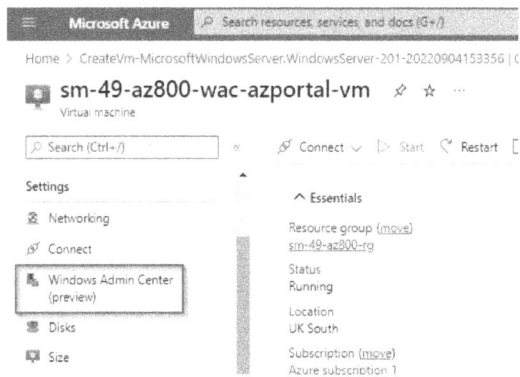

Figure 12.3 – Azure VM to be managed

2. From the **Windows Admin Center** (preview) blade, click **Install**:

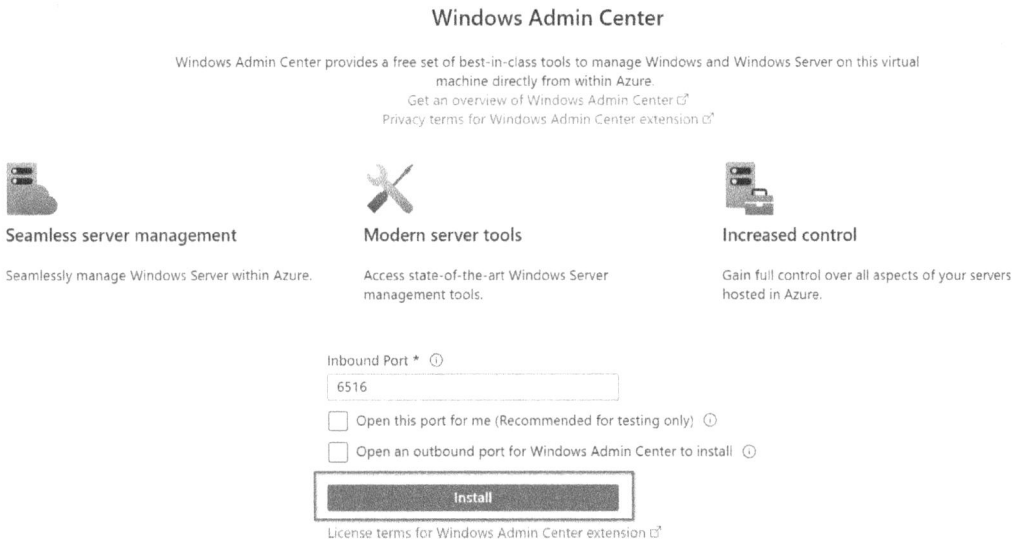

Windows Admin Center

Windows Admin Center provides a free set of best-in-class tools to manage Windows and Windows Server on this virtual machine directly from within Azure.
Get an overview of Windows Admin Center ✍
Privacy terms for Windows Admin Center extension ✍

Seamless server management

Seamlessly manage Windows Server within Azure.

Modern server tools

Access state-of-the-art Windows Server management tools.

Increased control

Gain full control over all aspects of your servers hosted in Azure.

Inbound Port * ⓘ

6516

☐ Open this port for me (Recommended for testing only) ⓘ

☐ Open an outbound port for Windows Admin Center to install ⓘ

Install

License terms for Windows Admin Center extension ✍

Figure 12.4 – Installing WAC extension to VM

3. You will receive a notification in the portal that the WAC extension was successfully installed on the VM:

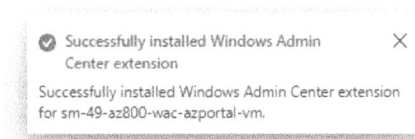

✓ Successfully installed Windows Admin Center extension   ✕

Successfully installed Windows Admin Center extension for sm-49-az800-wac-azportal-vm.

Figure 12.5 – WAC extension successfully installed on VM

4. Now that the WAC extension has been installed on the VM, we can use WAC within the Azure portal to manage the VM without the need for RDP or PowerShell.

5. To connect to the VM, navigate to the VM in the Azure portal, select an IP address to connect to the VM (public or private), and select **Connect**:

Windows Admin Center

Windows Admin Center provides a free set of best-in-class tools to manage Windows and Windows Server on this virtual machine directly from within Azure.

IP Address * ⓘ

| Public IP address (20.108.6.126) | ∨ |

| Connect |

Figure 12.6 – Connecting WAC to VM

6. You will be prompted to sign in; you should use a local admin account for the VM for this exercise (*the admin account set when creating the VM was used in this lab example*).

7. You will now have access to WAC running integrated within the Azure portal:

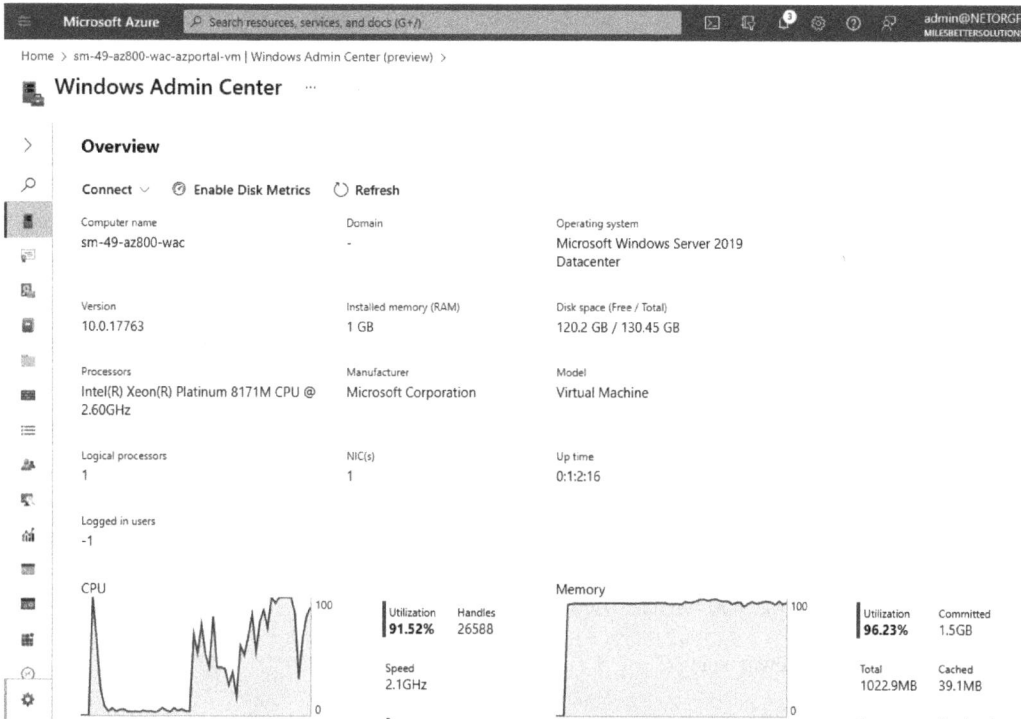

Figure 12.7 – WAC within Azure portal

With this, we have completed this exercise. This exercise taught us the skills needed to install WAC on an Azure VM and connect to it using WAC within the Azure portal. Now, let's summarize this chapter.

Summary

This chapter provided coverage for the *AZ-800 Administering Windows Server Hybrid Core Infrastructure: Manage Windows Servers and workloads in a hybrid environment* exam.

This chapter's content further developed your knowledge and skills for hybrid management services.

You learned about deploying a WAC gateway for the management of multiple VMs and how to install WAC so that a VM can be managed using WAC from within the Azure portal; we then concluded by looking at PowerShell remoting. We then finished the chapter with a hands-on exercise to develop your skills further.

You added new skills through the information provided, with the chapter's goal to take your knowledge beyond the exam objectives so that you are prepared for a real-world, day-to-day, hybrid environment-focused role.

The next chapter will teach you about implementing and managing Azure network infrastructure.

Further reading

This section provides links to additional study references and additional exam information:

- *Microsoft Certified: Windows Server Hybrid Administrator Associate*: `https://docs.microsoft.com/en-us/learn/certifications/windows-server-hybrid-administrator/`

- *Exam AZ-800: Administering Windows Server Hybrid Core Infrastructure*: `https://docs.microsoft.com/en-us/learn/certifications/exams/az-800`

- *Exam AZ-800*: skills outline: `https://query.prod.cms.rt.microsoft.com/cms/api/am/binary/RWKI0r`

- *Microsoft Learn*: `https://docs.microsoft.com/en-us/learn/paths/manage-windows-servers-workloads-hybrid-environment/`

Skills check

Challenge yourself with what you have learned in this chapter:

1. What is a WAC gateway server?

2. How does WAC work?

3. Which ports are required to be allowed for WAC deployment?

4. How is WAC accessed?

5. List the requirements for implementing WAC within the Azure portal.

6. What is PowerShell remoting?

7. How is PowerShell remoting enabled?

8. What is second-hop PowerShell remoting?

9. What is JEA?

10. What are the requirements for JEA?

13
Managing Windows Servers Using Azure Services

In the previous chapter, we learned about the skills for managing Windows Servers in a hybrid environment.

This chapter covers content from *AZ-800 Administering Windows Server Hybrid Core Infrastructure: Manage Windows servers and workloads in a hybrid environment.*

We will cover managing access for Azure-hosted Windows Server **virtual machines** (**VMs**), managing on-premises Windows servers using Azure Arc, and introducing Azure Automation's capabilities. We will then look at security and governance with Microsoft Defender for Cloud, Azure Policy, and Azure Sentinel.

We will then conclude with a hands-on exercise to develop your skills further.

This content does not aim to cover all aspects; the content is limited in scope to only the exam objectives and will cover the following topics:

- Managing access for Windows Server VMs in Azure
- Implementing Azure Arc for hybrid Windows servers
- Implementing Azure Automation for hybrid Windows servers
- Implementing Microsoft Defender for Cloud for hybrid Windows servers
- Hands-on exercises

In addition to the topics listed in this chapter, this chapter's goal is to take your knowledge beyond the exam objectives to prepare you for a real-world, day-to-day hybrid environment-focused role.

Managing access for Windows Server VMs in Azure

This section looks at the user access controls and network access methods for Windows Server Azure **infrastructure-as-a-service** (**IaaS**) VMs.

User access

User access to Azure IaaS VMs for management purposes is controlled through **Azure role-based access control** (**Azure RBAC**).

The following are the built-in default compute category roles:

- **Virtual Machine Administrator Login**: This lets you manage the VM only. You are unable to access the VM, the virtual network, and any storage account(s) they're connected to
- **Virtual Machine Contributor**: This lets you view VMs in the portal and log in as a regular user.
- **Virtual Machine User Login**: This lets you view VMs in the portal and log in as an administrator.

These roles are managed via the **Access Control** (**IAM**) blade for a VM resource; the following screenshot illustrates the **Compute** category roles shown in the Azure portal:

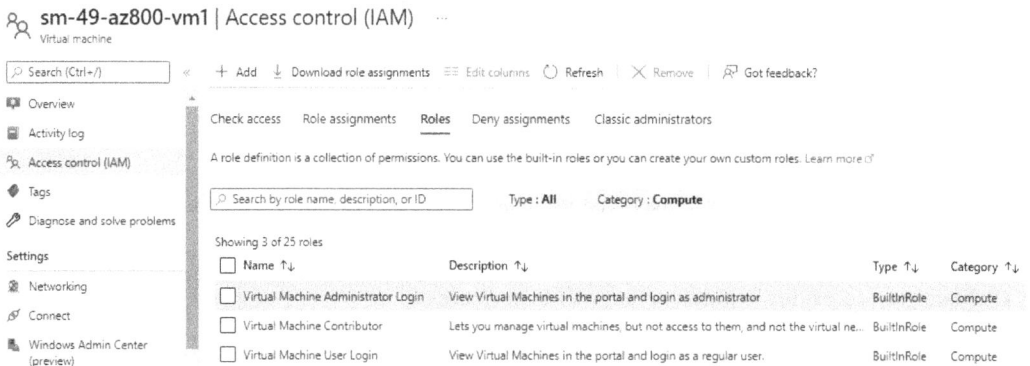

Figure 13.1 – Access control blade

We can use the PowerShell `Get-AzRoleDefinition` cmdlet to list all the roles:

```
Get-AzRoleDefinition | FT Name, Description | sort
```

The following screenshot example shows a listing of role assignments, formatted into a table showing the name and description in columns and listed alphabetically:

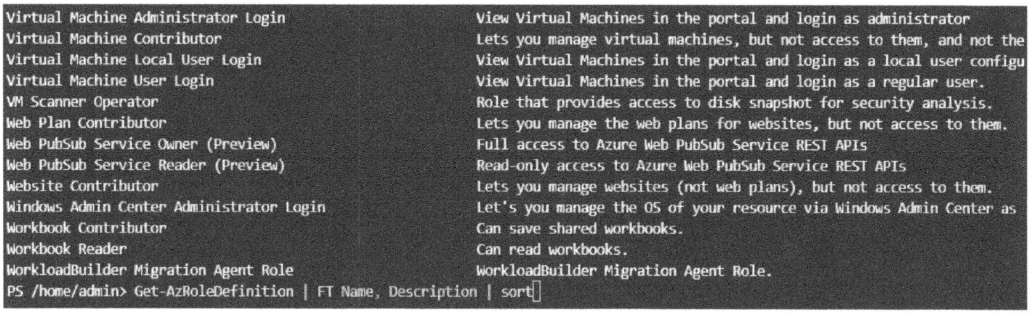

```
Virtual Machine Administrator Login            View Virtual Machines in the portal and login as administrator
Virtual Machine Contributor                    Lets you manage virtual machines, but not access to them, and not the
Virtual Machine Local User Login               View Virtual Machines in the portal and login as a local user configu
Virtual Machine User Login                     View Virtual Machines in the portal and login as a regular user.
VM Scanner Operator                            Role that provides access to disk snapshot for security analysis.
Web Plan Contributor                           Lets you manage the web plans for websites, but not access to them.
Web PubSub Service Owner (Preview)             Full access to Azure Web PubSub Service REST APIs
Web PubSub Service Reader (Preview)            Read-only access to Azure Web PubSub Service REST APIs
Website Contributor                            Lets you manage websites (not web plans), but not access to them.
Windows Admin Center Administrator Login       Let's you manage the OS of your resource via Windows Admin Center as
Workbook Contributor                           Can save shared workbooks.
Workbook Reader                                Can read workbooks.
WorkloadBuilder Migration Agent Role           WorkloadBuilder Migration Agent Role.
PS /home/admin> Get-AzRoleDefinition | FT Name, Description | sort
```

Figure 13.2 – Role definitions

The following Azure CLI command can also be used as an alternative to PowerShell:

```
az role definition create --role-definition
```

We can use the PowerShell Get-AzRoleDefinition <role_name> command to list the details of a specific role:

```
Get-AzRoleDefinition "Virtual Machine Administrator Login"
```

Here is a screenshot example:

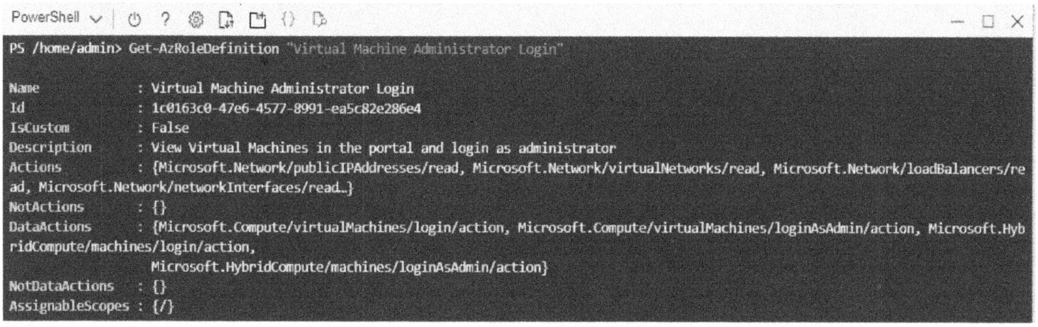

```
PowerShell ∨   ⏻  ?  ⚙  ⎗  ⬚  {}  ⬚                                                            —  ☐  ✕
PS /home/admin> Get-AzRoleDefinition "Virtual Machine Administrator Login"

Name             : Virtual Machine Administrator Login
Id               : 1c0163c0-47e6-4577-8991-ea5c82e286e4
IsCustom         : False
Description      : View Virtual Machines in the portal and login as administrator
Actions          : {Microsoft.Network/publicIPAddresses/read, Microsoft.Network/virtualNetworks/read, Microsoft.Network/loadBalancers/re
ad, Microsoft.Network/networkInterfaces/read…}
NotActions       : {}
DataActions      : {Microsoft.Compute/virtualMachines/login/action, Microsoft.Compute/virtualMachines/loginAsAdmin/action, Microsoft.Hyb
ridCompute/machines/login/action,
                   Microsoft.HybridCompute/machines/loginAsAdmin/action}
NotDataActions   : {}
AssignableScopes : {/}
```

Figure 13.3 – Role information

If we look back at the Azure portal view and click on one of these roles, we will see **Permissions** details that are granted with the role, as illustrated in the following screenshot:

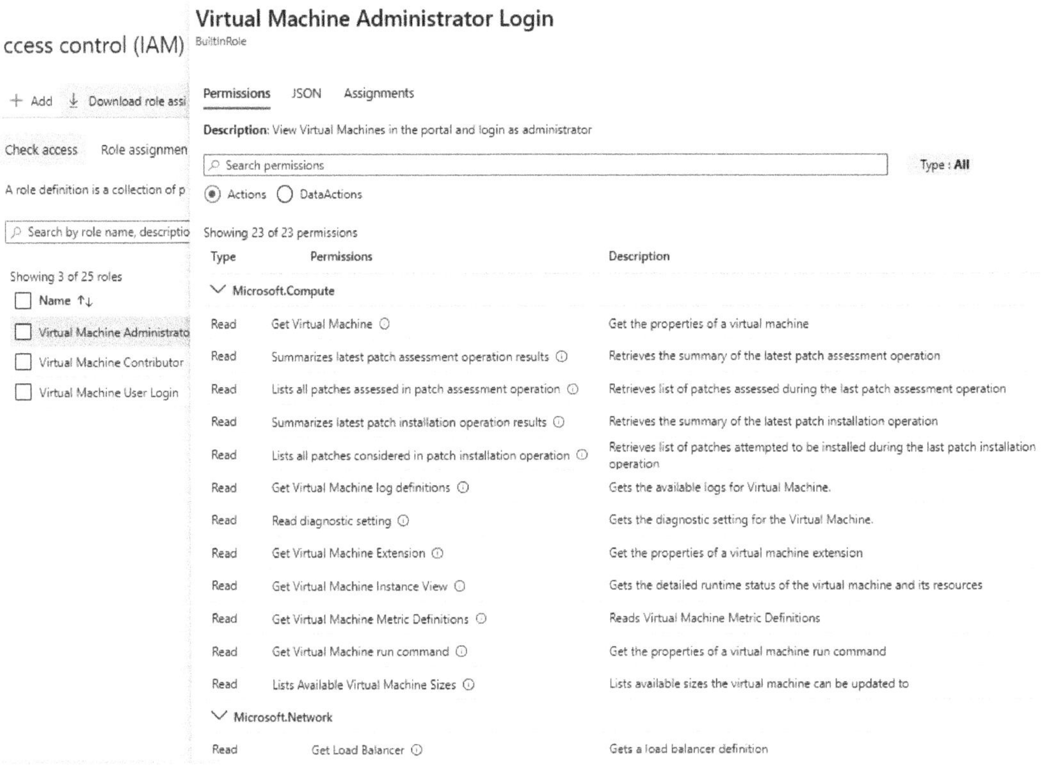

Figure 13.4 – Role permissions

We can use the following cmdlet to list permissions of a role definition, as shown in the following screenshot example:

```
(Get-AzRoleDefinition "Virtual Machine Contributor").Actions
```

The output is shown in the following screenshot example:

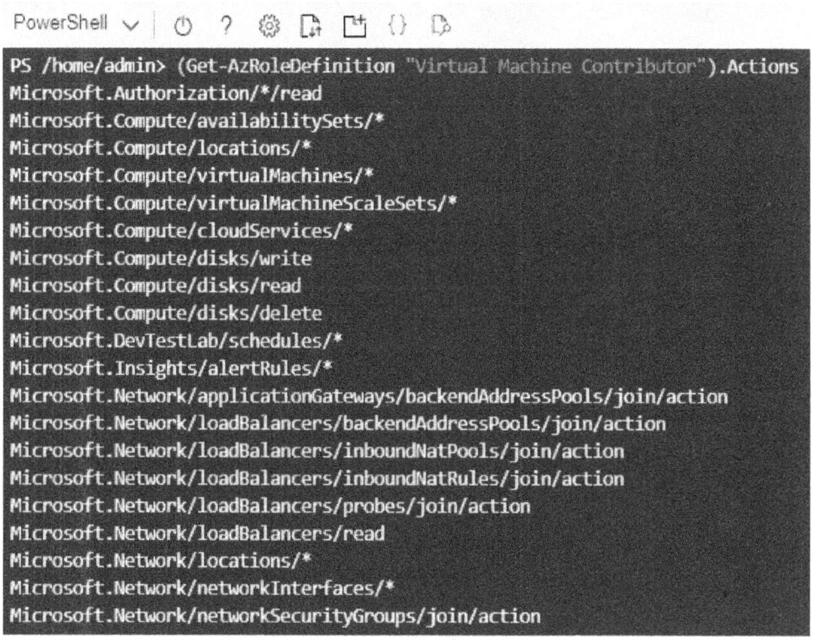
PowerShell
PS /home/admin> (Get-AzRoleDefinition "Virtual Machine Contributor").Actions
Microsoft.Authorization/*/read
Microsoft.Compute/availabilitySets/*
Microsoft.Compute/locations/*
Microsoft.Compute/virtualMachines/*
Microsoft.Compute/virtualMachineScaleSets/*
Microsoft.Compute/cloudServices/*
Microsoft.Compute/disks/write
Microsoft.Compute/disks/read
Microsoft.Compute/disks/delete
Microsoft.DevTestLab/schedules/*
Microsoft.Insights/alertRules/*
Microsoft.Network/applicationGateways/backendAddressPools/join/action
Microsoft.Network/loadBalancers/backendAddressPools/join/action
Microsoft.Network/loadBalancers/inboundNatPools/join/action
Microsoft.Network/loadBalancers/inboundNatRules/join/action
Microsoft.Network/loadBalancers/probes/join/action
Microsoft.Network/loadBalancers/read
Microsoft.Network/locations/*
Microsoft.Network/networkInterfaces/*
Microsoft.Network/networkSecurityGroups/join/action

Figure 13.5 – Role permissions (continued)

Returning to the Azure portal view, we can also see **Assignments** details for the role, as illustrated in the following screenshot:

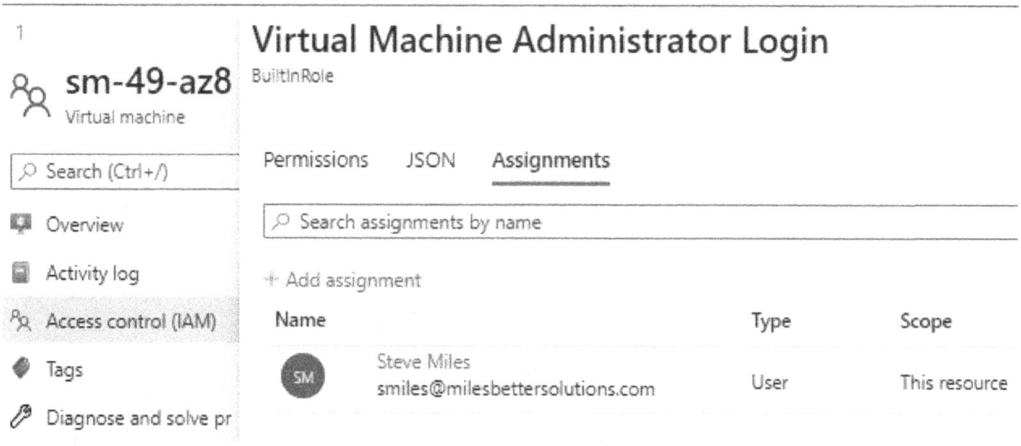
sm-49-az8
Virtual machine
Search (Ctrl+/)
Overview
Activity log
Access control (IAM)
Tags
Diagnose and solve pr
Virtual Machine Administrator Login
BuiltInRole
Permissions
JSON
Assignments
Search assignments by name
Add assignment
Name
Type
Scope
SM
Steve Miles
smiles@milesbettersolutions.com
User
This resource

Figure 13.6 – Role assignments

We can list *role assignments* using PowerShell from Cloud Shell using the `Get-AzRoleAssignment` cmdlet.

Here is an example listing the role assignments for a user:

```
Get-AzRoleAssignment -SignInName smiles@milesbettersolutions.
com
```

The preceding code is represented in the following figure:

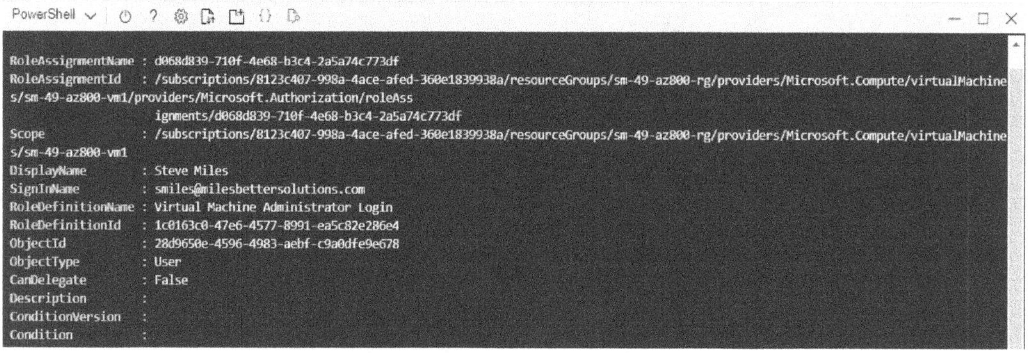

Figure 13.7 – User role assignments

We can also list user assignments at the resource level by using a `-Scope` parameter; here is an example:

```
Get-AzRoleAssignment -Scope /subscriptions/8123c407-998a-4ace-
afed-360e1839938a/resourceGroups/sm-49-az800-rg/providers/
Microsoft.Compute/virtualMachines/sm-49-az800-vm1
```

Here's the output:

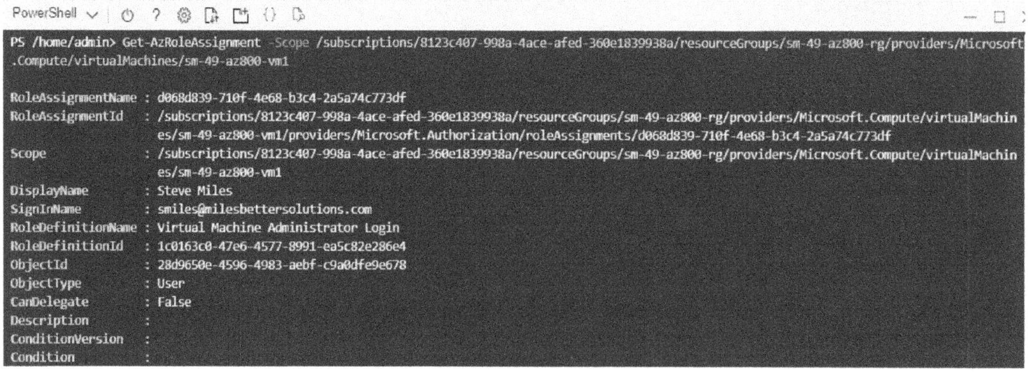

Figure 13.8 – Resource role assignments

We should always follow the principle of least privileged access when planning and configuring role assignments, ensuring a user is not assigned a role with more access than is required for their task.

Network access

The following are Azure first-party interfaces and tools that can be used to access Windows Server Azure VMs for management purposes:

- **Azure Bastion**: Provides a managed jump-host service from which you can make connections to VMs; no public IPs are required on VMs, and no RDP ports are required to be open. It uses HTTPS (443).

- **Azure Serial Console**: Provides a direct connection to an Azure VM without connection via the IP network.

- **Cloud Shell**: Provides access to PowerShell via a browser-based session.

- **RDP**: Makes a connection using a traditional RDP session; allows an incoming rule for RDP (3389), and if connecting over a public IP, restricts to known IP addresses.

- **Remote PowerShell**: Enables you to run commands on a remote computer from a local PowerShell session.

- **Just-Enough-Administration (JEA)** and **Just-in-Time (JIT)** VM access: Security controls to govern access.

This section looked at managing user access to Windows Server IaaS VMs in Azure and tools and interfaces for admin access. The next section will cover connecting hybrid Windows servers to Azure Arc.

Implementing Azure Arc for hybrid Windows servers

Azure Arc is a *hybrid management and governance tool* that supports physical and virtual Windows servers. These hybrid servers can be on-premises, in provider edge locations, or hosted on other cloud provider's platforms. This is represented in the following diagram:

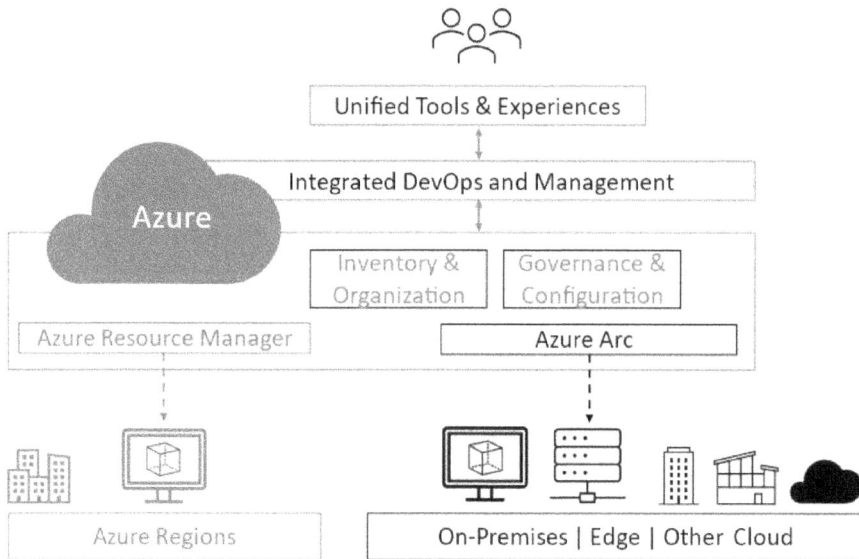

Figure 13.9 – Azure Arc for servers

When connected in this way, a hybrid server becomes an *Azure resource* that can be *controlled*, *secured*, and *managed* the same as an Azure native VM.

Each hybrid machine is given an *Azure resource ID* allowing the machine to be added to a resource group and be managed by the **Azure Resource Manager** (**ARM**); we class these Azure Arc-managed servers as **Arc-enabled servers**.

To connect a Windows server to Azure Arc, an **Azure Connected Machine agent** is deployed and configured on the server. It should be noted that this does not replace the **Azure Monitor Agent** (**AMA**) for Windows servers; *both agents* are required for *Arc-enabled servers*.

The following hybrid functionality and cloud operations can be performed when a hybrid server is Azure Arc-enabled:

- **Configuring**: **Azure Automation** can be used to speed up routine management tasks. **Change Tracking and Inventory** capabilities are used for configuration changes, software installation, services, and registry updates.

- **Governing**: Azure Policy guest configurations can be assigned to perform machine settings audits.

- **Monitoring**: **VM insights** is used to monitor the OS, components, processes, services, and software applications. Log data collection and analysis is carried out through **Log Analytics**.

- **Protecting**: **Microsoft Defender for Cloud** provides endpoint protection through **Microsoft Defender for Endpoint**. Microsoft Sentinel can provide **security information and event management** (**SIEM**) and orchestrated and automated responses.

Azure Arc can be enabled manually on servers or automated through **Windows Admin Center** (**WAC**). This can be used to deploy the **Connected Machine agent** and perform the *Azure registration* in a single place.

Deployment prerequisites

The following requirements should be met:

- The hybrid machines must have a supported OS for the **Connected Machine agent**; these are as follows: *Windows Server 2008 R2 SP1, 2012 R2, 2016, 2019, and 2022*; Desktop and Server Core experiences. *Azure Stack HCI* supports Windows Server *Azure editions*.

- The hybrid machines must access Azure resources directly or via a proxy.

- You must have an administrator account on the local machine to install and configure the **Azure Connected Machine** agent.

- The **Azure Connected Machine Onboarding** Azure built-in role is required.

- The **Azure Connected Machine Resource Administrator** Azure built-in role is required to *read*, *modify*, and *delete* a VM.

Further information on the prerequisites can be found in the following Microsoft documentation article: `https://docs.microsoft.com/en-gb/azure/azure-arc/servers/prerequisites`.

Network connectivity

Azure Arc provides secure outbound connections to *public endpoints* using HTTPS (TCP/443). You can configure the agent to use a proxy for this communication if your scenario requires this.

You can additionally use **Azure Private Link** to ensure the connections are private so that you do not need to open any public network access, with a **virtual private network** (**VPN**) or ExpressRoute for private network access.

If your outbound traffic is filtered, will need to allow the URLs required by the **Connected Machine agent**; these can be found in the following Microsoft documentation article: `https://docs.microsoft.com/en-gb/azure/azure-arc/servers/network-requirements?tabs=azure-cloud#urls`.

Azure Policy guest configuration

Azure Policy is a *governance tool* that helps determine compliance and identify non-compliance of environments. Using **guest configuration policies**, we can audit the state of Azure Arc-enabled servers and apply settings within the machines.

For any of the provided policies that are assigned, resources can be marked as non-compliant with the policy when a condition is found `true` when evaluated.

In the *Hands-on exercises* section of this chapter, we will walk through creating a policy assignment. We will look at the **Log Analytics extension should be installed on your Windows Azure Arc machines** policy. This policy audits Windows Azure Arc machines if the Log Analytics extension is not installed; if the result is `false`—that is, the extension is not installed—then the machine will be marked as non-compliant with the policy.

A list of all the Azure Policy definitions can be found in the following Microsoft documentation article: `https://docs.microsoft.com/en-gb/azure/governance/policy/samples/built-in-policies`.

Log Analytics agent for hybrid Windows servers

The **Log Analytics agent** is used for sending data to a **Log Analytics workspace**. It is supported for Windows and Linux servers that can be Azure and non-Azure VMs; it is also referred to as the **Microsoft Monitoring Agent** (**MMA**).

You should be aware that the Log Analytics agent is being deprecated and will not be supported after August 31, 2024; you should migrate to the AMA.

The primary scenario for deploying the Log Analytics agent is when you need to do any of the following:

- Collect log and performance data from any Azure or non-Azure machines, physical or virtual
- Send data to a Log Analytics workspace; for use with features such as log queries in Azure Monitor
- Monitor machines with VM Insights
- Manage protection of machines with Microsoft Defender for Cloud or Microsoft Sentinel
- Use the Azure Automation features of **Update Management**, **State Configuration**, or **Change Tracking and Inventory**

The Log Analytics agent has the following limitations:

- Data cannot be sent to Azure Monitor Metrics, Azure Storage, or Azure Event Hubs
- Unique monitoring definitions for individual agents are not possible
- Scale limitations are due to each machine having a unique configuration

A list of the supported OSes can be found in the following Microsoft documentation article: `https://docs.microsoft.com/en-gb/azure/azure-monitor/agents/agents-overview#supported-operating-systems`.

There are several deployment options for deploying the Log Analytics agent to hybrid Windows servers; these are as follows:

- Use Azure Arc for servers, which deploys the VM extension to Azure Arc-enabled servers
- Manually install the agent
- Integrate **System Center Operations Manager** (**SCOM**) with Azure Monitor

The following types of data can be collected with the Log Analytics agent for Windows servers:

- Windows event logs
- Performance
- **Internet Information Services** (**IIS**) logs
- Custom logs

The Log Analytics agent has a wider scope of usage scenarios than just collecting data for ingestion into a Log Analytics workspace for Azure Monitor consumption. It is also used as a core component in Microsoft Defender for Cloud and Microsoft Sentinel, which also use it in conjunction with a connected Log Analytics workspace.

The agent for Windows Server requires outbound communication over TCP 443, and the following URLs will need to be allowed:

- `*.ods.opinsights.azure.com`
- `*.oms.opinsights.azure.com`
- `*.blob.core.windows.net`
- `*.azure-automation.net`

If no direct internet connection is allowed, a Log Analytics gateway will need to be set up, which then acts as a proxy for the agent and the workspace.

The Windows agent uses **SHA-2** signing. To ensure data in transit to Azure is secure, it is recommended the agent should use **Transport Layer Security** (**TLS**) **1.2**.

In this section, we looked at Azure Arc for hybrid servers. In the next section, we will cover the content for implementing Azure Automation for hybrid Windows servers.

Implementing Azure Automation for hybrid Windows servers

Process automation is the automation of frequent management tasks that can be error-prone, risk configuration drift, and consume an unequal amount of time to the value the task delivers. In this next section, we will introduce this topic.

Introduction

Automation can be provided through the **Azure Automation** service and is available to hybrid workloads.

To implement Azure Automation in this manner for non-Azure hybrid workloads, an **Azure Automation account** is required. Azure Automation is an Azure cloud service responsible for automating the configuration and management of Azure and non-Azure platforms.

The foundation of Azure Automation is **runbooks**, a set of tasks that execute on the target machine; these runbooks are executed as jobs. A process called a **worker** is responsible for running the jobs when the runbook is executed.

Azure Automation relies on a **Hybrid Runbook Worker** feature that allows runbooks to be executed directly on Azure Arc-enabled servers. Hybrid runbook workers can be deployed with two approaches, as follows:

- **Agent-based** (**v1**): This deploys **AMA/Log Analytics agent for Windows** and uses a **Log Analytics** workspace to report to.
- **Extension-based** (**v2**): This is deployed as an Azure VM extension: the **Hybrid Runbook Worker VM extension**. It does not rely on **AMA** for Windows servers and does not use a **Log Analytics** workspace to report to.

The extension-based *Hybrid Runbook Worker* is the recommended approach for an integrated and seamless experience.

There are two types of runbook worker types: **System** and **User**; only the **System** type is supported for the extension-based *Hybrid Runbook Worker*.

Azure Automation Update Management

Azure Automation includes **Update Management**, which provides management of OS updates for Azure and non-Azure machines that can be virtual and physical, located on-premises, at edge locations, or on other cloud platforms.

Update Management uses an automation account and stores update assessments and deployment results data in a Log Analytics workspace through integration with Azure Monitor. The Log Analytics agent must be deployed and configured for each machine that will use **Update Management**. A machine registered for **Update Management** *cannot* report to *more than one* Log Analytics workspace. System information updates from SCOM agents in a management group connected to the workspace can be used to collect information.

Windows machines must use Microsoft Update to report to **Windows Server Update Services (WSUS)**.

When **Update Management** is enabled, a *Hybrid Runbook Worker group* is automatically added to the automation account. All machines connected directly to the Log Analytics workspace have the *System Hybrid Runbook Worker* configured to support the **Update Management** runbooks.

The following are the supported Windows OSs for **Update Management**:

- Windows Server 2008 R2 (*RTM and SP1 standard*)

- Windows Server 2012

- Windows Server 2012 R2 (*Standard/Datacenter*)

- Windows Server 2016 (*Standard/Datacenter, not Server Core*)

- Windows Server 2019 (*Standard/Datacenter, including Server Core*)

At the time of publishing, the *Windows Client OS* (Windows 7/10/11) is not supported for **Update Management**. Microsoft Endpoint Configuration Manager is the recommended patch management solution for **Azure Virtual Desktop (AVD)**.

This section introduced Azure Automation for hybrid servers and covered the Azure **Update Management** capability. The next section will look at Microsoft Defender for Cloud for hybrid Windows servers.

Implementing Microsoft Defender for Cloud for hybrid Windows servers

Microsoft Defender for Cloud is Microsoft's **Cloud Security Posture Management (CSPM)** and **Cloud Workload Protection Platform (CWPP)** solution. We can use it to monitor and manage the security posture of hybrid Windows servers in on-premises environments.

In addition, we can implement Microsoft Sentinel, which provides a cloud-native SIEM and **security orchestration, automation, and response (SOAR)** solution for onboarded hybrid Windows servers.

When Defender for Cloud is enabled, the Log Analytics agent needs to be deployed to the machines; the agents then send data to a connected Log Analytics workspace. Defender for Cloud can then determine the security posture of the machines, applications, data, and networks and provide a secure score.

For Azure VMs, the Log Analytics agent is automatically provisioned by Defender for Cloud. For on-premises hybrid machines, the agent can be manually installed or automated through a solution such as Microsoft Endpoint Protection Manager.

The following are Defender for Cloud-specific roles:

- **Security Reader**: This role has read-only rights; it can view recommendations, alerts, security policies, and security states.

- **Security Admin**: This role has the same rights as the **Security Reader** role, plus it can update security policies and dismiss alerts and recommendations.

Microsoft Sentinel's role is the ingestion of security operations data from many different sources to allow correlation and response.

Onboarding Microsoft Sentinel starts with enabling and connecting security data sources with many built-in data connectors that allow data ingestion.

Microsoft Sentinel needs access to a Log Analytics workspace and should be a separate and dedicated instance and not shared with the Defender for Cloud workspace; this is referred to as a *Sentinel-enabled Log Analytics workspace*. A Sentinel-enabled Log Analytics workspace has its data retention period changed to 90 days.

To enable Microsoft Sentinel, the following access is required:

- **Contributor** access to the Azure subscription where the Sentinel workspace is created

- **Contributor** or **Reader** access to the same resource group where the Sentinel workspace is created should be inherited from the subscription **contributor role** access

This section covered implementing Defender for Cloud and Microsoft Sentinel.

Next, we will complete a hands-on exercise section to re-enforce some of the concepts covered in this chapter.

Hands-on exercises

To support your learning with some practical skills, we will utilize the concepts and understanding gained from the earlier sections of this chapter and put them into practical application.

We will look at the following exercises:

- Exercise—assigning access to an Azure VM

- Exercise—creating an Azure Policy assignment

- Exercise—setting up Microsoft Defender for Cloud

- Exercise—setting up Microsoft Sentinel

Getting started

To get started with this section, if you do not already have an Azure subscription, you can create a free Azure account at `https://azure.microsoft.com/free`. This free Azure account provides the following:

- 12 months of free services

- USD $200 credit to explore Azure for 30 days

- 25+ services that are always free

Let's move on to the exercises for this chapter.

Exercise – assigning access to an Azure VM

This exercise will look at assigning access to an Azure VM using the Azure portal.

You will need the following in place for this exercise:

- Azure portal access with an **Owner** or **Contributor** subscription role

- An Azure VM resource to assign access

Follow these steps to get started:

1. Log in to the Azure portal: `https://portal.azure.com`. You can alternatively use the Azure desktop app: `https://portal.azure.com/App/Download`.

2. In the search bar, type in `virtual machines`; click **Virtual machines** from the list of services:

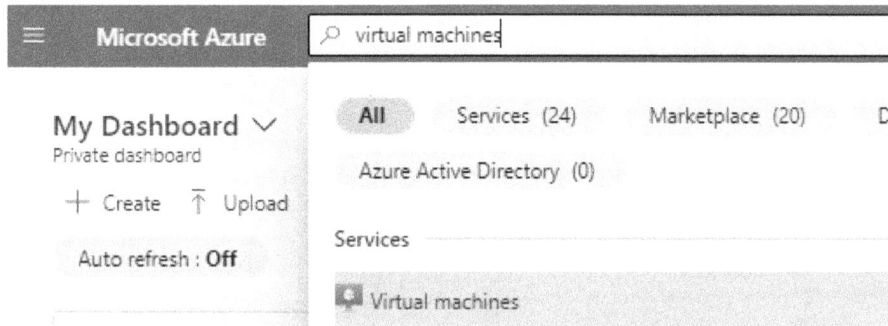

Figure 13.10 – Searching for the Virtual machines service

3. Locate your VM for this exercise from the **Virtual machines** blade and click on the VM *name* to open the **Virtual machines** blade:

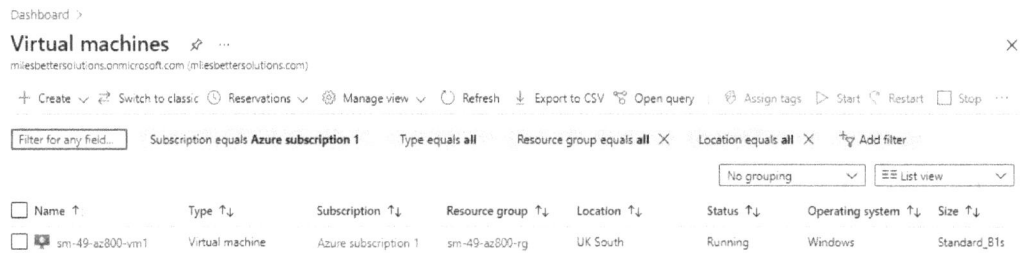

Figure 13.11 – Selecting your VM

4.   From the **Virtual machines** blade, click **Access control (IAM)**:

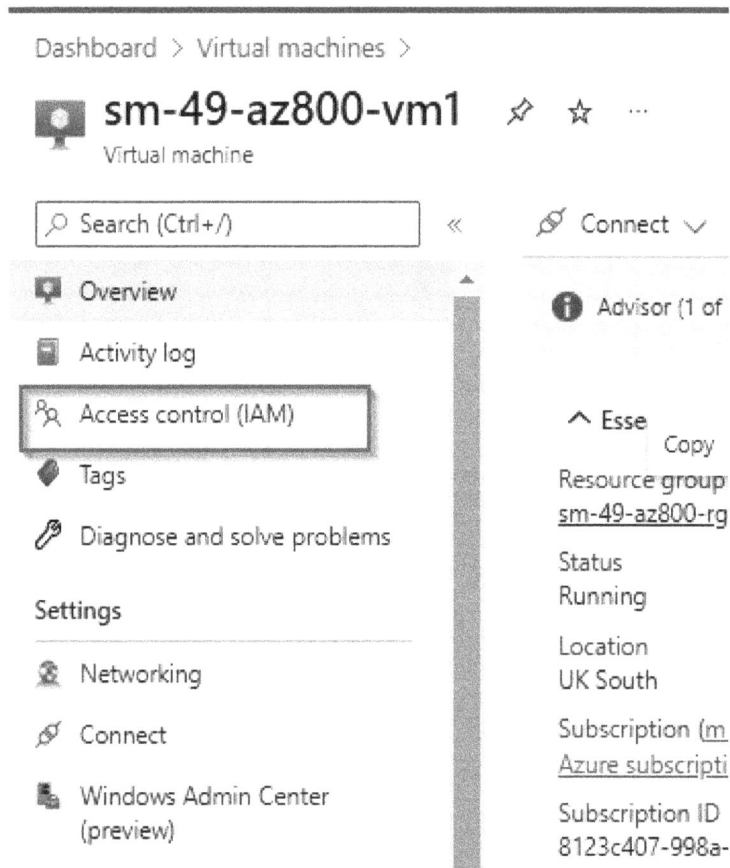

Figure 13.12 – Entering access control for VM

5. From the **Access control (IAM)** blade, from the options at the top of the blade, click + **Add** and then **Add role assignment**:

Dashboard > Virtual machines > sm-49-az800-vm1

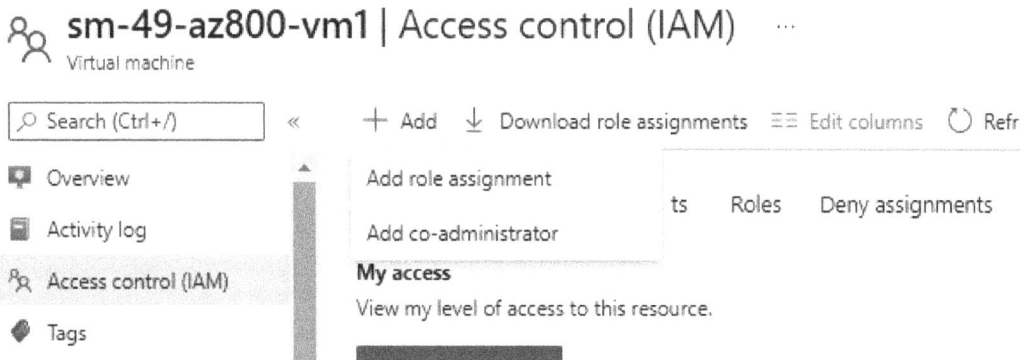

sm-49-az800-vm1 | Access control (IAM)   ...
Virtual machine

| Search (Ctrl+/) | « | + Add | ↓ Download role assignments | ☰☰ Edit columns | ◯ Refr |

- 🖵 Overview
- 🗎 Activity log
- Access control (IAM)
- 🏷 Tags

Add role assignment

Add co-administrator

ts   Roles   Deny assignments

My access
View my level of access to this resource.

Figure 13.13 – Adding a role assignment

6. From the **Add role assignment** blade, in the **Category** filter, select **Compute**:

Dashboard > Virtual machines > sm-49-az800-vm1 | Access control (IAM) >

Add role assignment   ...

🗩 Got feedback?

Role   Members   Review + assign

A role definition is a collection of permissions. You can use the built-in roles or you can create your own custom roles. Learn more ☐

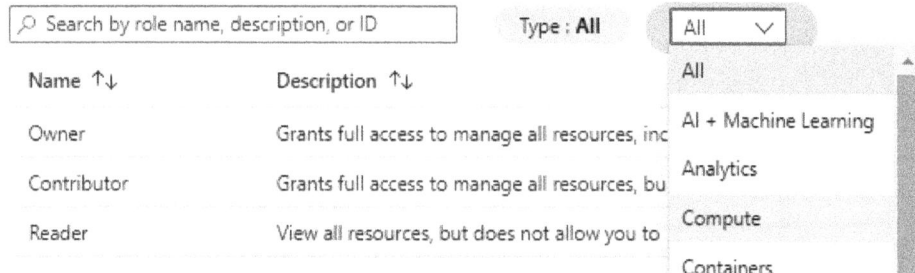

| Search by role name, description, or ID | Type : **All** | All ∨ |

Name ↑↓	Description ↑↓
Owner	Grants full access to manage all resources, inc
Contributor	Grants full access to manage all resources, bu
Reader	View all resources, but does not allow you to

All

AI + Machine Learning

Analytics

Compute

Containers

Figure 13.14 – Selecting an assignment category

7. From the **Role** tab, select the *name of the role* to assign and click **Next**; we will use the **Virtual Machine Administrator Login** role for this exercise:

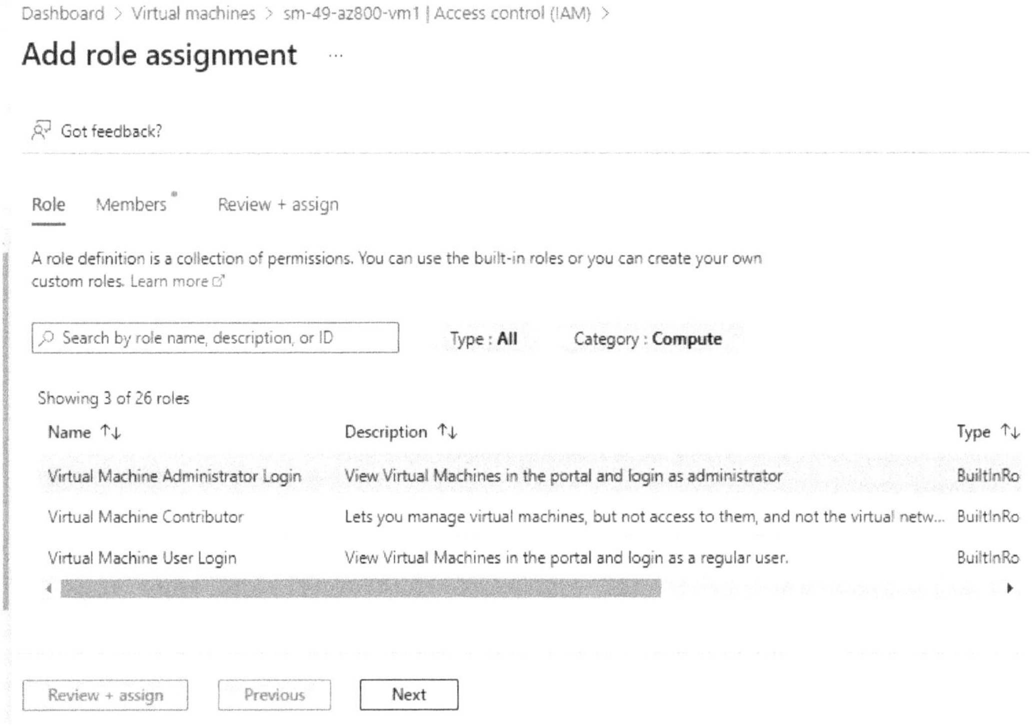

Dashboard > Virtual machines > sm-49-az800-vm1 | Access control (IAM) >

Add role assignment   ...

🖉 Got feedback?

Role   Members*   Review + assign

A role definition is a collection of permissions. You can use the built-in roles or you can create your own custom roles. Learn more ☑

🔎 Search by role name, description, or ID		Type : **All**	Category : **Compute**

Showing 3 of 26 roles

Name ↑↓	Description ↑↓	Type ↑↓
Virtual Machine Administrator Login	View Virtual Machines in the portal and login as administrator	BuiltInRo
Virtual Machine Contributor	Lets you manage virtual machines, but not access to them, and not the virtual netw...	BuiltInRo
Virtual Machine User Login	View Virtual Machines in the portal and login as a regular user.	BuiltInRo

Review + assign	Previous	Next

Figure 13.15 – Selecting a member to assign

8. From the **Members** tab, click + **Select members** under the **Members** section. From the **Select members** blade, search for a user to *assign*, click the user, then click **Select**:

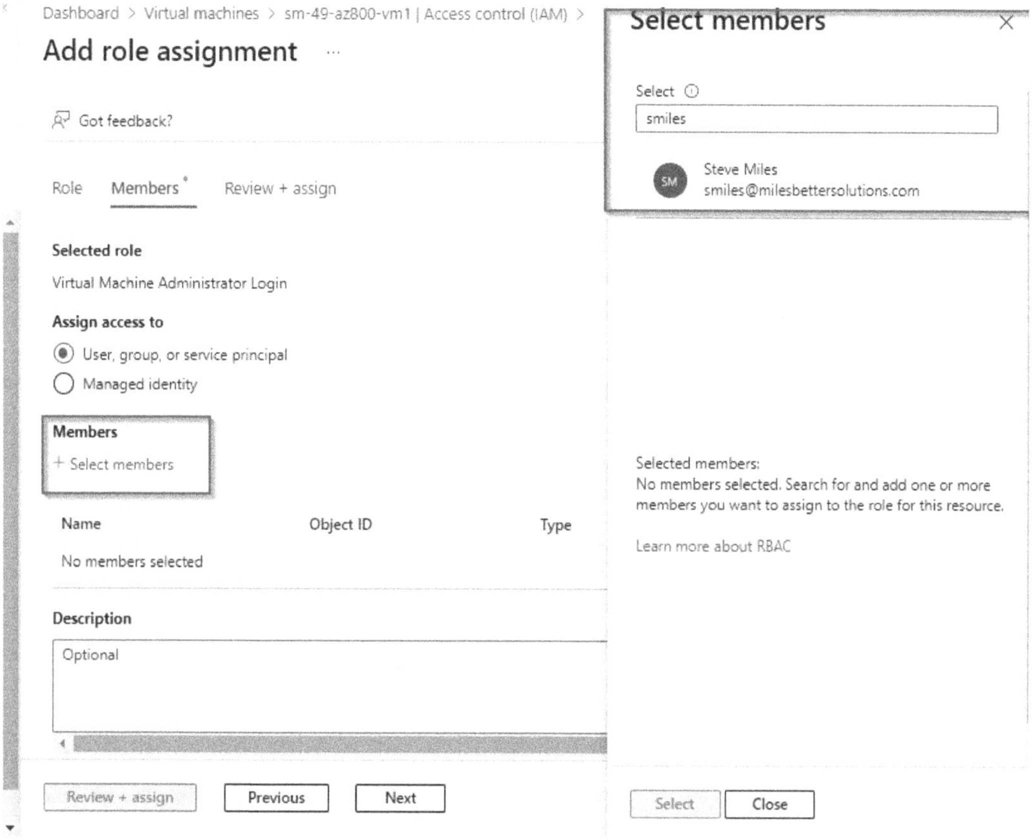

Dashboard > Virtual machines > sm-49-az800-vm1 | Access control (IAM) >

Add role assignment ···

Select members   ✕

🗨 Got feedback?

Select ⓘ

smiles

Role   Members˙   Review + assign

SM   Steve Miles
     smiles@milesbettersolutions.com

Selected role

Virtual Machine Administrator Login

Assign access to

◉ User, group, or service principal

◯ Managed identity

Members

+ Select members

Selected members:
No members selected. Search for and add one or more
members you want to assign to the role for this resource.

Learn more about RBAC

Name   Object ID   Type

No members selected

Description

Optional

Review + assign   Previous   Next

Select   Close

Figure 13.16 – Selecting a member to assign (continued)

9. From the **Review + assign** blade, click **Review + assign**:

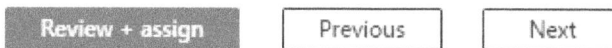

Review + assign   Previous   Next

Figure 13.17 – Creating an assignment

10. You will receive a notification that the role assignment has been added:

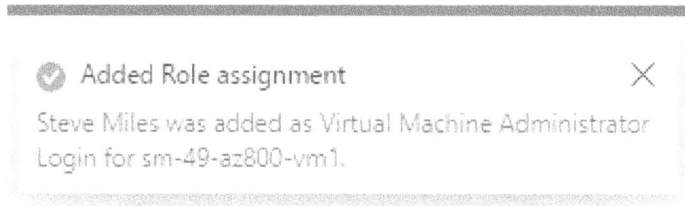

Figure 13.18 – Assignment notification

11. You will now see this new role assignment in the resource's **Access control** (**IAM**) blade on the **Role assignments** tab:

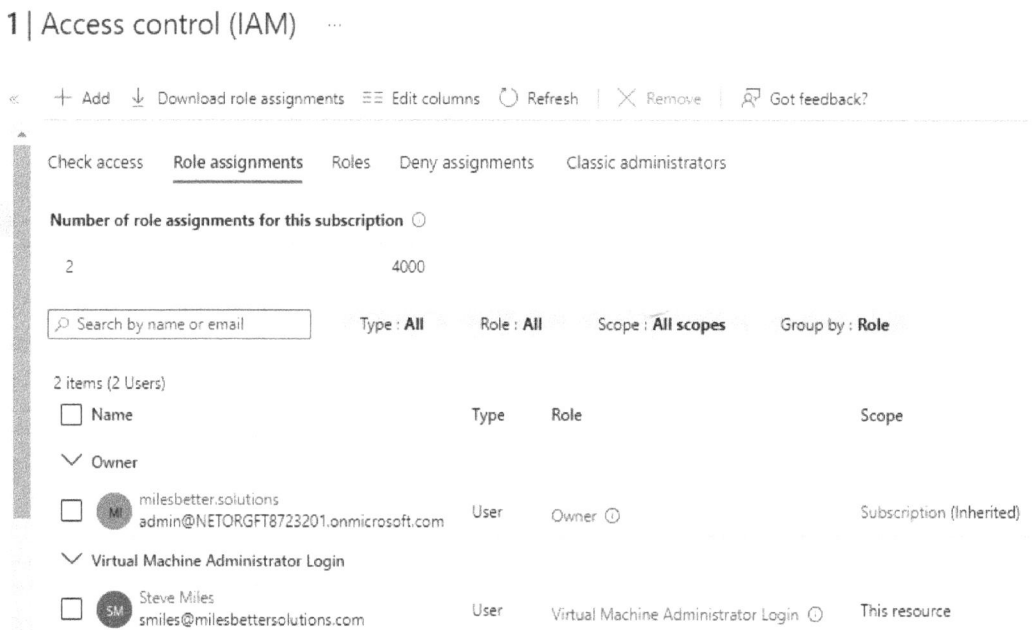

Figure 13.19 – Assignment created

12. If you wish to validate the level of access assigned, from the **Check access** tab, you can click **View my access**; if you wish to check the access for another identity, you can click **Check access**.

13. From **Check access**, click on the **Check access** button, and from the **Current role assignments** tab, click the **Virtual Machine Administrator Login** role entry:

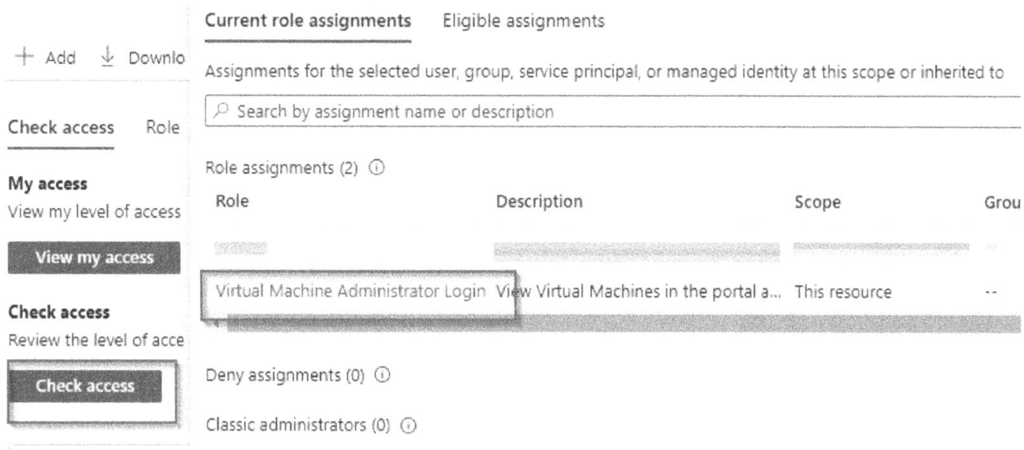

Figure 13.20 – Checking access

14. You will now see the *assigned permissions* for this role and a *description* for each role.

15. You should now *repeat the exercise steps* starting from *step 4*, but this time, for *step 7*, you should select **Virtual Machine User Login** as the role.

16. You should now *repeat steps 12 to 14* to view the *differences* in assigned permissions for this role compared to the **Virtual Machine Administrator Login** role.

With this, we have completed this exercise. This exercise taught us the skills needed to assign access to a VM. In the next exercise, we will look at creating an Azure Policy assignment.

Exercise – creating an Azure Policy assignment

This exercise will look at creating an Azure Policy assignment to audit the existence of the Azure Log Analytics agent required for Azure Arc-enabled Windows servers in a hybrid environment.

You will need the following in place for this exercise:

- Azure portal access with an **Owner** or **Contributor** subscription role

Follow these steps to get started:

1. Log in to the Azure portal: `https://portal.azure.com`. You can alternatively use the Azure desktop app: `https://portal.azure.com/App/Download`.

2. In the search bar, type in `policy`; click **Policy** from the list of services:

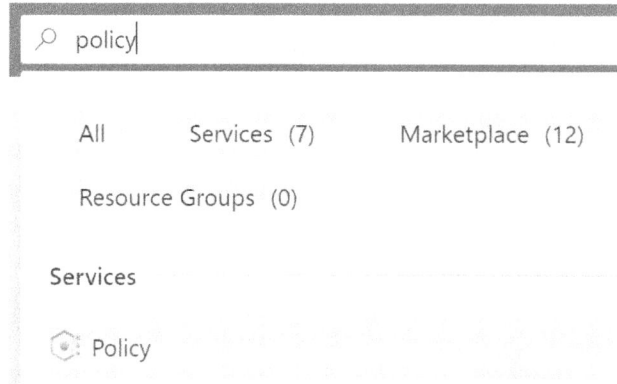

Figure 13.21 – Searching for policy

3.   From the **Policy** blade, click **Assignments** under the **Authoring** section:

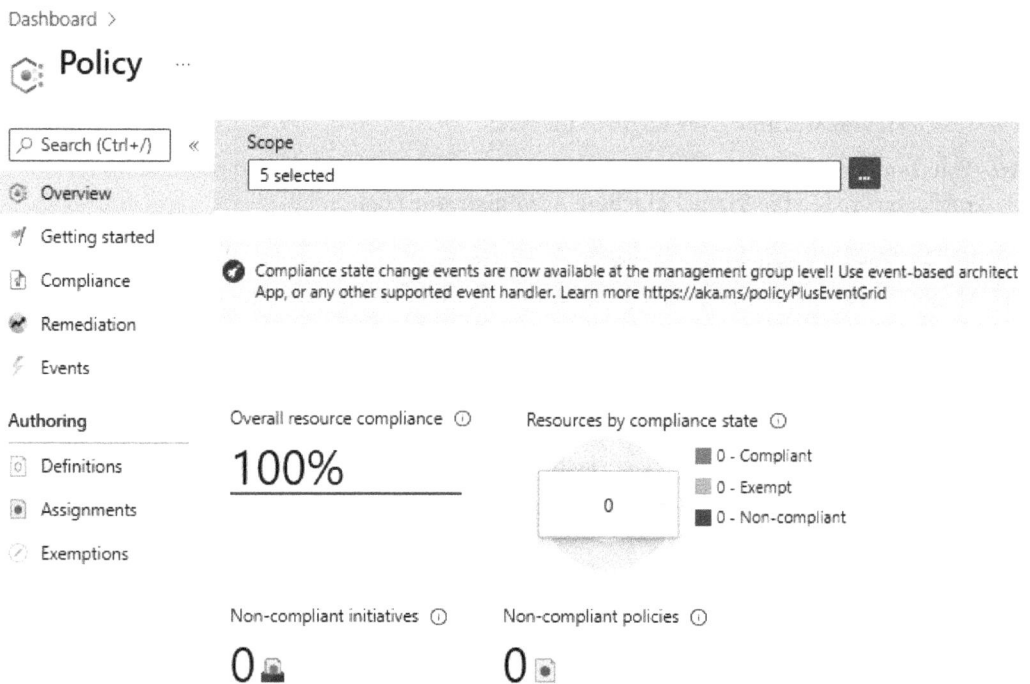

Figure 13.22 – Policy assignment

4. Select **Assign policy** from the top options on the **Assignments** blade:

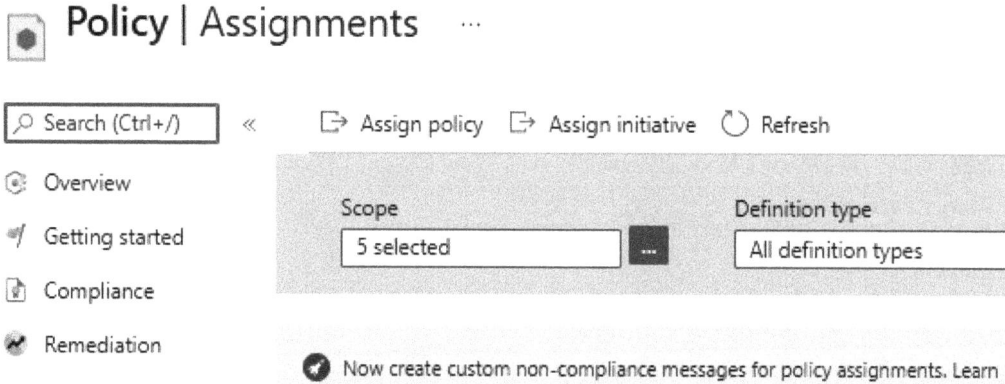

Figure 13.23 – Assign policy option

5. From the **Assign policy** blade, from the **Basics** tab, set a **Scope** type for the subscription and select the required subscription:

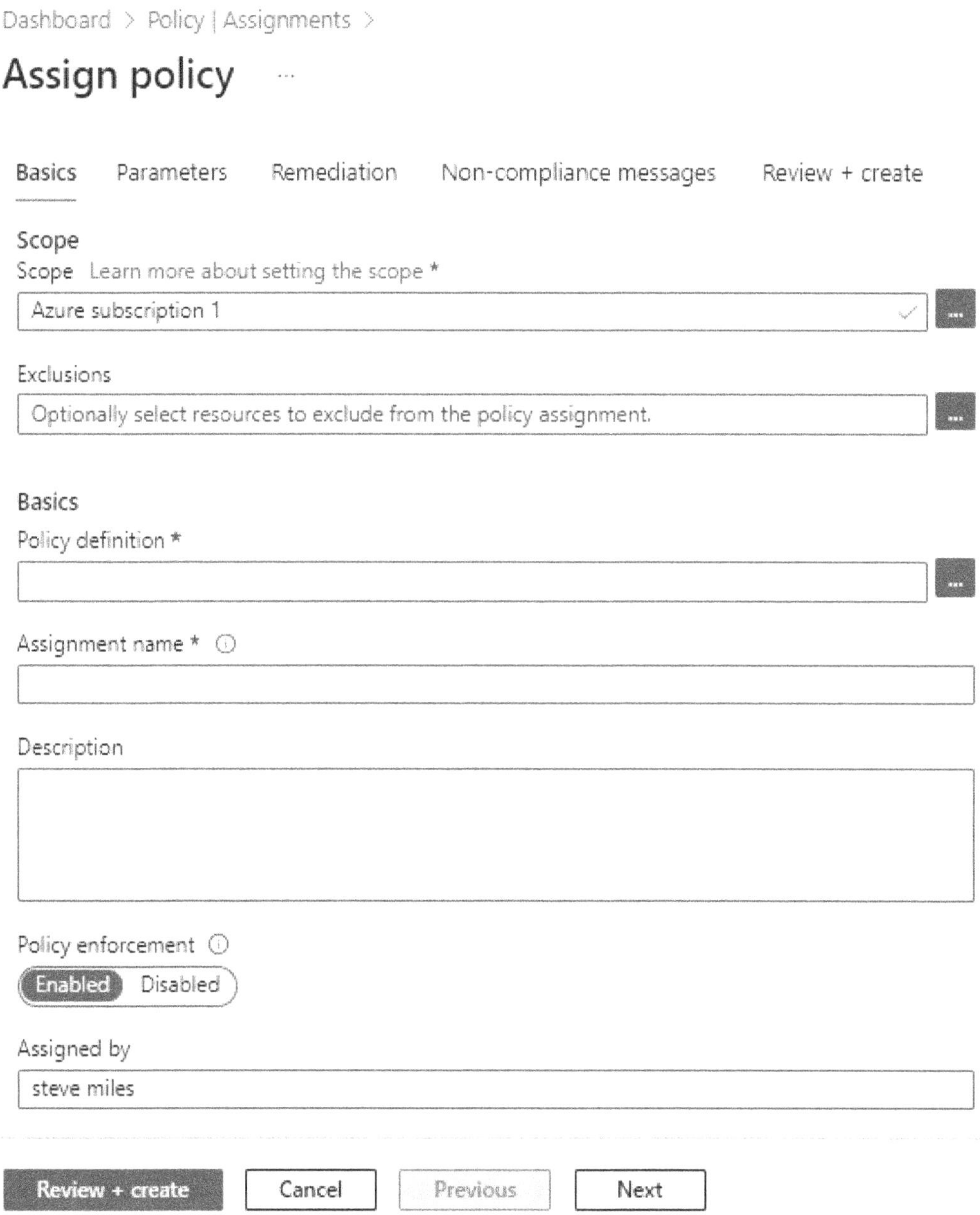

Dashboard > Policy | Assignments >

Assign policy   ...

Basics   Parameters   Remediation   Non-compliance messages   Review + create

Scope
Scope   Learn more about setting the scope *

| Azure subscription 1 | ✓ | ... |

Exclusions

| Optionally select resources to exclude from the policy assignment. | ... |

Basics
Policy definition *

| | ... |

Assignment name *   ⓘ

| |

Description

| |

Policy enforcement ⓘ

(**Enabled**   Disabled)

Assigned by

| steve miles |

Review + create   Cancel   Previous   Next

Figure 13.24 – Assign policy wizard

6. From the **Policy definition** setting under the **Basics** section, search for the following policy definition and click **Select**:

[Preview]: Log Analytics extension should be installed on your Windows Azure Arc machines:

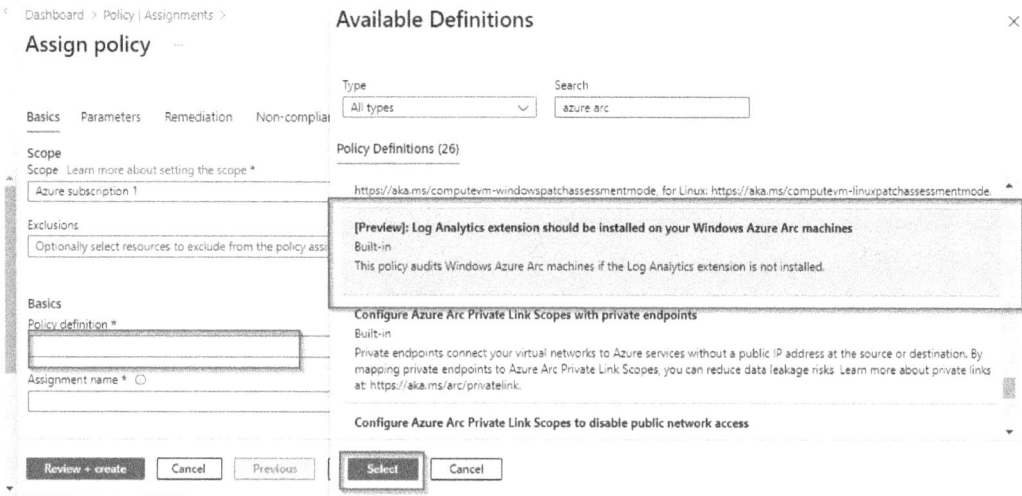

Figure 13.25 – Selecting a policy definition

7. The assignment's name will have been automatically pre-filled:

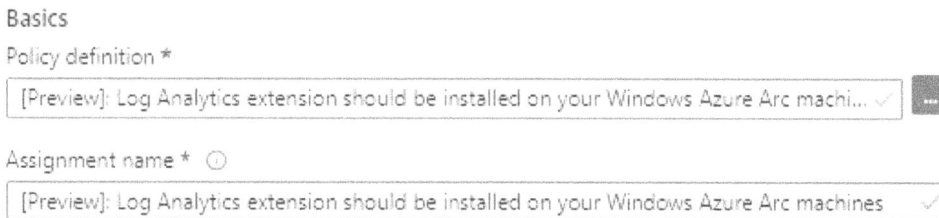

Figure 13.26 – Assignment name setting

8. No other settings are completed on the other tabs for this exercise; click **Review + Create**, and then **Create**:

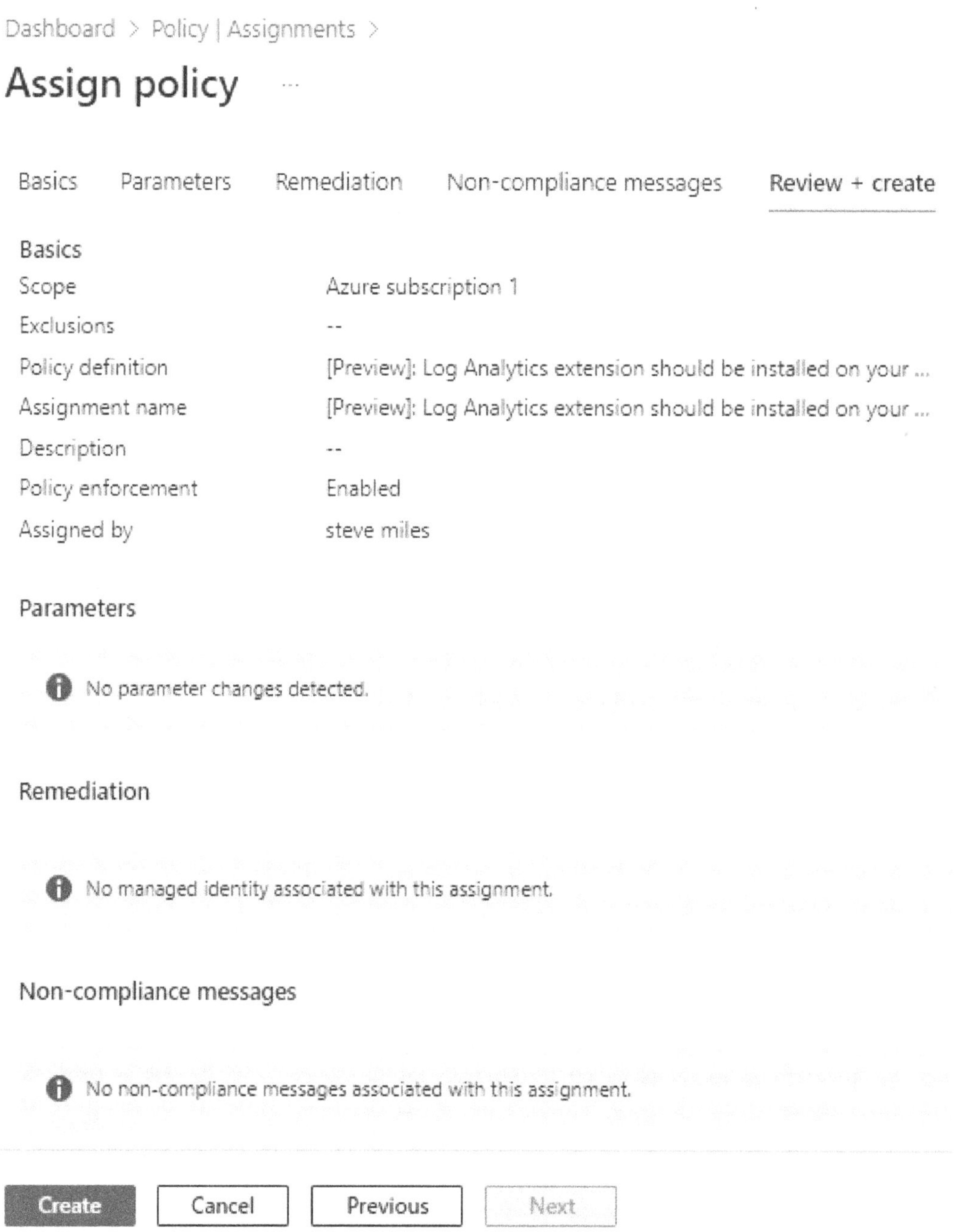

Dashboard > Policy | Assignments >

Assign policy ...

Basics   Parameters   Remediation   Non-compliance messages   Review + create

Basics

Scope   Azure subscription 1
Exclusions   --
Policy definition   [Preview]: Log Analytics extension should be installed on your ...
Assignment name   [Preview]: Log Analytics extension should be installed on your ...
Description   --
Policy enforcement   Enabled
Assigned by   steve miles

Parameters

ⓘ No parameter changes detected.

Remediation

ⓘ No managed identity associated with this assignment.

Non-compliance messages

ⓘ No non-compliance messages associated with this assignment.

| Create | Cancel | Previous | Next |

Figure 13.27 – Creating a policy assignment

9. Click **Compliance** on the left-hand menu, then locate and click on the policy assignment created:

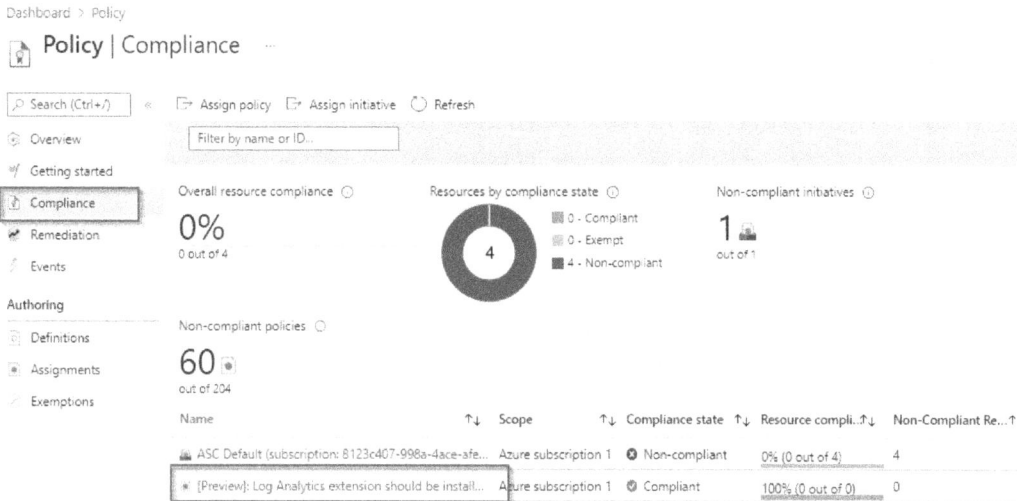

Dashboard > Policy

Policy | Compliance   ···

Search (Ctrl+/) «	Assign policy Assign initiative Refresh
Overview	Filter by name or ID...
Getting started	
Compliance	Overall resource compliance ⓘ
Remediation	**0%**
Events	0 out of 4

Resources by compliance state ⓘ

- 0 - Compliant
- 0 - Exempt
- 4 - Non-compliant

4

Non-compliant initiatives ⓘ

1
out of 1

Authoring

Definitions	Non-compliant policies ⓘ
Assignments	**60**
Exemptions	out of 204

Name ↑↓	Scope ↑↓	Compliance state ↑↓	Resource compli...↑↓	Non-Compliant Re...↑
ASC Default (subscription: 8123c407-998a-4ace-afe...	Azure subscription 1	⊗ Non-compliant	0% (0 out of 4)	4
[Preview]: Log Analytics extension should be install...	Azure subscription 1	✓ Compliant	100% (0 out of 0)	0

Figure 13.28 – Policy compliance dashboard

10. You will see the status of all resources affected by the policy and whether their compliance state is **Compliant** or **Non-compliant**:

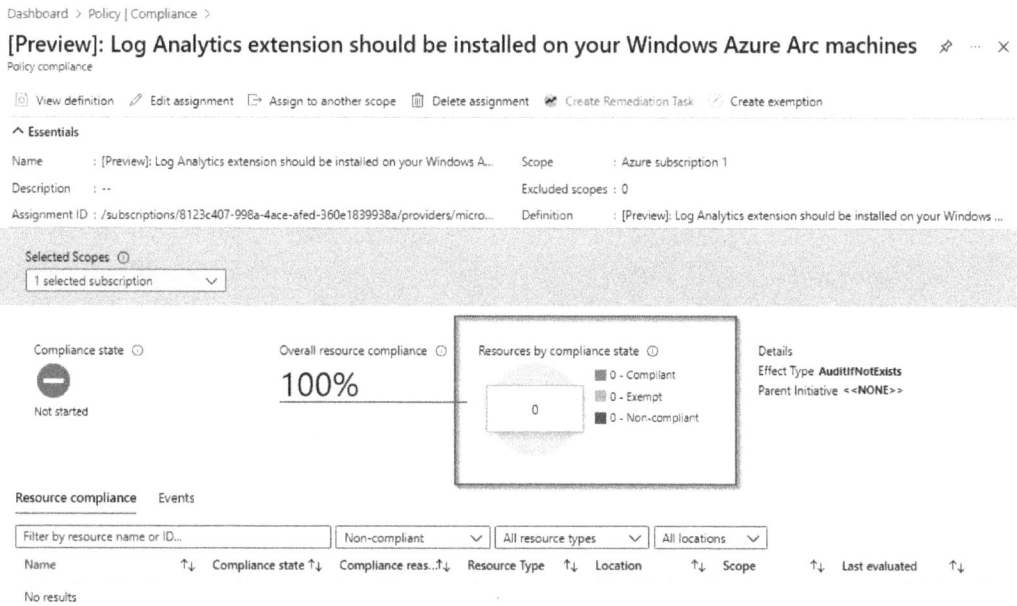

Dashboard > Policy | Compliance >

[Preview]: Log Analytics extension should be installed on your Windows Azure Arc machines   📌 ··· ✕
Policy compliance

View definition   Edit assignment   Assign to another scope   🗑 Delete assignment   Create Remediation Task   Create exemption

∧ Essentials

Name	: [Preview]: Log Analytics extension should be installed on your Windows A...	Scope	: Azure subscription 1
Description	: --	Excluded scopes : 0	
Assignment ID	: /subscriptions/8123c407-998a-4ace-afed-360e1839938a/providers/micro...	Definition	: [Preview]: Log Analytics extension should be installed on your Windows ...

Selected Scopes ⓘ

1 selected subscription ∨

Compliance state ⓘ	Overall resource compliance ⓘ	Resources by compliance state ⓘ	Details
⊖ Not started	**100%**	0	Effect Type **AuditIfNotExists**
		- 0 - Compliant	Parent Initiative <<NONE>>
		- 0 - Exempt	
		- 0 - Non-compliant	

Resource compliance   Events

Filter by resource name or ID...	Non-compliant ∨	All resource types ∨	All locations ∨

Name ↑↓	Compliance state ↑↓	Compliance reas...↑↓	Resource Type ↑↓	Location ↑↓	Scope ↑↓	Last evaluated ↑↓
No results						

Figure 13.29 – Policy dashboard

With this, we have completed this exercise. This exercise taught us the skills needed to create an Azure Policy assignment. The next exercise will look at setting up Microsoft Defender for Cloud.

Exercise – setting up Microsoft Defender for Cloud

This exercise will look at setting up Microsoft Defender for Cloud.

You will need the following in place for this exercise:

- Azure portal access with an **Owner** or **Contributor** subscription role
- An Azure VM resource to assign access

In the following sub-sections, you can see the procedure to complete the exercise, segregated into tasks for a better understanding.

Task 1 – accessing the Azure portal

1. Log in to the Azure portal: `https://portal.azure.com`. You can alternatively use the Azure desktop app: `https://portal.azure.com/App/Download`.

Task 2 – creating a Log Analytics workspace for Defender for Cloud

2. In the search bar, type in `log analytics workspaces`; click **Log Analytics workspaces** from the list of services:

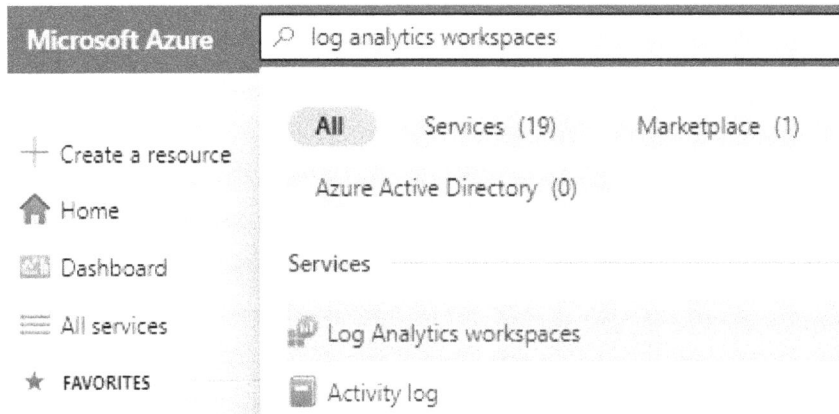

Figure 13.30 – Searching for Log Analytics workspaces

3. From the **Log Analytics workspaces** blade, click **Create**:

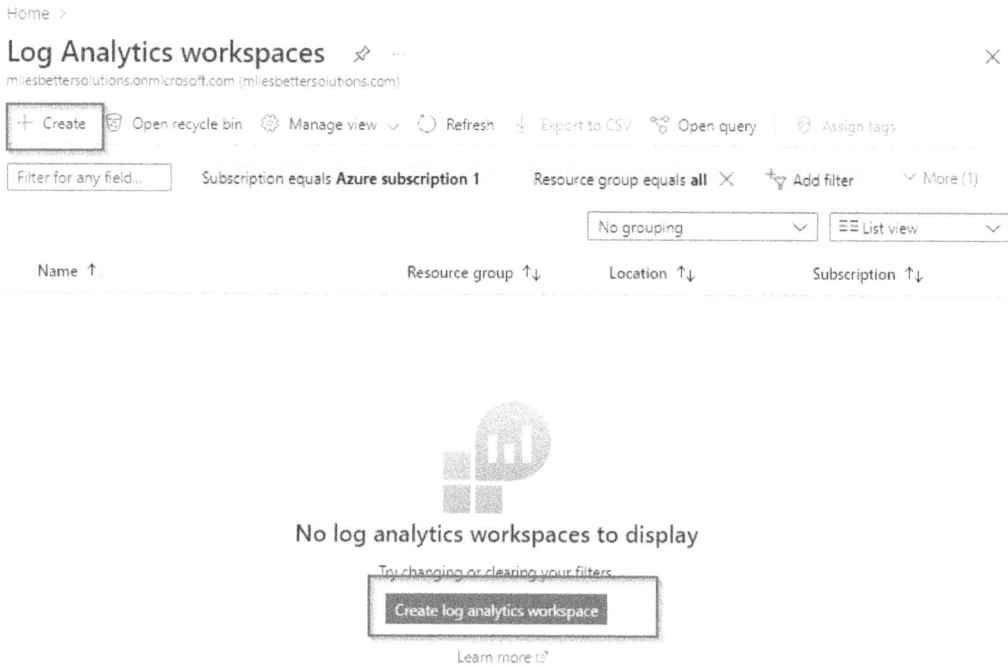

Home >

Log Analytics workspaces

milesbettersolutions.onmicrosoft.com (milesbettersolutions.com)

+ Create   Open recycle bin   Manage view ∨   ○ Refresh   Export to CSV   Open query   Assign tags

| Filter for any field... | Subscription equals **Azure subscription 1** | Resource group equals **all** ✕ | Add filter | More (1) |

No grouping ∨   List view ∨

| Name ↑ | Resource group ↑↓ | Location ↑↓ | Subscription ↑↓ |

No log analytics workspaces to display

Try changing or clearing your filters

Create log analytics workspace

Learn more

Figure 13.31 – Creating a Log Analytics workspace

4.   Fill out the **Project details** and **Instance details** settings as required, then click **Review + Create**:

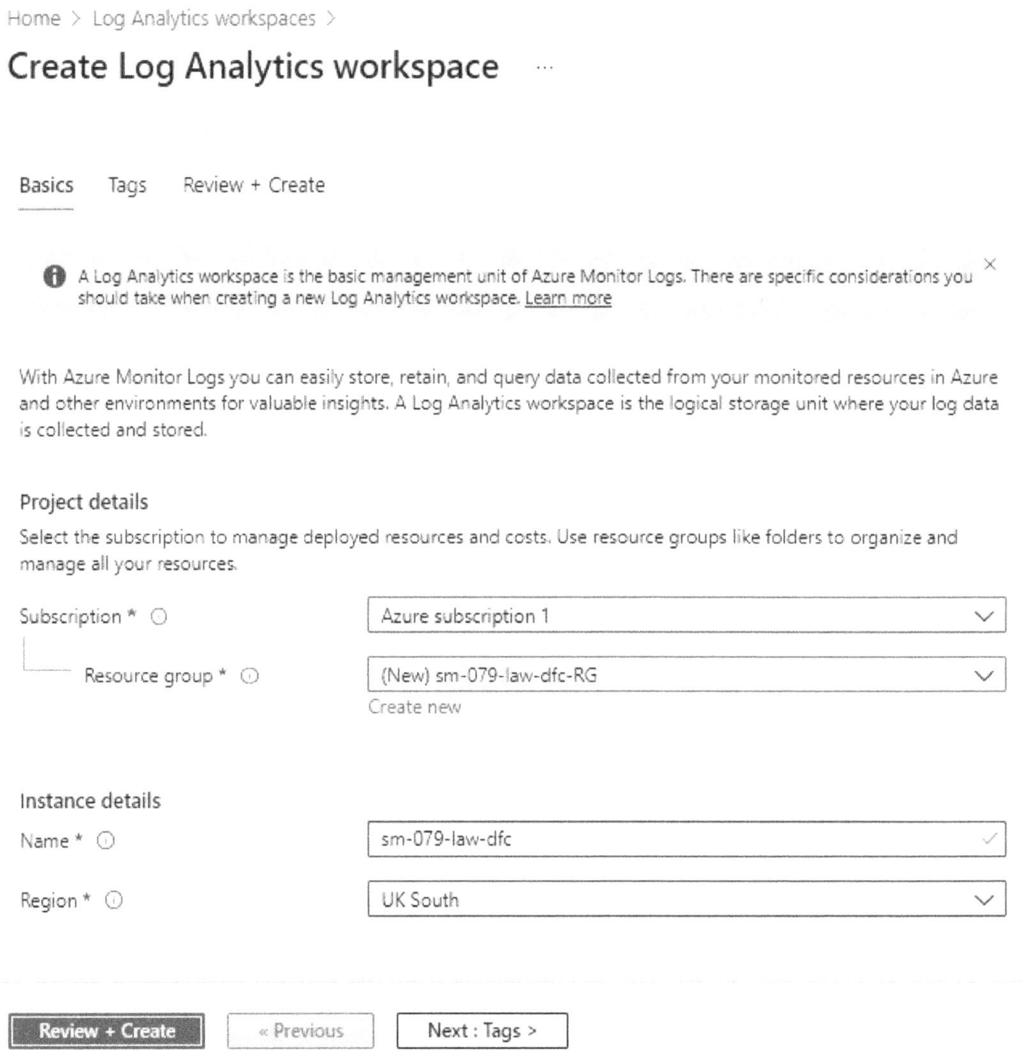

Home > Log Analytics workspaces >

Create Log Analytics workspace   ...

Basics   Tags   Review + Create

ⓘ A Log Analytics workspace is the basic management unit of Azure Monitor Logs. There are specific considerations you should take when creating a new Log Analytics workspace. Learn more

With Azure Monitor Logs you can easily store, retain, and query data collected from your monitored resources in Azure and other environments for valuable insights. A Log Analytics workspace is the logical storage unit where your log data is collected and stored.

Project details

Select the subscription to manage deployed resources and costs. Use resource groups like folders to organize and manage all your resources.

Subscription * ○   Azure subscription 1   ⌄

   └─ Resource group * ⓘ   (New) sm-079-law-dfc-RG   ⌄
                            Create new

Instance details

Name * ⓘ   sm-079-law-dfc   ✓

Region * ⓘ   UK South   ⌄

[Review + Create]   « Previous   [Next : Tags >]

Figure 13.32 – Create Log Analytics workspace wizard

5. Review the settings, then click **Create**:

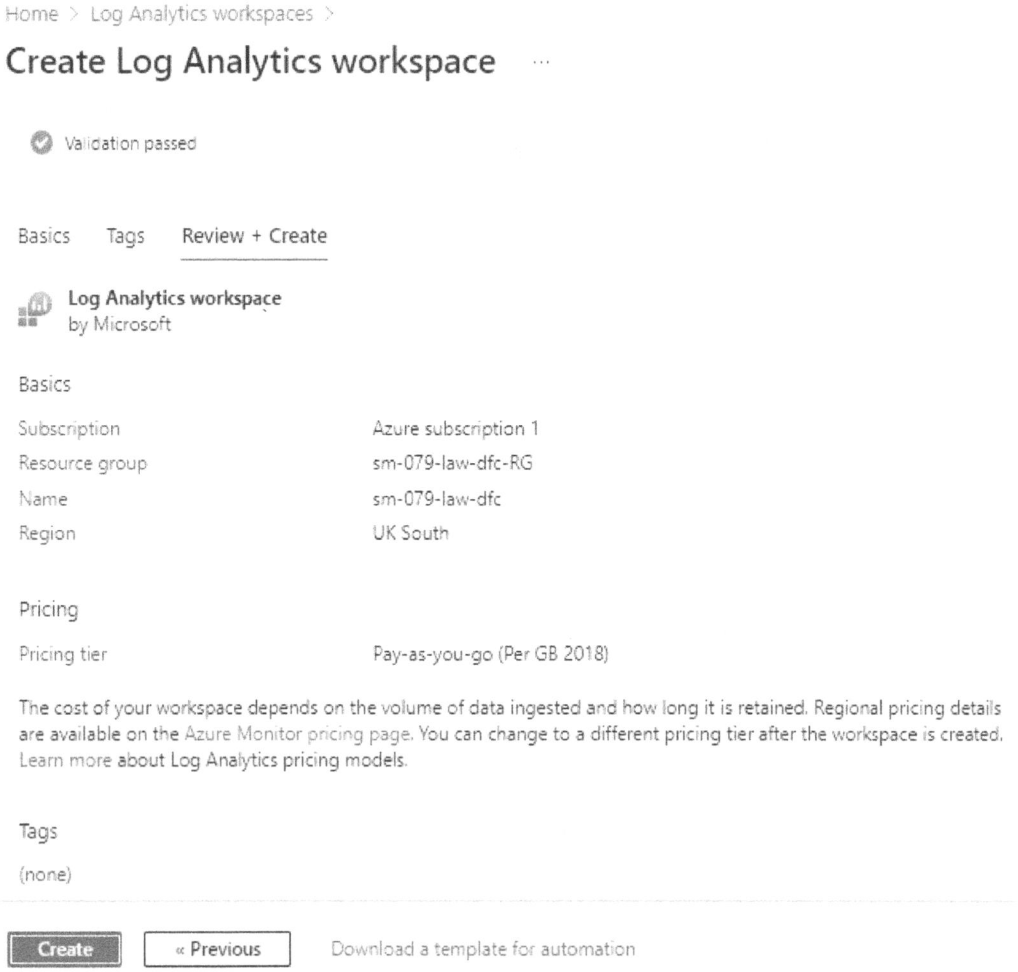

Home > Log Analytics workspaces >

Create Log Analytics workspace   ...

✅ Validation passed

Basics   Tags   Review + Create

Log Analytics workspace
by Microsoft

Basics

Subscription	Azure subscription 1
Resource group	sm-079-law-dfc-RG
Name	sm-079-law-dfc
Region	UK South

Pricing

Pricing tier	Pay-as-you-go (Per GB 2018)

The cost of your workspace depends on the volume of data ingested and how long it is retained. Regional pricing details are available on the Azure Monitor pricing page. You can change to a different pricing tier after the workspace is created. Learn more about Log Analytics pricing models.

Tags

(none)

| Create | « Previous | Download a template for automation |

Figure 13.33 – Creating a Log Analytics workspace

6.   You will receive information that the deployment is complete:

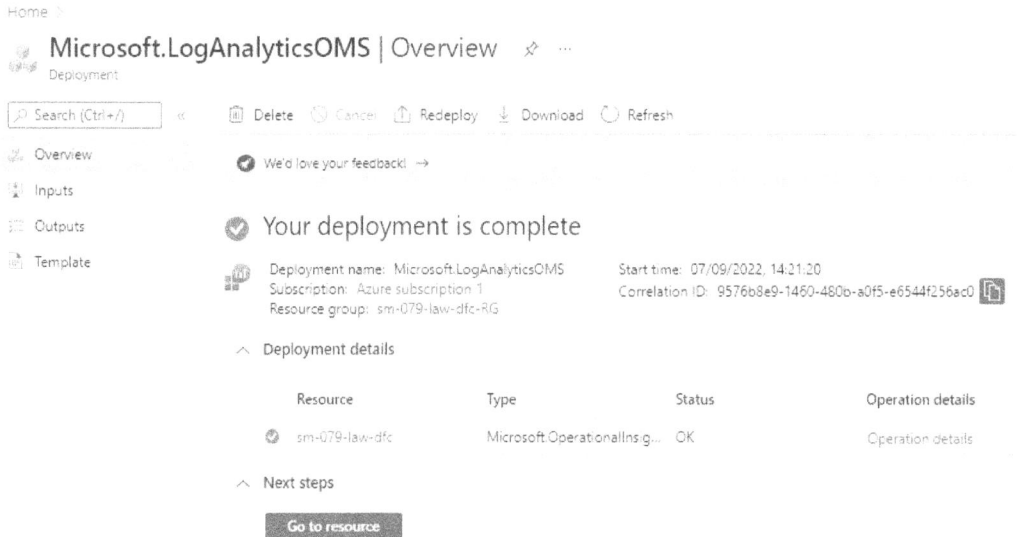

Figure 13.34 – Log Analytics workspace deployment completed

Task 3 – enabling advanced features and upgrading Defender for Cloud

We will enable advanced features and upgrade Defender for Cloud in the subscription. Follow these steps:

7.   In the search bar, type in `defender for cloud`; click **Microsoft Defender for Cloud** from the list of services:

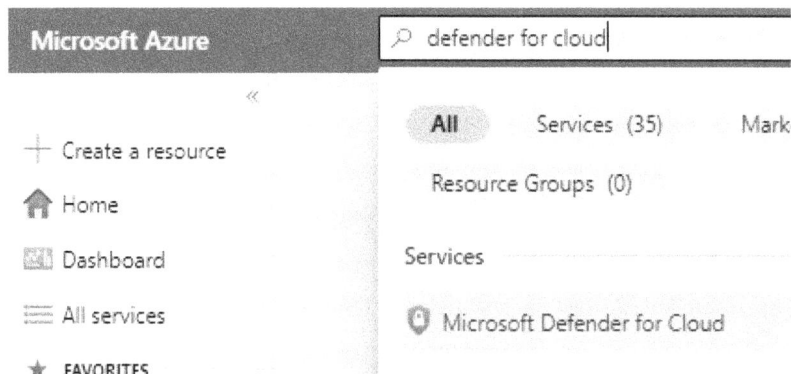

Figure 13.35 – Searching for Defender for Cloud

8. From the **Getting started** blade, ensure that your subscription and created workspace are selected, then click **Upgrade**:

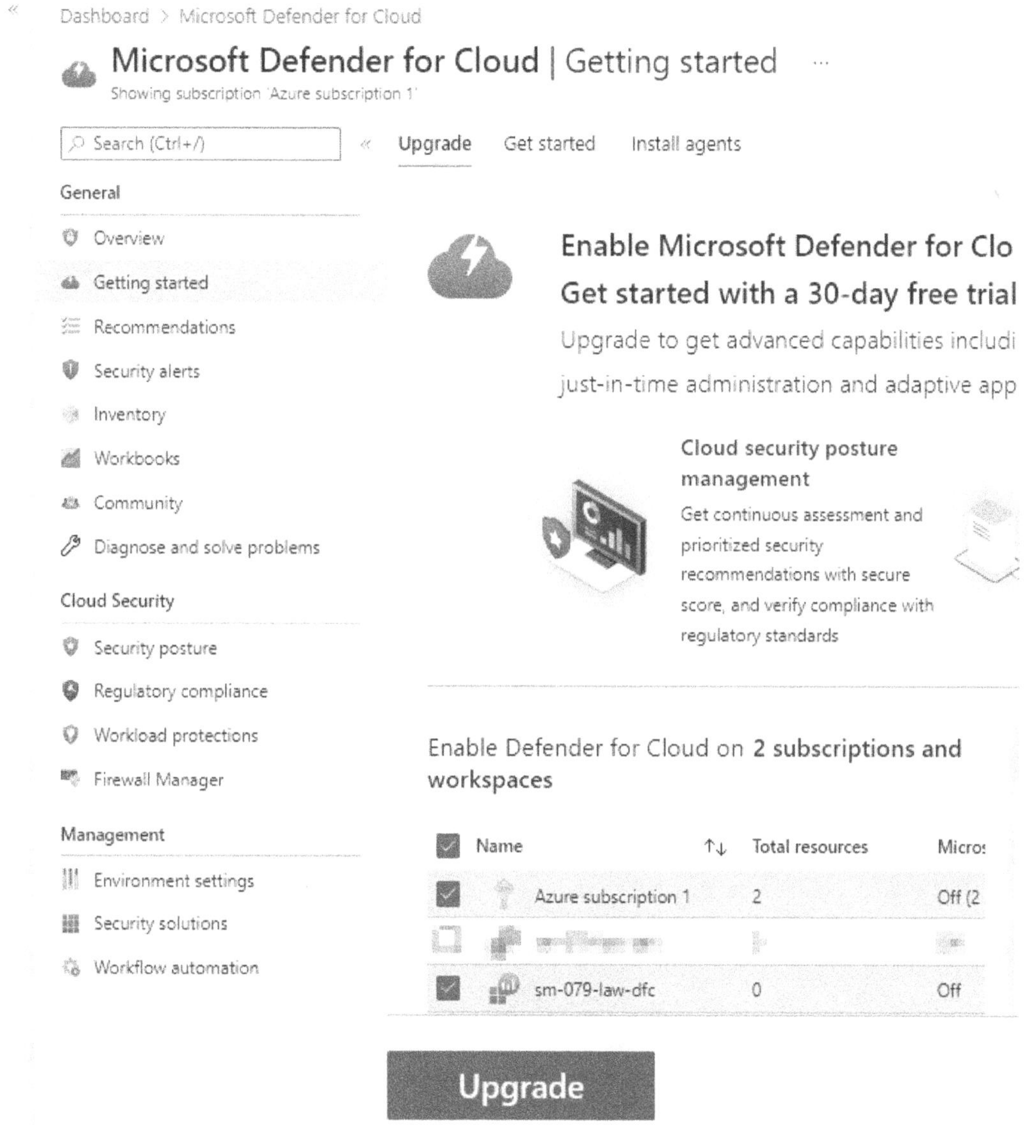

Figure 13.36 – Upgrading Defender for Cloud

9. You will now see the **Install agents** blade. Click **Install agents**:

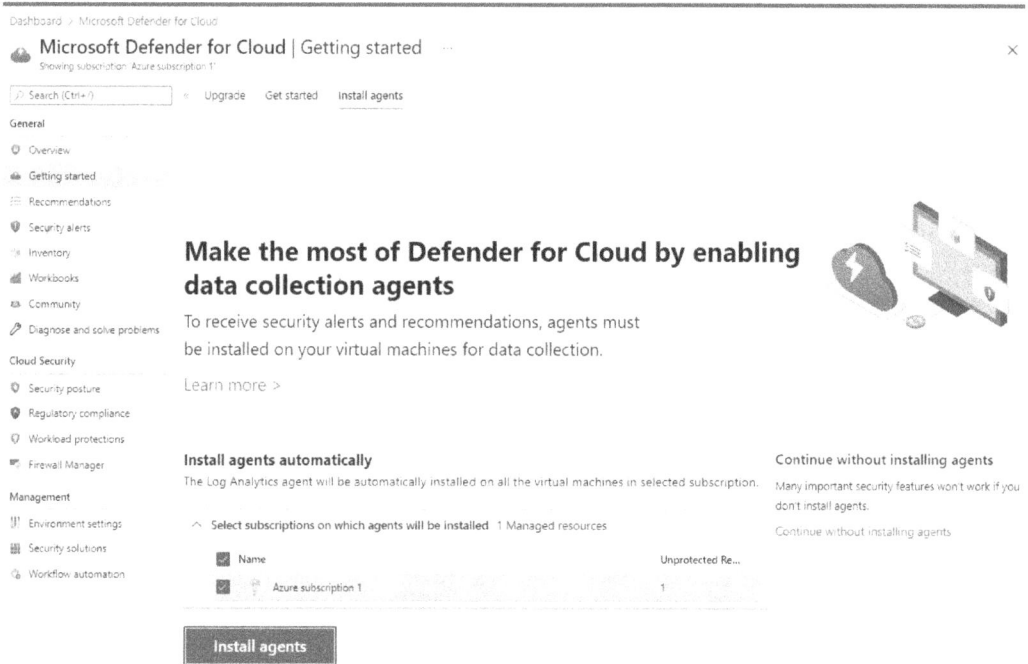

Figure 13.37 – Install agents blade

10. You will now have access to the CSPM and CWPP capabilities:

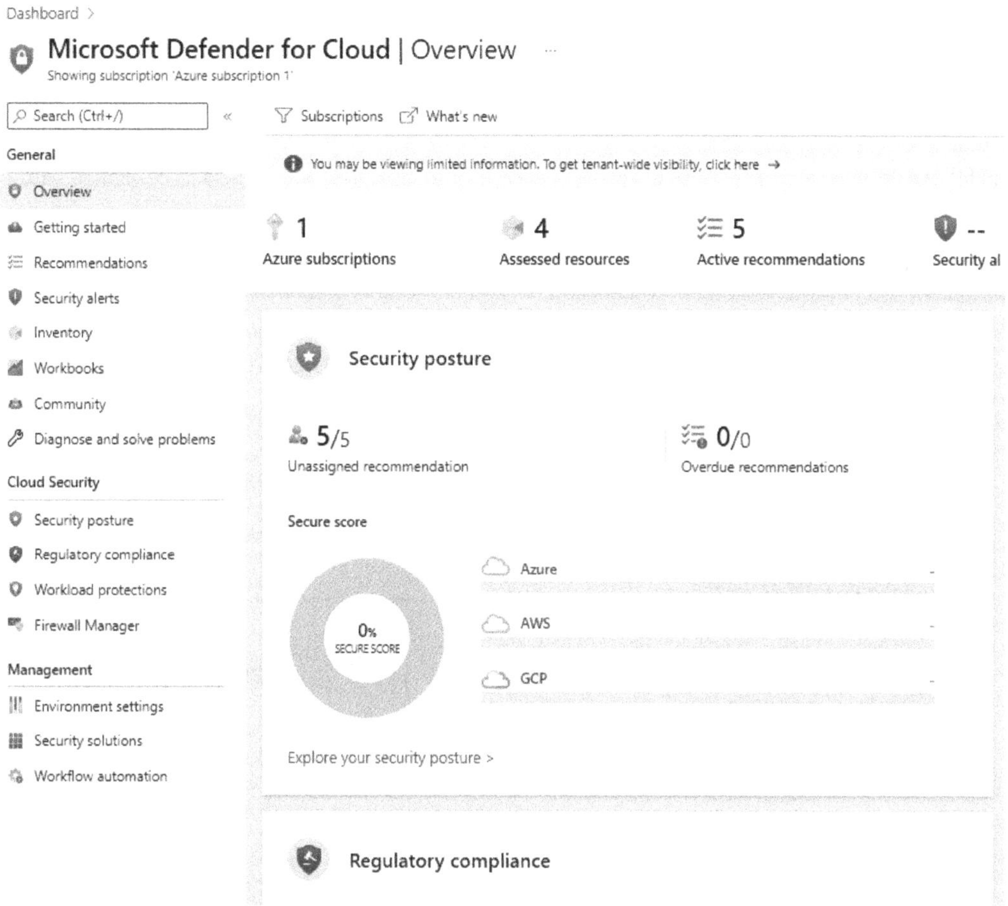

Figure 13.38 – Defender for Cloud setup

In this task, we enabled advanced features and upgraded Defender for Cloud.

With this, we have completed this exercise. This exercise taught us the skills to set up Microsoft Defender for Cloud. The next exercise will look at setting up Microsoft Sentinel and integrating Microsoft Defender for Cloud.

Exercise – setting up Microsoft Sentinel

You will need the following in place for this exercise:

- Azure portal access with an **Owner** or **Contributor** subscription role

- **Contributor** role access to the resource group where the Sentinel workspace is created

In the following sub-sections, you will see the procedure to complete the exercise, segregated into tasks for a better understanding:

Task 1 – accessing the Azure portal

1. Log in to the Azure portal: `https://portal.azure.com`. You can alternatively use the Azure desktop app: `https://portal.azure.com/App/Download`.

Task 2 – enabling Microsoft Sentinel

2. In the search bar, type in `sentinel`; click **Microsoft Sentinel** from the list of services:

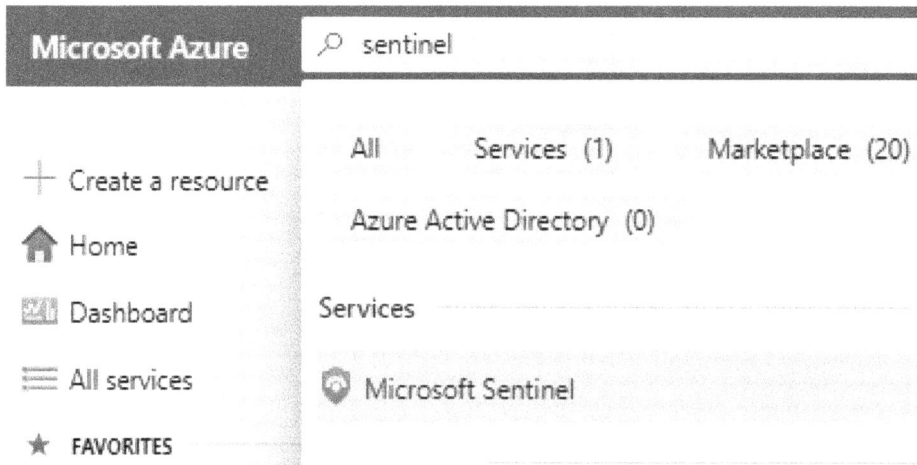

Figure 13.39 – Searching for Sentinel

3. From the **Microsoft Sentinel** blade, click **+ Create** from the top menu:

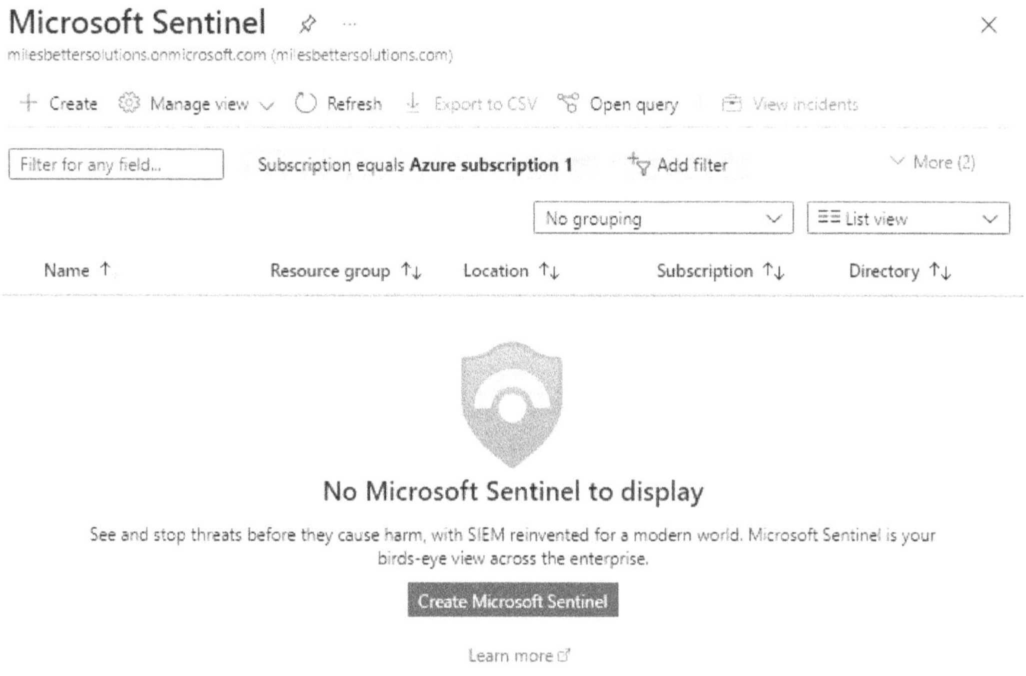

Microsoft Sentinel 📌 ⋯

milesbettersolutions.onmicrosoft.com (milesbettersolutions.com)

+ Create   ⚙ Manage view ⌄   ○ Refresh   ⤓ Export to CSV   ⚷ Open query   🗒 View incidents

| Filter for any field... | Subscription equals **Azure subscription 1** | ⁺▽ Add filter | ⌄ More (2) |

| | No grouping ⌄ | ☰ List view ⌄ |

| Name ↑ | Resource group ↑↓ | Location ↑↓ | Subscription ↑↓ | Directory ↑↓ |

No Microsoft Sentinel to display

See and stop threats before they cause harm, with SIEM reinvented for a modern world. Microsoft Sentinel is your birds-eye view across the enterprise.

Create Microsoft Sentinel

Learn more ⌕

Figure 13.40 – Creating a Sentinel instance

4. From the **Add Microsoft Sentinel to a workspace** blade, click **+ Create a new workspace**:

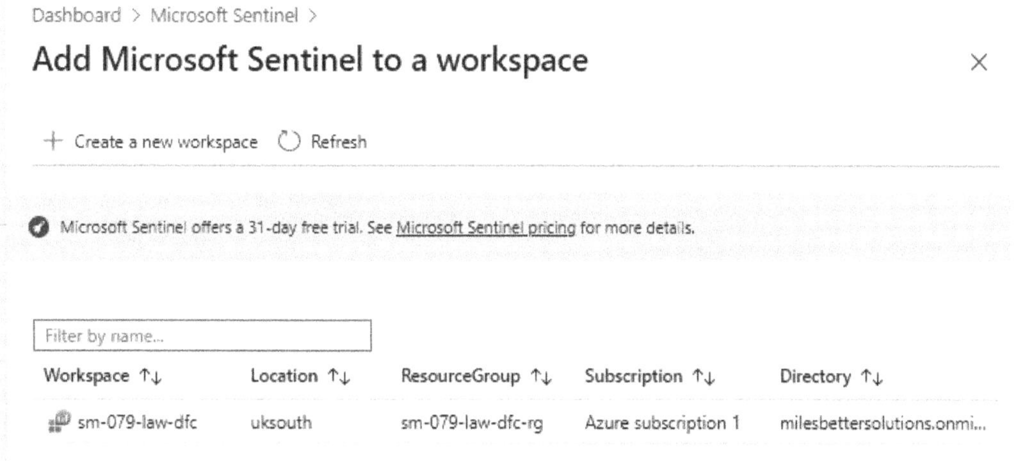

Dashboard > Microsoft Sentinel >

Add Microsoft Sentinel to a workspace   ×

+ Create a new workspace   ○ Refresh

🛈 Microsoft Sentinel offers a 31-day free trial. See Microsoft Sentinel pricing for more details.

| Filter by name... |

Workspace ↑↓	Location ↑↓	ResourceGroup ↑↓	Subscription ↑↓	Directory ↑↓
sm-079-law-dfc	uksouth	sm-079-law-dfc-rg	Azure subscription 1	milesbettersolutions.onmi...

Figure 13.41 – Creating a workspace

5. On the **Create Log Analytics workspace** blade, fill out the **Project details** and **Instance details** settings as required, then click **Review + Create**:

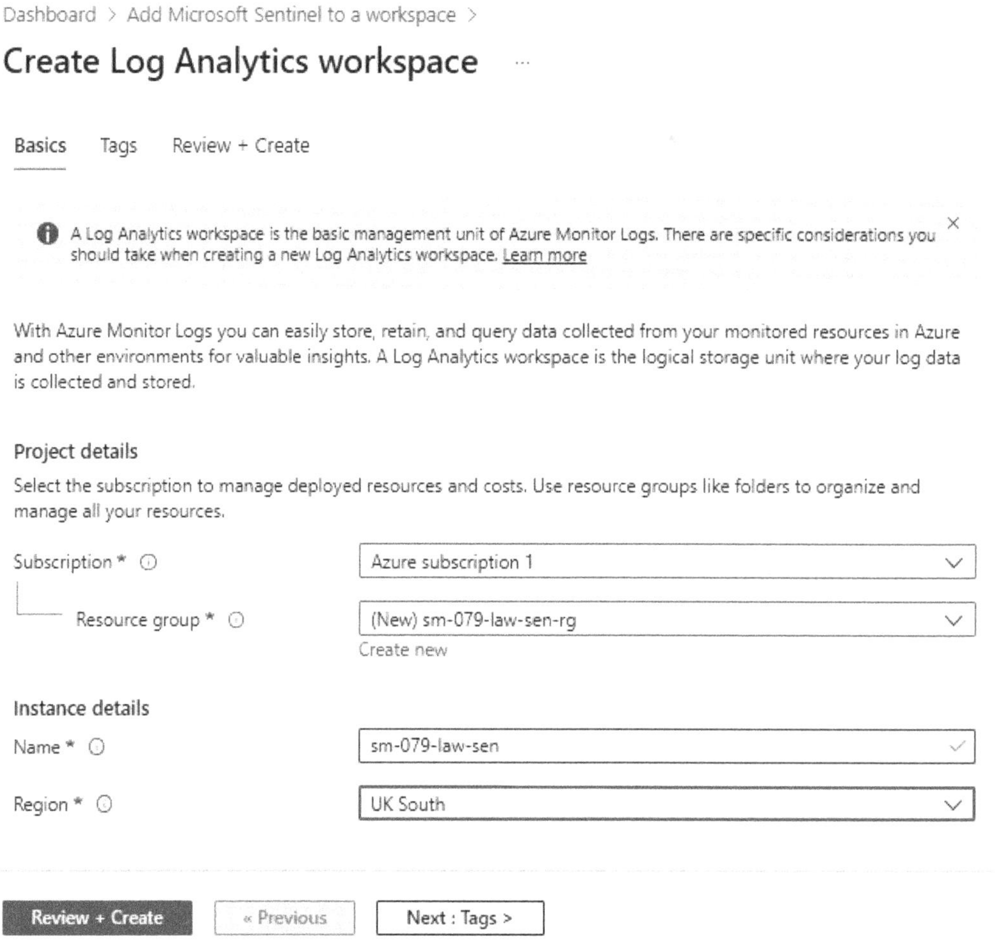

Dashboard > Add Microsoft Sentinel to a workspace >

Create Log Analytics workspace   ...

Basics   Tags   Review + Create

> ℹ️ A Log Analytics workspace is the basic management unit of Azure Monitor Logs. There are specific considerations you ✕ should take when creating a new Log Analytics workspace. Learn more

With Azure Monitor Logs you can easily store, retain, and query data collected from your monitored resources in Azure and other environments for valuable insights. A Log Analytics workspace is the logical storage unit where your log data is collected and stored.

Project details

Select the subscription to manage deployed resources and costs. Use resource groups like folders to organize and manage all your resources.

Subscription * ⓘ	Azure subscription 1 ⌄
Resource group * ⓘ	(New) sm-079-law-sen-rg ⌄
	Create new

Instance details

Name * ⓘ	sm-079-law-sen ✓
Region * ⓘ	UK South ⌄

[Review + Create]   [« Previous]   [Next : Tags >]

Figure 13.42 – Create Log Analytics workspace wizard

6. Review the settings and then click **Create**:

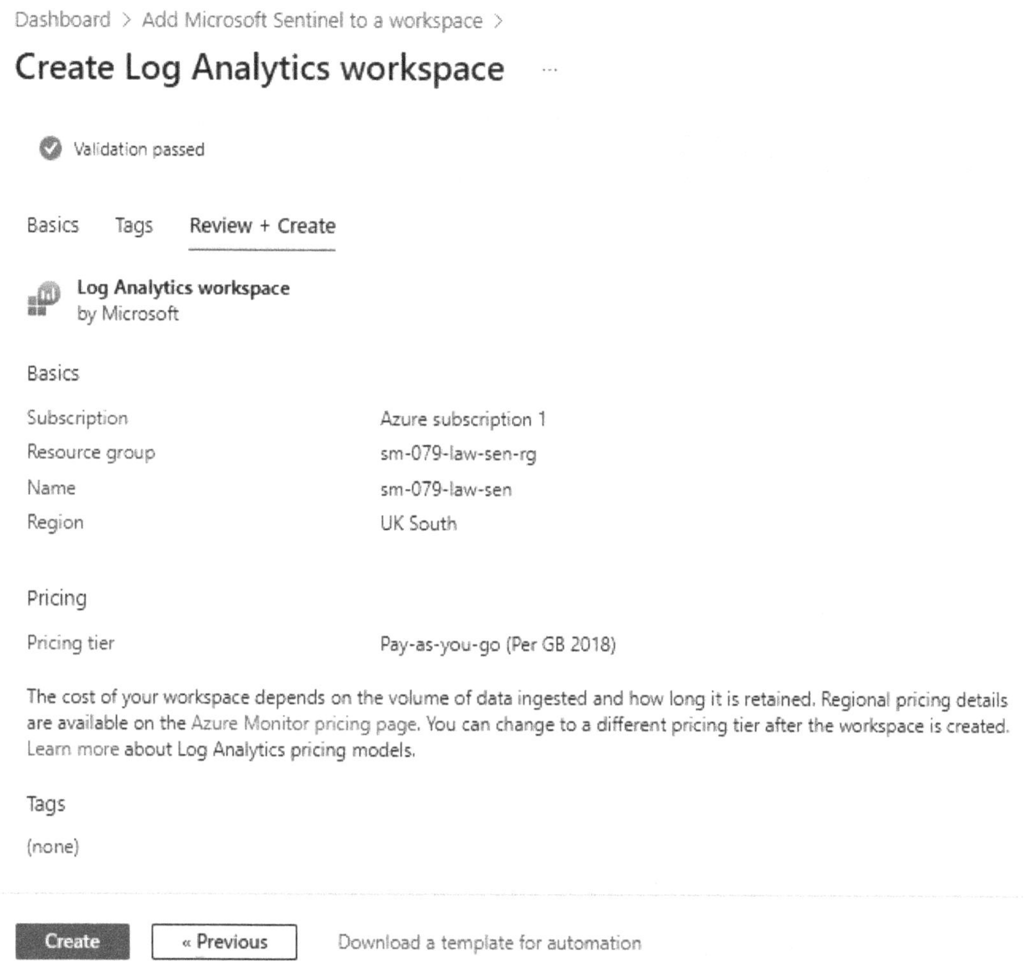

Dashboard > Add Microsoft Sentinel to a workspace >

Create Log Analytics workspace   ...

✓ Validation passed

Basics   Tags   Review + Create
_____

Log Analytics workspace
by Microsoft

Basics

Subscription	Azure subscription 1
Resource group	sm-079-law-sen-rg
Name	sm-079-law-sen
Region	UK South

Pricing

Pricing tier	Pay-as-you-go (Per GB 2018)

The cost of your workspace depends on the volume of data ingested and how long it is retained. Regional pricing details are available on the Azure Monitor pricing page. You can change to a different pricing tier after the workspace is created. Learn more about Log Analytics pricing models.

Tags

(none)

| Create | « Previous   Download a template for automation

Figure 13.43 – Reviewing the workspace

7. You will receive information that the deployment is complete.

8. Click **Refresh** to see your *created workspace*, select the workspace to be Sentinel-enabled, and click **Add**:

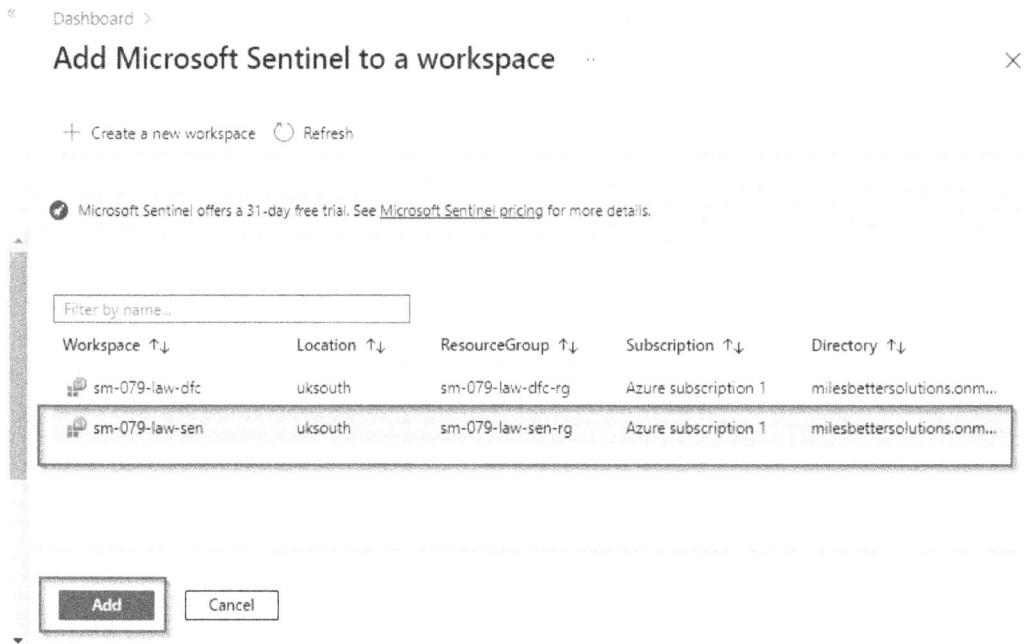

Figure 13.44 – Adding a workspace to Sentinel

9.   You will receive a notification that *Sentinel was added*:

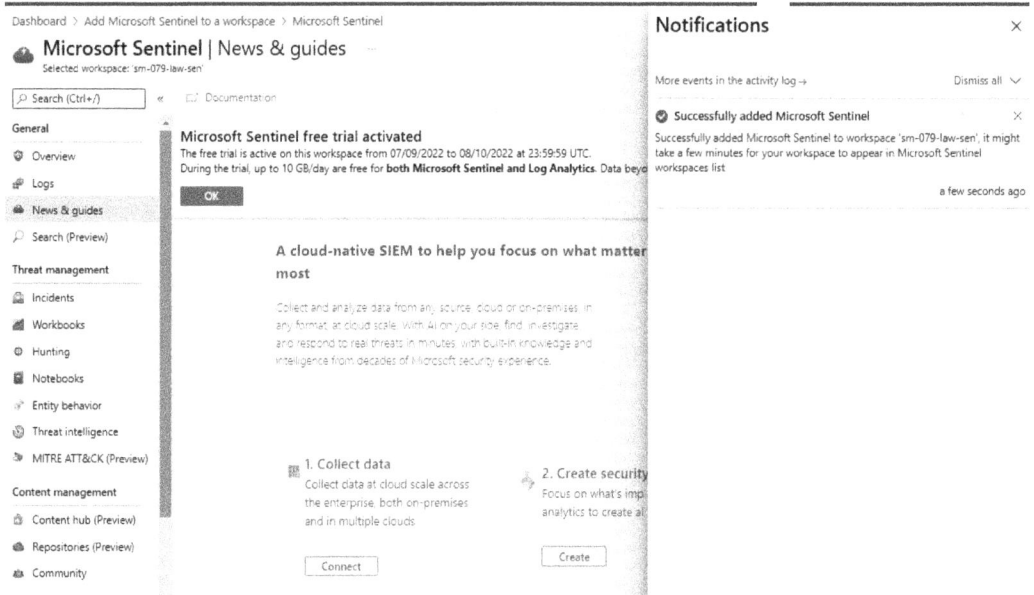

Figure 13.45 – Sentinel added

10. Sentinel is now *enabled*.

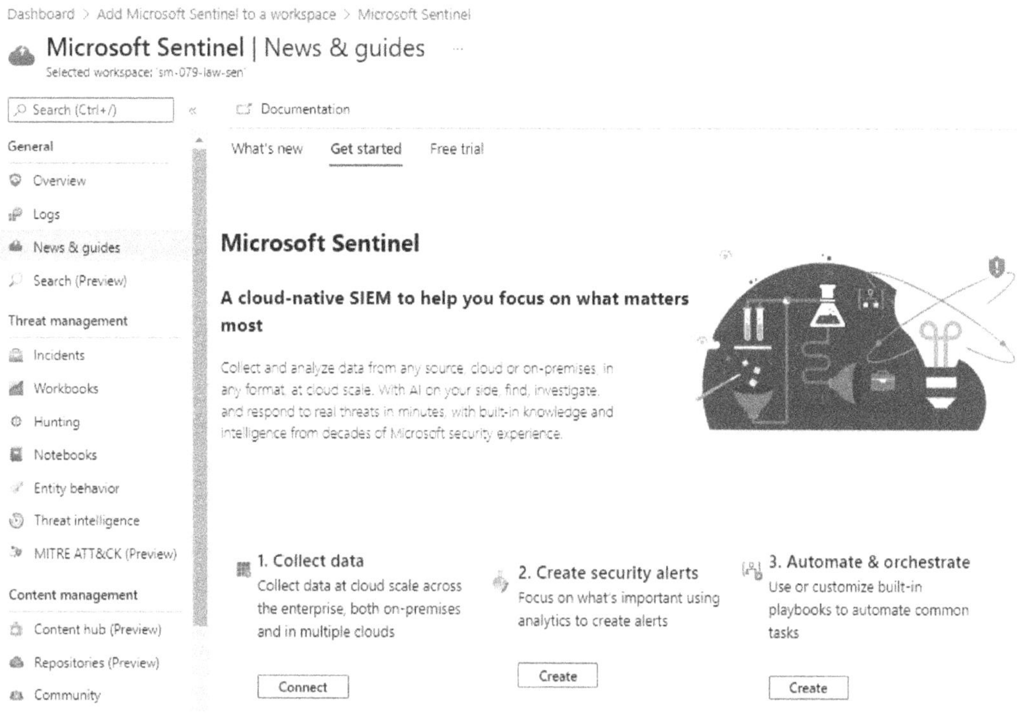

Figure 13.46 – Sentinel enabled

Task 3 – integrating Sentinel with Defender for Cloud

We will integrate Sentinel with Defender for Cloud in this task. Follow these steps:

1.   Navigate to the **Data connectors** blade under the **Configuration** section:

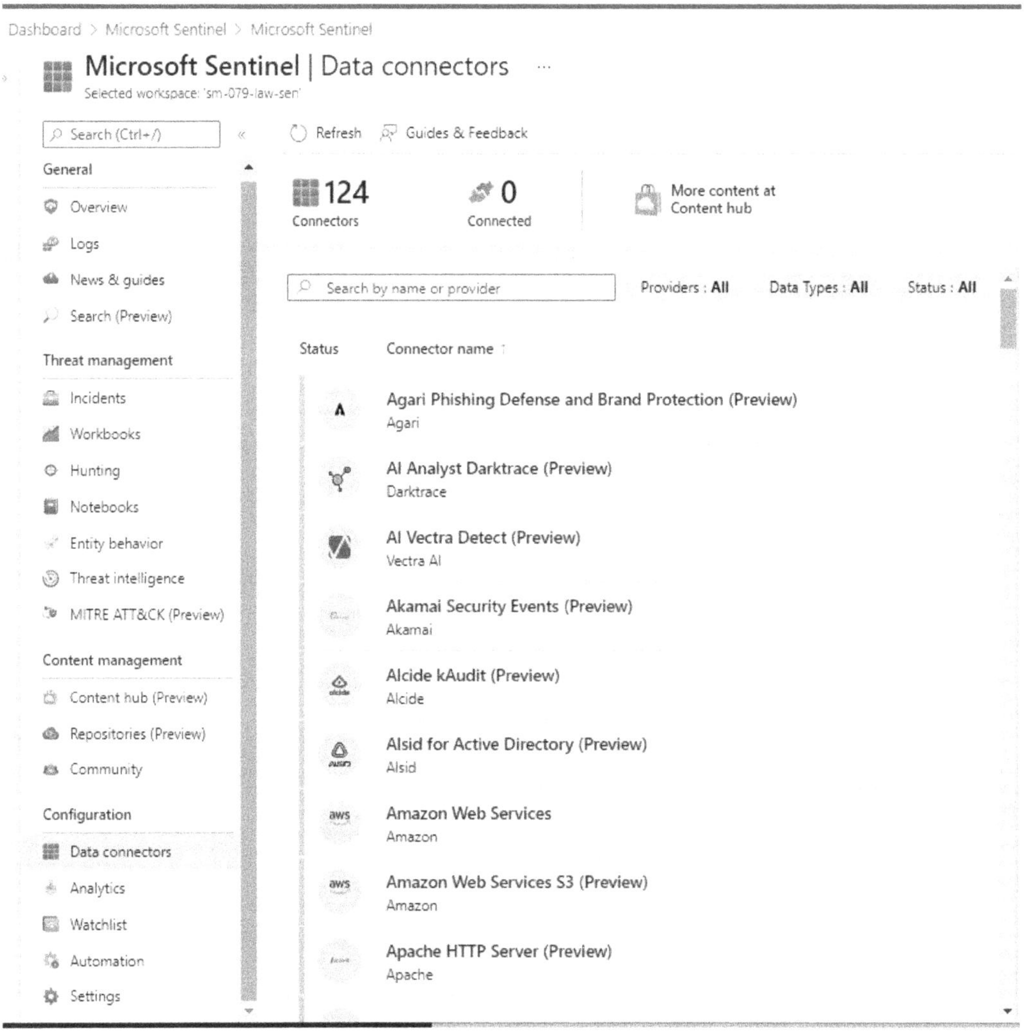

Figure 13.47 – Sentinel data connectors

2. Search for the **Microsoft Defender for Cloud** connector and click to open the **Data connectors** blade; then, select **Open connector page**:

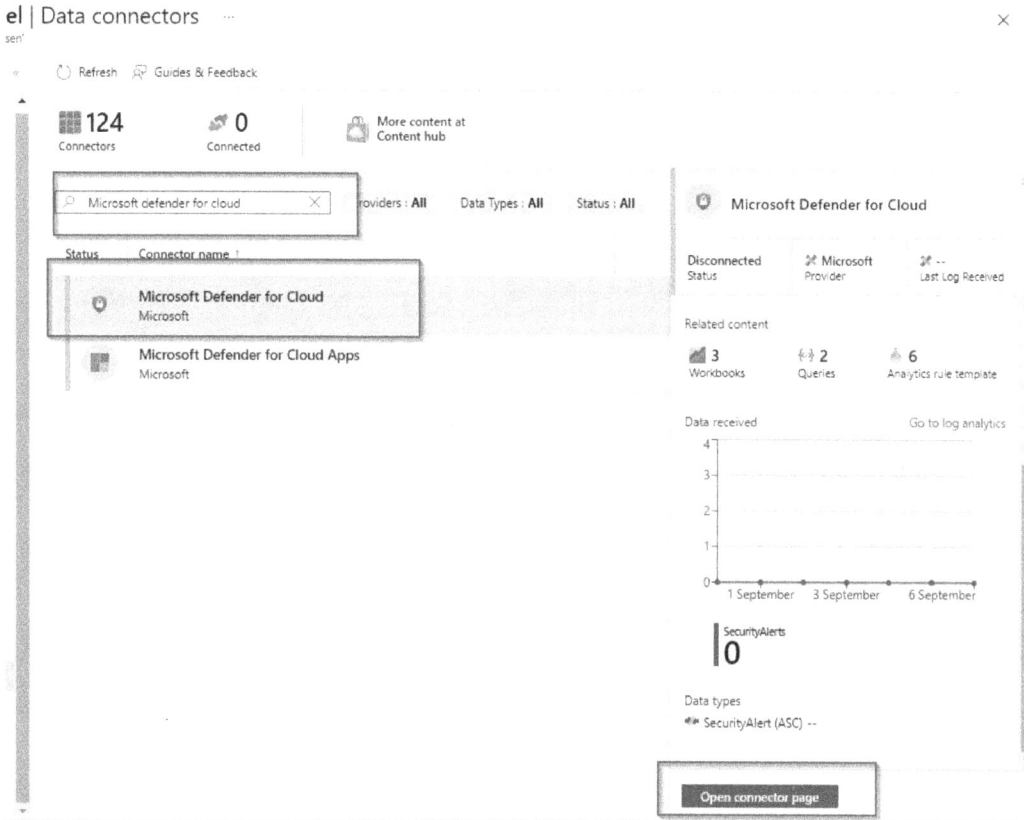

Figure 13.48 – Defender for Cloud data connector

3. On the connector page, select your subscription; at the bottom of the screen, click **Enable** for **Create incidents – Recommended!**. Then, click **Connect** and accept the pop-up dialog box:

Instructions   Next steps

🛡 **Subscription:** read security data

Configuration

Connect Microsoft Defender for Cloud to Microsoft Sentinel

Mark the check box of each Azure subscription whose alerts you want to import into Microsoft Sentinel, then select **Connect** above the list.

The connector can be enabled only on subscriptions that have at least one Microsoft Defender plan enabled in Microsoft Defender for Cloud, and only by users with Security Reader permissions on the subscription.

Connect   Disconnect   Enable bi-directional sync   Disable bi-directional sync   🛡 Enable Microsoft Defender for all subscriptions >

🔍 Search

☑ Subscription ↑↓	Status	Bi-directional sync ○	Microsoft Defender plans
☑ Azure subscription 1	● Disconnected	⊘ Disabled ∨	Some enabled Enable all >

Create incidents - Recommended!
Create incidents automatically from all alerts generated in this connected service.   **Enable**

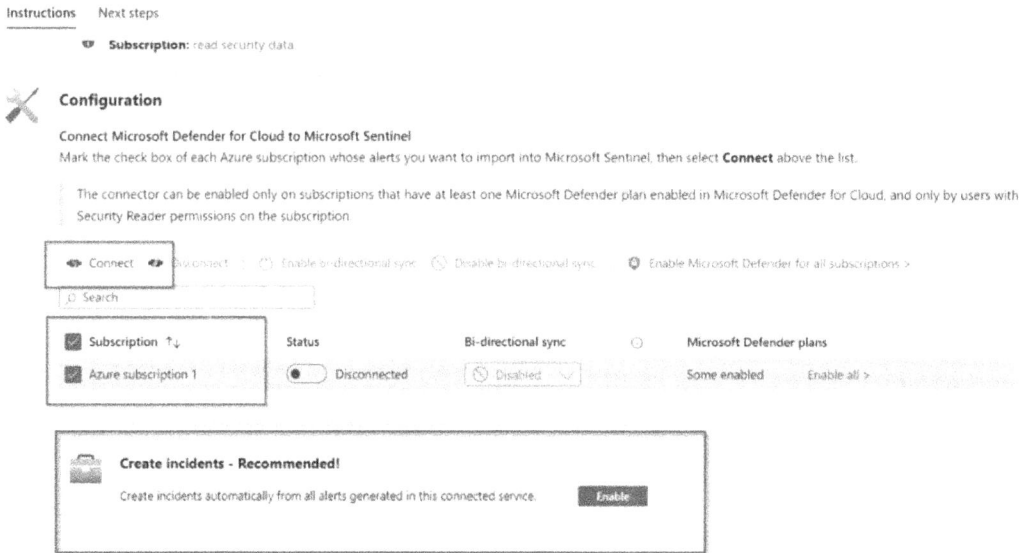

Figure 13.49 – Connecting the data connector

4. You will now see the Defender for Cloud data connector marked as **Connected**:

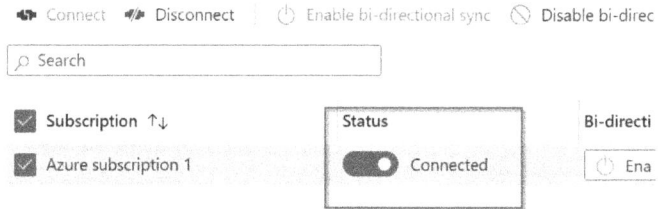

Connect   Disconnect   Enable bi-directional sync   ⊘ Disable bi-direc

🔍 Search

☑ Subscription ↑↓	Status	Bi-directi
☑ Azure subscription 1	⬤ Connected	Ena

Figure 13.50 – Connected data connector

5. The **Data connectors** blade confirms Microsoft Defender for Cloud is *connected* to Sentinel:

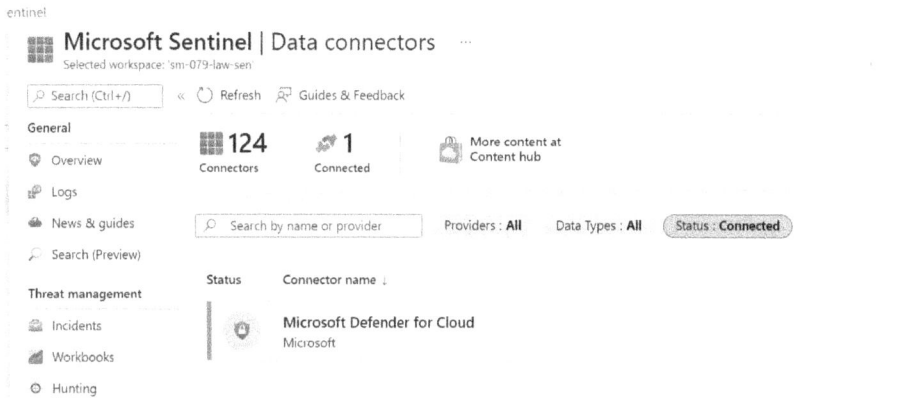

Figure 13.51 – Connected data connector (continued)

6. Defender for Cloud is now integrated with Sentinel and ready for you to start security operations:

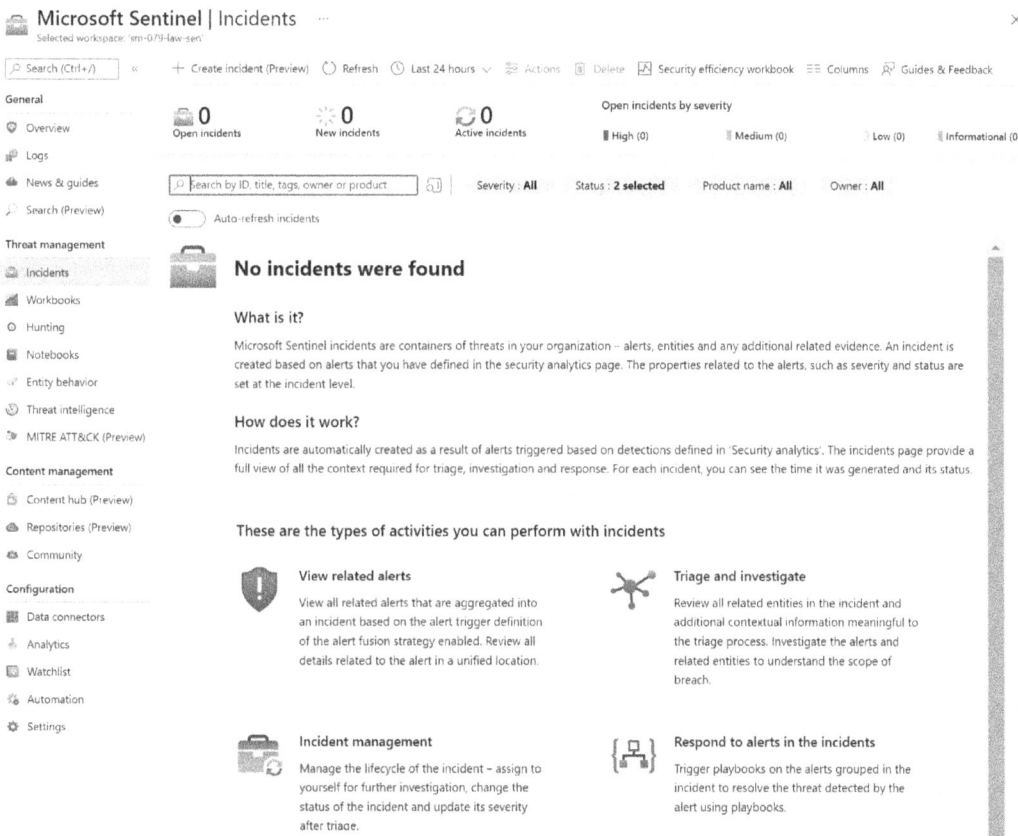

Figure 13.52 – Defender for Cloud integrated with Sentinel

With this, we have completed this exercise. This exercise taught us the skills to set up Microsoft Sentinel and integrate Microsoft Defender for Cloud. Now, let's summarize this chapter.

Summary

This chapter provided coverage for the *AZ-800 Administering Windows Server Hybrid Core Infrastructure: Manage Windows servers and workloads in a hybrid environment* exam.

This chapter's content further developed your knowledge and skills for on-premises network infrastructure services.

You learned about managing Windows Server user and network access, implementing Azure Arc and Azure Automation for hybrid Windows servers, setting up Microsoft Defender for Cloud, and integrating it with Microsoft Sentinel. We then finished the chapter with hands-on exercises to develop your skills further.

You added new skills through the information provided, with the chapter's goal to take your knowledge beyond the exam objectives so that you are prepared for a real-world, day-to-day hybrid environment-focused role.

The next chapter will teach you about implementing and managing Azure network infrastructure.

Further reading

This section provides links to additional study references and additional exam information:

- *Microsoft Certified: Windows Server Hybrid Administrator Associate*: `https://docs.microsoft.com/en-us/learn/certifications/windows-server-hybrid-administrator/`

- *Exam AZ-800: Administering Windows Server Hybrid Core Infrastructure*: `https://docs.microsoft.com/en-us/learn/certifications/exams/az-800`

- *Exam AZ-800*: skills outline: `https://query.prod.cms.rt.microsoft.com/cms/api/am/binary/RWKI0r`

- *Microsoft Learn*: `https://docs.microsoft.com/en-us/training/paths/manage-windows-servers-workloads-hybrid-environment/`

Skills check

Challenge yourself with what you have learned in this chapter:

1. What are the built-in compute category access roles?

2. Which PowerShell cmdlet can be used to manage roles?

3. Which interfaces and tools can be used to access Windows Server Azure VMs?

4. What is Azure Arc?

5. What are Azure Arc-enabled servers?

6. What are the deployment roles required for Azure Arc?

7. What are the network connectivity options?

8. What are Azure Policy and guest configuration policies?

9. What is the Log Analytics agent? Name some use case scenarios.

10. What are the two components of Log Analytics?

11. What is the relationship between Azure Log Analytics and Azure Arc?

12. What is Azure Automation Update Management?

13. What is Microsoft Defender for Cloud?

14. What is Microsoft Sentinel?

15. How does Defender for Cloud integrate with Microsoft Sentinel?

Part 5:
Exam Prep

This part will provide practice tests for each of the *Skills measured* sections for the *AZ-800: Administering Windows Server Hybrid Core Infrastructure* exam.

This part of the book comprises the following chapter:

- *Chapter 14, Exam Preparation Practice Tests*

14
Exam Preparation Practice Tests

This chapter will cover exam preparation tests for the exam objectives from each key *Skills measured* section. Combined with the hands-on exercises throughout this chapter, you will be come confident and ready to apply the knowledge and skills that have prepared you for a real-world, day-to-day, hybrid Windows Server role.

Test questions

Each of the following tests maps to the *Exam AZ-800: Administering Windows Server Hybrid Core Infrastructure Skills measured* exam outline: `https://docs.microsoft.com/en-us/certifications/exams/az-800`.

For questions that have more than one answer, you may select multiple choices as required.

Practice test 1 – deploying and managing AD DS in on-premises and cloud environments

1. Which of the following components are AD logical components?

    A. Domain and Forest

    B. OU

    C. Partition

    D. Datastore

2. Which of the following components are AD physical components?

    A. Global catalog

    B. Domain controller

    C. Container

    D. Site and subnet

3. Which of the following is a correct UPN suffix?

    A. `MILESBETTER\smiles`

    B. `smiles@milesbetter.solutions`

    C. `@milesbetter.solutions`

    D. `milesbetter.solutions`

    E. `milesbetter.onmicrosoft.com`

4. Which FSMO role can add/remove a domain and create partitions?

    A. Schema master

    B. Domain naming master

    C. RID master

    D. Infrastructure master

    E. PDC emulator master

5. The **Key Distribution Services (KDS)** root key allows you to create which of the following?

    A. Standard managed service accounts (sMSAs)

    B. Group-managed service accounts (gMSAs)

    C. System-assigned managed identities

    D. User-assigned managed identities

6. You need an account that can be shared across the nodes of a cluster; you can use an sMSA to do this.

    A. True

    B. False

7. Which PowerShell cmdlet is used to create a gMSA?

    A. `Install-ADServiceAccount`

    B. `New-ADServiceAccount`

   C. `New-ADGroupServiceAccount`

   D. `Set-ADServiceAccount`

8. A service that runs in the context of the Network Service does not need to specify a user account for authentication.

   A. True

   B. False

9. A replica AD domain controller running in Azure requires which type of resources to be created?

   A. An Azure AD instance

   B. Azure AD Domain Services

   C. AD Domain Services deployed to an Azure IaaS VM

   D. Azure AD Connect

   E. Azure VPN Gateway or NVA

10. Modifying AD site link costs will minimize AD change convergence time.

   A. True

   B. False

Practice test 2 – managing Windows servers and workloads in a hybrid environment

1. A JEA role capability file uses which file extension?

   A. `.pssc`

   B. `.psrc`

   C. `.psm1`

   D. `.ps1`

2. A JEA configuration file can be created with which PowerShell command?

   A. `New-PSSessionConfigurationFile -SessionType RestrictedRemoteServer -Path .\Az800JEAEndpoint.pssc`

   B. `New-PSSessionConfigurationFile -SessionType RestrictedRemoteServer -Path .\Az800JEAEndpoint.psrc`

C. `Add-PSSessionConfigurationFile -SessionType RestrictedRemoteServer -Path .\Az800JEAEndpoint.psm1`

D. `Add-PSSessionConfigurationFile -SessionType RestrictedRemoteServer -Path .\Az800JEAEndpoint.ps1`

3. A JEA endpoint is registered with which PowerShell cmdlet?

A. `Get-PSSessionConfiguration`

B. `New-PSSessionConfiguration`

C. `New-PSSessionRegistration`

D. `Import-DscResource`

4. What PowerShell cmdlet is used to enable PowerShell Remoting?

A. `Enable-ServerManagerStandardUserRemoting`

B. `Enable-PSServerRemoting`

C. `Enable-PSRemoting`

D. `Set-PSRemoting`

5. Which agents are required to connect a Windows Server to Azure Arc and collect VM Insights?

A. Azure Log Analytics agent

B. Azure Monitor agent

C. Azure Connected Machine agent

D. Azure VM agent

6. All machines have to be hybrid Azure AD-joined for Azure Arc-enabled servers.

A. True

B. False

7. Which protocols/ports must be allowed on the network for the WAC gateway?

A. HTTP – `80`

B. HTTPS – `443`

C. `6516`

D. RDP – `3389`

8. Which of the following are supported configuration methods for second-hop PowerShell Remoting?

   A. CredSSP

   B. JEA

   C. Kerberos-constrained delegation

   D. Resource-based Kerberos-constrained delegation

   E. All of the above

9. What version of PowerShell is required for JEA?

   A. 4.0 and later

   B. 5.0 and later

   C. 6.0 and later

   D. 7.0 and later

10. Which roles are required for implementing Azure Arc for hybrid Windows Servers?

   A. Azure Subscription Contributor

   B. Virtual Machine Contributor

   C. Azure Connected Machine Onboarding

   D. Azure Connected Machine Resource Administrator

Practice test 3 – managing virtual machines and containers

1. Which of the following tasks will require a VM to be restarted?

   A. Resizing a VM

   B. Attaching a data disk

   C. Resizing a data disk

   D. Changing a data disk's type from HDD to SSD

2. For a VM that will have the AD DS role installed as a replica domain controller, which of the following disks should host ntds.dit and SYSVOL?

   A. OS disk with caching set to *read-only*

   B. Temp disk

   C. Data disk with caching set to *none*

   D. Data disk with caching set to *read-write*

3. Which virtual processor scheduling types are supported for Hyper-V?

    A. Classic Scheduler

    B. Core Scheduler

    C. Root Scheduler

4. Which virtual network switch types are supported with Hyper-V?

    A. Public

    B. Private

    C. Internal

    D. External

5. Enhanced session mode provides access to which of the following local resources?

    A. Drives

    B. USB devices

    C. Smart cards

    D. Printers

    E. All of the above

6. What are the remote access methods for managing a Hyper-V Windows Server VM?

    A. PowerShell Remoting

    B. PowerShell Direct

    C. RDP

    D. HVC.exe

7. Which characteristics are correct for the Hyper-V isolation mode in Windows Server containers?

    A. There is a per-container user mode

    B. There is a shared kernel between the containers and the host

    C. Hardware-level isolation is provided between each container and the host

    D. Each container runs inside a VM

8. Which of the following are networking drivers for Windows Server containers?

    A. L2bridge network driver

    B. L2tunnel network driver

C.   NAT network driver

D.   Overlay network driver

E.   All of the above

9.   Which Windows Server container base image is best for lift and shift scenarios?

A.   Nano Server

B.   Server Core

C.   Windows

D.   Windows Server

10.   Which command(s) make a device available to a VM via DDA?

A.   `Mount-VMHostAssignableDevice -LocationPath $locationPath`

B.   `Add-VMAssignableDevice -LocationPath $locationPath -VMName VMName`

C.   `Dismount-VMHostAssignableDevice -LocationPath $locationPath`

Then, use the following command:

```
Add-VMAssignableDevice -LocationPath $locationPath
-VMName VMName
```

D.   `Remove-VMAssignableDevice -LocationPath $locationPath -VMName VMName`

Then, use the following command:

```
Set-VMAssignableDevice -LocationPath $locationPath
-VMName VMName
```

Practice test 4 – implementing and managing an on-premises and hybrid networking infrastructure

1.   Which hybrid network connectivity solution requires your IP address space to *NOT* overlap with on-premises?

A.   VNet peering

B.   Azure VPN Gateway

C.   Azure NAT gateway

D.   Azure Extended Network

2. Which hybrid remote access solutions provide time-limited access?

    A. Just enough access (JEA)

    B. Just-in-time (JIT) access

    C. Azure Bastion Service

    D. Remote Desktop Services (RDS)

3. Which of the following capabilities are *NOT* provided by Azure DNS?

    A. Reverse DNS lookup

    B. Conditional forwarding

    C. DNSSEC

    D. Zone transfers

    E. All of the above

4. A DNS zone must have a trust anchor to be recognized as supporting DNSSEC.

    A. True

    B. False

5. When a server receives a DHCP lease of 20 days, on which day will a renewal be attempted?

    A. Day 20

    B. Day 19

    C. Day 10

    D. Day 15

6. Which PowerShell cmdlets can be used to manage DHCP scopes?

    A. `Add-DhcpServerv4Scope`

    B. `Get-DhcpServerv4Scope`

    C. `New-DhcpServerv4Scope`

    D. `Get-DhcpServerv4ScopeStatistics`

7. Which of the following is the least secure authentication protocol?

    A. Password Authentication Protocol (PAP)

    B. Challenge Handshake Authentication Protocol (CHAP)

C. Microsoft Challenge Handshake Authentication Protocol version 2 (MS-CHAPv2)

D. Extensible Authentication Protocol (EAP)

8. An Azure AVD session host needs an AD replica domain controller as an Azure VM. To resolve names on-premises, what should you implement/configure?

A. Azure AD Domain Services

B. A VPN gateway or NVA

C. A VM as an AD Domain Controller with AD-integrated DNS

D. The VNet to use the custom DNS settings of the AD domain controllers

9. What is the default lookup method if a DNS server does not hold a primary or secondary zone for a domain?

A. Forwarding

B. Recursive lookup

C. Root hints

D. Conditional forwarding

10. Which of the following will be required when creating a DHCP reservation?

A. MAC address

B. IP address

C. FQDN

D. Computer name

Practice test 5 – managing storage and file services

1. You need to change an Azure storage account performance tier after it has been created. Can this be done?

A. Yes

B. No

2. You need to change an Azure VM disk type from HDD to SSD after it has been created. Is this possible?

A. Yes

B. No

3. You need to change a VM type to support premium disks after it has been created. Is this possible?

   A. Yes

   B. No

4. Which of the following storage authentication methods supports NTFS permissions?

   A. Access key

   B. Shared Access Signature (SAS)

   C. Identity-based using Kerberos

5. Which of the following are features of FSRM?

   A. Quota management

   B. File screening management

   C. Storage reports

   D. File classification infrastructure

   E. All of the above

6. Which of the following are iSCSI logical components?

   A. Internet Storage Name Service (iSNS)

   B. iSCSI initiator

   C. iSCSI Qualified Name (IQN)

   D. iSCSI target

   E. All of the above

7. Which authentication method is used for iSCSI initiator connections?

   A. Password Authentication Protocol (PAP)

   B. Challenge Handshake Authentication Protocol (CHAP)

   C. Microsoft Challenge Handshake Authentication Protocol version 2 (MS-CHAPv2)

   D. Extensible Authentication Protocol (EAP)

8. What resiliency levels are provided by storage spaces?

   A. Simple

   B. Basic

C.   Parity

D.   Mirror

9.   Which data volumes cannot be processed for deduplication?

A.   System state files

B.   Encrypted files

C.   Extended attribute files

D.   32 KB or smaller files

E.   All of the above

10. Which communications need to be allowed for storage replicas?

A.   ICMP

B.   SMB (port 445)

C.   SMB Direct (port 4445)

D.   WS-MAN (port 5985)

E.   All of the above

Test answers

Practice test 1 – deploying and managing AD DS in on-premises and cloud environments

1.   A, B, C.

2.   A, B, D.

3.   D.

4.   B.

5.   B.

6.   False. The correct answer is gMSAs.

7.   B.

8.   False. The correct answer is that a user account must be specified.

9.   C, E.

10. False. The correct answer is that you must modify the replication schedule.

Practice test 2 – managing Windows servers and workloads in a hybrid environment

1. B.
2. A.
3. A.
4. C.
5. B, C.
6. False. *Note*: servers can be standalone workgroup servers or domain-joined.
7. B. Note: port 6516 is required when connecting to WAC deployed within the Azure portal.
8. E.
9. B.
10. C, D.

Practice test 3 – managing virtual machines and containers

1. A.
2. C.
3. A, B.
4. B, C, D.
5. E.
6. A, B, C. Note: Linux VMs use HVC.exe.
7. C, D.
8. E.
9. B.
10. C.

Practice test 4 – implementing and managing an on-premises and hybrid networking infrastructure

1. Azure VPN Gateway. *Note*: Azure Extended Network does not require you to re-IP resources in Azure.
2. B.
3. E.
4. A.

5. C.

6. A, B, D.

7. A.

8. B, C, D.

9. C.

10. A, B.

Practice test 5 – managing storage and file services

1. B.

2. A.

3. A.

4. C.

5. E.

6. A, C.

7. B.

8. A, C, D.

9. E.

10. E.

Summary

This chapter covered exam preparation tests in each key *Skills measured* section from the exam objectives.

With the knowledge and skills you've learned in this book, you will now not only be ready to take the *Administering Windows Server Hybrid Core Infrastructure* exam but, through the hands-on exercises throughout this book, you will also be confident and ready to apply the knowledge and skills that have prepared you for a real-world, day-to-day, Azure-focused role.

With this being the last chapter in this book, I would like to close off by saying a big thank you to those of you who have given your time and commitment to broadening your skills and knowledge in Azure and hybrid Windows Server through this content. I hope it may act as a springboard and spark your interest in pursuing further training and education.

Index

‹packt›

Packt.com

Subscribe to our online digital library for full access to over 7,000 books and videos, as well as industry leading tools to help you plan your personal development and advance your career. For more information, please visit our website.

Why subscribe?

- Spend less time learning and more time coding with practical eBooks and Videos from over 4,000 industry professionals

- Improve your learning with Skill Plans built especially for you

- Get a free eBook or video every month

- Fully searchable for easy access to vital information

- Copy and paste, print, and bookmark content

Did you know that Packt offers eBook versions of every book published, with PDF and ePub files available? You can upgrade to the eBook version at packt.com and as a print book customer, you are entitled to a discount on the eBook copy. Get in touch with us at customercare@packtpub.com for more details.

At www.packt.com, you can also read a collection of free technical articles, sign up for a range of free newsletters, and receive exclusive discounts and offers on Packt books and eBooks.

Other Books You May Enjoy

If you enjoyed this book, you may be interested in these other books by Packt:

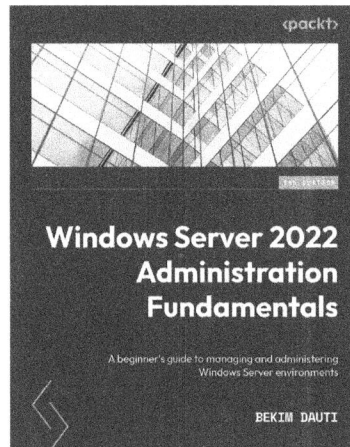

Windows Server 2022 Administration Fundamentals - Third Edition

Bekim Dauti

ISBN: 978-1-80323-215-7

- Grasp the fundamentals of Windows Server 2022
- Understand how to deploy Windows Server 2022
- Discover Windows Server post-installation tasks
- Add roles to your Windows Server environment
- Apply Windows Server 2022 GPOs to your network
- Delve into virtualization and Hyper-V concepts
- Tune, maintain, update, and troubleshoot Windows Server 2022
- Get familiar with Microsoft's role-based certifications

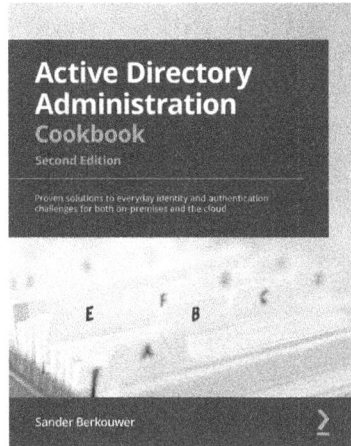

Active Directory Administration Cookbook - Second Edition

Sander Berkouwer

ISBN: 978-1-80324-250-7

- Manage the Recycle Bin, gMSAs, and fine-grained password policies
- Work with Active Directory from both the graphical user interface (GUI) and command line
- Use Windows PowerShell to automate tasks
- Create and remove forests, domains, domain controllers, and trusts
- Create groups, modify group scope and type, and manage memberships
- Delegate, view, and modify permissions
- Set up, manage, and optionally decommission certificate authorities
- Optimize Active Directory and Azure AD for security

Packt is searching for authors like you

If you're interested in becoming an author for Packt, please visit `authors.packtpub.com` and apply today. We have worked with thousands of developers and tech professionals, just like you, to help them share their insight with the global tech community. You can make a general application, apply for a specific hot topic that we are recruiting an author for, or submit your own idea.

Share Your Thoughts

Now you've finished *Administering Windows Server Hybrid Core Infrastructure AZ-800 Exam Guide*, we'd love to hear your thoughts! Scan the QR code below to go straight to the Amazon review page for this book and share your feedback or leave a review on the site that you purchased it from.

`https://packt.link/r/1803239204`

Your review is important to us and the tech community and will help us make sure we're delivering excellent quality content.

Download a free PDF copy of this book

Thanks for purchasing this book!

Do you like to read on the go but are unable to carry your print books everywhere? Is your eBook purchase not compatible with the device of your choice?

Don't worry, now with every Packt book you get a DRM-free PDF version of that book at no cost.

Read anywhere, any place, on any device. Search, copy, and paste code from your favorite technical books directly into your application.

The perks don't stop there, you can get exclusive access to discounts, newsletters, and great free content in your inbox daily

Follow these simple steps to get the benefits:

1. Scan the QR code or visit the link below

https://packt.link/free-ebook/9781803239200

2. Submit your proof of purchase
3. That's it! We'll send your free PDF and other benefits to your email directly

www.ingramcontent.com/pod-product-compliance
Lightning Source LLC
Chambersburg PA
CBHW080121220326
41598CB00032B/4913